Statistics in Criminology and Criminal Justice
Analysis and Interpretation

Third Edition

Jeffery T. Walker, PhD
Professor
Department of Criminal Justice
University of Arkansas–Little Rock

Sean Maddan, PhD
Assistant Professor
Department of Criminology and Criminal Justice
University of Tampa

JONES AND BARTLETT PUBLISHERS
Sudbury, Massachusetts
BOSTON   TORONTO   LONDON   SINGAPORE

World Headquarters

Jones and Bartlett Publishers
40 Tall Pine Drive
Sudbury, MA 01776
978-443-5000
info@jbpub.com
www.jbpub.com

Jones and Bartlett Publishers
Canada
6339 Ormindale Way
Mississauga, Ontario L5V 1J2
Canada

Jones and Bartlett Publishers
International
Barb House, Barb Mews
London W6 7PA
United Kingdom

Jones and Bartlett's books and products are available through most bookstores and online booksellers. To contact Jones and Bartlett Publishers directly, call 800-832-0034, fax 978-443-8000, or visit our website www.jbpub.com.

Substantial discounts on bulk quantities of Jones and Bartlett's publications are available to corporations, professional associations, and other qualified organizations. For details and specific discount information, contact the special sales department at Jones and Bartlett via the above contact information or send an email to specialsales@jbpub.com.

This publication is designed to provide accurate and authoritative information in regard to the Subject Matter covered. It is sold with the understanding that the publisher is not engaged in rendering legal, accounting, or other professional service. If legal advice or other expert assistance is required, the service of a competent professional person should be sought.

Production Credits

Acquisitions Editor: Jeremy Spiegel
Editorial Assistant: Maro Asadoorian
Production Director: Amy Rose
Senior Production Editor: Renée Sekerak
Production Assistant: Julia Waugaman
Associate Marketing Manager: Lisa Gordon
Manufacturing and Inventory Control Supervisor: Amy Bacus
Cover Design: Kristin E. Ohlin
Composition: Northeast Compositors, Inc.
Cover Image: © Joseph/ShutterStock, Inc.
Chapter Opener Image: © Stephen Sweet/ShutterStock, Inc.
Printing and Binding: Malloy Incorporated
Cover Printing: Malloy Incorporated

Library of Congress Cataloging-in-Publication Data

Walker, Jeffery T.
    Statistics in criminology and criminal justice / by Jeffery T. Walker and Sean Maddan. — 3rd ed.
        p. cm.
    Includes bibliographical references and index.
    ISBN 978-0-7637-5548-5 (hardcover)
    1. Criminal justice, Administration of—Statistical methods. 2. Criminal statistics—Research. I. Maddan, Sean. II. Title.
    HV7415.W32 2009
    364.01′5195—dc22

                                 2008018116

6048
Printed in the United States of America
12 11 10 09 08   10 9 8 7 6 5 4 3 2 1

Contents

11 Introduction to Multivariate Statistics 254

12 Multiple Regression I: Ordinary Least Squares Regression 269

13 Multiple Regression II: Limited Dependent Variables 301

Use of SPSS and Data Sets with This Book

Although there are some formulas that can be worked by hand, this book is centered around determining the proper statistical analysis procedure to use with particular data and how to interpret the analyses. Throughout the book, SPSS output is displayed showing analyses. If you purchased a book bundled with SPSS, you may also follow the procedures in the book to conduct your own analyses. Sample data sets associated with the book are available at the Jones and Bartlett website associated with the book. The version of SPSS bundled with this book (or a copy of SPSS Grad Pack purchased separately) can be used to conduct most of the analyses in this book. A few of the multivariate analyses (Chapters 12, 13, and 14) require either a complete copy of SPSS or another program altogether, like SAS or Stata (for poisson/negative binomial regression).

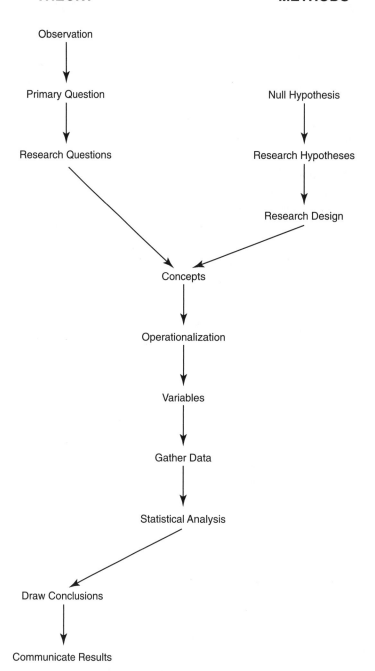

THEORY   **METHODS**

Observation

Primary Question   Null Hypothesis

Research Questions   Research Hypotheses

                            Research Design

                Concepts

                Operationalization

                Variables

                Gather Data

                Statistical Analysis

Draw Conclusions

Communicate Results

The Logic of Comparisons and Analysis

CHAPTER

1

Statistical thinking will one day be as necessary for efficient citizenship as the ability to read and write.

—H. G. Wells

■ Introduction: Why Analyze Data?

Discovery and innovation may be the distinguishing characteristics between modern human activity and that of our ancestors. The Renaissance period brought forth an emphasis on learning and advancing our way of doing things that has prevailed to the present. Scientists, inventors, and others involved in the process of scientific inquiry have often been held in awe for their works. Galileo, Einstein, Madam Curie, and others are singled out in grade school books for their works and discoveries. As you will learn, so too should be people such as Pearson, Kendall, and Yule.

Statistical analysis is all about **discovery**. The process of **scientific inquiry** provides a method of examining things that interest us in a systematic manner. This process generally requires evidence to support an argument. One of the clearest methods of establishing evidence is by examining numbers associated with the objects being studied. That examination takes place through statistical analysis. As such, statistical analysis is the linchpin of discovery, and mastery of it draws us closer to Einstein and Galileo.

■ Some Statistical History

The earliest form of what is now considered **statistical analysis** was developed by Pythagoras in the 6th century BC. This was the forerunner of **descriptive statistics** (what would eventually be known as the *mean*, or what is commonly known as an *average*). The other type of statistical analysis (**inferential statistics**) is thought to have first developed in the Orient around 200 BC (Dudycha and Dudycha, 1972). This was a form of probability analysis used in assessing whether an expected child was likely to be male or female. Probability theory, as it would come to be known, continued in the form of gambling mathematics in the works of Blaise Pascal (1623–1662) and Lord Christianus Huygens (1629–1695) (David, 1962). Many of the other descriptive statistics were developed in the late 1800s and early 1900s by mathematicians and scientists such as Galton (1883) and Pearson (1895).

Statistics moved beyond gambling and purely mathematical concepts through what was called *political arithmetic,* a term coined because of its close association with those studying political topics, including economics. (This probably began the close association between political lying and statistical lying.) The first known use of this political arithmetic was by John Graunt (1662), who used what is now called descriptive statistics to study London's death rates. Although there is fierce debate concerning the original use of the term **statistics** (Yule, 1905), the greatest support is that it was coined by Eberhard August Wilhelm von Zimmerman in the preface of *A Political Survey of the Present State of Europe* (1787). Modern use of the term *statistics* (as opposed to *mathematics*) is often attributed to R. A. Fisher and his work *Statistical Methods for Research Workers* (1925), wherein he stated that "a statistic is a value calculated from an observed sample with a view to characterizing the population from which it is drawn." Since that time, statisticians have added to the techniques available to analyze data, many adding their names to the procedures; and the addition of statistical techniques continues today. Analysis procedures have been added to the statistical repertoire in the past few years that have greatly increased the ability of researchers in criminology, criminal justice, and other fields to examine the relationship between variables more accurately.

A single death is a tragedy, a million deaths is a statistic.

—Joseph Stalin

■ Uses of Statistics

The term *statistics* is often misunderstood because there are actually two practical applications of it. The first, reflecting the history of the term, is a collection of data—often expressed in summary form—that is collected and preserved. The best example of these are census statistics or mortality statistics, which depict the characteristics of the living or the causes of death, respectively. The second application is the subject of this book: a method of analyzing data. Statistics as you will come to know them are methods used to examine data collected in the process of scientific inquiry. These methods allow researchers to think logically about the data and to do one of two things: to come to some succinct and meaningful conclusions about the data *(descriptive statistics)*, or to determine—or infer—characteristics of large groups based on the data collected on small parts (samples) of the group *(inferential statistics)*. For example, data could be gathered on all correctional officers in Arkansas for a research project to determine the sex and race breakdown of the officers. This would be a descriptive analysis that could be used to examine the employment patterns for the Arkansas Department of Corrections. Alternatively, a sample of correctional officers from each state could be collected and the data from the sample used to make statements about all correctional officers in the nation. This would be drawing conclusions (inferences) about a large group based on information about a sample of the group.

Statistical analysis is the workhorse of discovery and knowledge. The scientific process, using research to test theory, requires that empirical evidence (data) drawn from the research subjects be examined systematically. The use of mathematics in general and statistical analysis in particular allows researchers to make these comparisons and to discover new information that will provide a better understanding of their subject.

In the scientific process, the purpose is usually to discover something that was previously unknown or to prove something true or false that was previously thought to be true but was never supported by hard evidence. The way to obtain that evidence is by gathering information (data) and subjecting it to statistical analysis.

■ Theory Construction at a Glance

Three elements in social science research, or any research for that matter, are essential to sound investigation: theory, research methods, and statistical analysis. Although these elements are intimately linked, there is debate—even among those most supportive of the research process—on their ordering, importance, and what should be included from each element in a textbook. It is not possible to cover all of these elements adequately in one course or in one textbook, so it becomes an issue of how much of each element should be included in a discussion of the other. In this book, theory is covered primarily in this chapter, research spans this chapter and several that follow, and statistical analysis prevails thereafter.

What Is Theory?

At the most basic level, **theory** consists of statements concerning the relationship or association among *social phenomena* such as events and characteristics of people or things. For example, in criminology, there are theories addressing how people learn to be criminal. In these theories, statements are constructed dealing with the role of peers in a person's learning criminal behavior, how the rewards from a crime can influence behavior, and what influence punishment can have on the decision to commit a crime.

The goal of these statements is to develop explanations of why things are as they appear and to try to explain their meaning. From an early age, humans have ideas about the causes of events and why things work the way that they do. The problem with these explanations, however, is that they are often too simplistic to be of any real value. Theory attempts to provide a stronger foundation for these ideas by asking questions about them, such as:

- What is the point of all of this?
- What does it mean?
- Why are things this way?

Without theory, there is often only conjecture and war stories. With theory, we may begin to develop statements or ideas that are based on sound observation and thought.

Theory and Research

Theory may be developed in several ways. Researchers may look at the world around them, find the **social phenomena** that pique their interest, and begin to develop statements concerning why these phenomena work the way they do. This is called **induction**. An example could be a researcher who follows crime trends in a city for a number of years. She may begin to see that the crimes follow a definite pattern of movement in the city, moving from east to west across the city. From this, she might set out to determine what the cause of this movement could be, ultimately developing a theory of crime movement in urban areas. This is a process of moving from data to theory and attempting to make sense of the data with the theory.

Alternatively, researchers may become curious about something and set out to develop statements and then to test them. This is called **deduction**. The process of deduction begins with an idea and an attempt to test the idea with data and analysis. For example, a researcher might believe that increased supervision of probationers would prevent them from becoming involved in subsequent crimes. This researcher might create an experiment where a random sample of probationers are put under intensive supervision while another random sample receives a normal amount of supervision. The results of this experiment could either support or refute the researcher's initial beliefs. This is a process of moving from theory to data, where the data tests the theory. It should be noted that Sherlock Holmes was not exactly correct in his understanding of the difference between induction and deduction. When Holmes made his famous statement, "brilliant deduction, Watson!", he should actually have been commending Watson on his inductive reasoning. Watson was drawing conclusions based on what he had observed, not testing previously developed conclusions, as discussed later.

Finally, and probably most often the case, a researcher may start with either induction or deduction, but by the time a project is finished, he or she has used both induction and deduction. This is called **retroduction**. With this process, the researcher investigating supervision of probationers might conduct the intensive supervision experiment as a deductive process. After examining the data, however, it might be obvious that the experiment could be done better or that there was something in the data that needed further explanation. For example, those probationers who received the most supervision were successful, whereas those who received intensive, but less than the most intensive, supervision were not successful. The researcher might then rethink part of the theory and set out to retest it. This process might continue until the theory was supported or disproven. This is a process of moving from theory to data to theory and so on; or data to theory to data and so on. The key here is that it is an alternating process between induction and deduction.

■ The Process of Scientific Inquiry

The process of scientific inquiry (using a deductive method) is shown in **Figure 1-1**. As shown in this diagram, theory is at the starting point of the process. Theory is driven by observations and leads researchers to initiate the research process through primary questions and research questions. It is from this process of theory building that researchers follow the process from developing a null hypothesis to communicating results. The process of scientific inquiry and its individual parts are discussed further in the remainder of the chapter.

Observation and Inquisitiveness

The first step in the process of scientific inquiry, and one of the most important, is often overlooked: **observation** and *inquisitiveness*. Many research projects are never begun because the researcher was not aware of his or her surroundings or did not recognize something as a topic worthy of research.

It is often theory that stimulates observation and scientific inquiry. As you go through school and read research and material you find interesting, you will sometimes think that you have a better way to do something, or what you read may stimulate you in other areas. By using a structured scientific process to evaluate your observations and formulate statements of why these phenomena are behaving the way they are, you are developing theory.

deduction

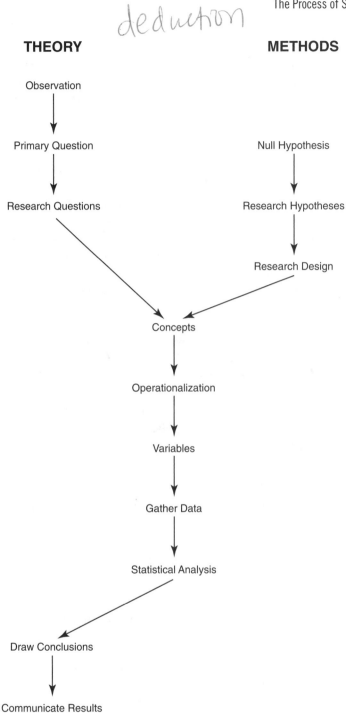

Figure 1-1 Process of Scientific Inquiry: Theory, Research Methods, and Statistical Analysis

An example of inductive theory development can be shown in Robert Burgess's Zonal Hypothesis. Students at the University of Chicago were making maps of Chicago showing different characteristics of neighborhoods, such as welfare, infant mortality, and housing. Burgess observed that these maps followed very similar patterns throughout the city. His observations led him to develop a theory about how cities grow and change. The theory he developed from these maps proposed that cities grow in rings similar to when a rock is thrown into the water. In this configuration, the rings closest to the center of the city will be the most run down, have the highest level of infant mortality, and be characterized by other social ills that are not present in the outer rings. All of this was developed simply by examining students' maps and by using inductive theory building.

Primary Questions

A **primary question** is the one driving thought behind a research project. It should represent the entire reason for the study. Primary questions are important because how well a researcher meets the goals of the primary question will often be the criteria by which the research will be evaluated. The primary question should be a carefully worded phrase that states exactly the focus of the study. For example, in research on police use of deadly force, a possible primary question might be: What factors most influence police use of deadly force?"[1] This question is very broad and somewhat vague, but it can easily represent the goal of a research project.

Research Questions

Often, the primary question will be theoretical, vague, and quite possibly not directly addressable through research. **Research questions** break down the primary question into subproblems that are more manageable and make the primary question testable through research. If the primary question establishes the goal of the research, the research questions suggest ways of achieving that goal.

In our earlier example concerning police use of deadly force, some possible research questions might include the following:

- What is the relationship between an officer's shift and the likelihood that an officer will use deadly force?
- What is the relationship between the violent crime rate of an area and the likelihood that an officer will use deadly force?
- What is the relationship between an officer's level of education and the likelihood that an officer will use deadly force?

These research questions break down the primary question into smaller parts that can be examined more easily. The answers to these questions are derived from the research process and statistical analysis and allow the researcher to answer the primary question.

Doing research is like defusing a bomb. When you begin, you are all excited and focused on the end result. If you have a good plan, know the layout, and work the plan, you will typically get the results you sought. If you run in and start cutting without a plan, it is likely to blow up in your face.

■ Research: Movement from Theory to Data and Back

Theory cannot stand alone; nor can research or statistics. Theory without research and statistical analyses to back it up is little more than fable. Research without theory is like building a house without plans, and research without statistical analysis is like building a house without nails: It is possible and it is done, but it would be more effective with them. Statistical analysis without theory and research methodology to guide the research is like having your own intercontinental ballistic missile: nice to have and may impress your neighbors, but not really useful.

Whatever the technique of developing theory, **research** is the method used to test and validate the theory. In its purest form research is a scientific, systematic study undertaken to discover new information or to test the validity of theories developed previously. The primary goal of research is discovery. Depending on whether an inductive or a deductive process is used, research is a systematic way of turning observation and statistical analysis into theory (induction) or testing theory through statistical analysis (deduction).

Although there are no exact, cookbook steps that must be followed in conducting a research project, there are some general guidelines that should be followed to ensure nothing is left out of the study. These steps are included in this chapter and provide a kind of map of where research and statistics fit into the overall process of scientific inquiry.

Formulating Hypotheses

Once the research questions have been developed, you must decide what the research is attempting to determine. Hypotheses are questions or statements whose answers will support or refute the theoretical propositions of the research. Hypotheses are generally broken down into research hypotheses and null hypotheses. Although these are covered in more detail in Chapters 14 through 17, a brief definition is provided here.

A **research hypothesis** is a statement, similar to a research question, that indicates the expected outcome of a part of a research project. If a research question in a project asked "What is the relationship between an officer's shift and the likelihood that he or she will use deadly force?", the research hypothesis might be: "There is a statistically significant correlation between an officer's shift and the likelihood that he or she will use deadly force." In using research hypotheses, the researcher turns the relatively abstract wording of theory development into a more concrete, testable form appropriate for statistical analysis.

One of the often difficult to understand but vital elements of statistical analysis and hypothesis testing is that research alone cannot prove anything. Even when researchers find a great deal of support for an association between two variables, it may be that these results are occurring because information is missing or the model is somehow flawed. Other researchers may very well be able to disprove these findings by conducting additional research. If research cannot prove anything, what can it do? It can be used to disprove something or eliminate alternatives. For example, even though research cannot prove that officers on night shifts use deadly force more than their daytime counterparts, it may disprove that there is no relationship between shift and the use of deadly force. This is accomplished with a **null hypothesis**, which generally takes the form of one of the following examples:

- There is no statistically significant difference between the groups being compared.
- There is no statistically significant difference between a group being studied and the general population.
- The differences between the groups are due to random errors.

An example of a null hypothesis is: "There is no statistically significant difference between an officer's shift and the likelihood that he or she will use deadly force." This null hypothesis contains several important components. The first is the phrase *statistically significant.* This is a specific type of relationship between the shift of duty and deadly force. The null hypothesis is not stating that there is no difference at all, just that there is no statistically significant difference. How will you know if there is no statistically significant difference? A test of the null hypothesis (test of significance) will determine statistical significance. This is discussed further in Chapter 6. The wording here is chosen carefully and should be followed carefully. Stating that "there is no difference between the shifts" is very different from stating that "there is no statistically significant difference." There probably is a difference in the use of deadly force between the shifts. As discussed in Chapter 5, though, this difference may not be sufficient to make a statistical or theoretical stand. There is also a small but tangible difference between the statements "there is no statistical difference" and "there is no statistically significant difference." It is possible to have a statistical difference (a difference in means, a difference in standard deviations, a difference in the epsilon) and not have a statistically significant difference (a difference between the variables at a particular level of confidence based on Chi-square, t-test, or another measure of statistical significance).

Another important part of the null hypothesis consists of the **variables** or items being compared. Most null hypotheses will contain two groups that are being examined (night shift, day shift, and mid shift in this example) and it will contain what is being measured (deadly force). If these items are absent from the null hypothesis, it makes it difficult to determine exactly what is being compared.

These examples show that proper wording is important in the null hypothesis. Although this phrasing does not have to be followed verbatim, it is a good example of the proper language for a null hypothesis, and any hypothesis presented should generally follow this format.

The goal of a null hypothesis is to disprove it. Disproving that there is no relationship (disproving the null hypothesis) helps support a conclusion that there is some relationship between the phenomena being studied.

Constructing the Research Design

Once a decision is made of exactly what will be studied, planning the actual research may begin. As a researcher, you should be cautious not to jump ahead of this step to other steps in the research process. You would not start building a house without first looking at other houses and thinking about what you want your house to look like; why would you start a research project without thorough consideration of what you want to do and find?

Activities in this step include determining the method to be used (experiment, survey, or other method) and generally how to approach the research. If the researcher has to collect the data, decisions must be made concerning how to collect it, what group it will be collected from, and other parameters. The decisions made here will drive the rest of the project, so they should be made carefully. This step in the research process is also dictated by the type of data gathered, which in turn dictates the statistical analyses that will be used.

Conceptualization

Once research questions and hypotheses have been developed, they must be broken down into more manageable parts. This is accomplished by drawing out the concepts from the questions and hypotheses. **Concepts** are terms that are generally agreed upon as representing a characteristic, a phenomenon, or a group of interrelated phenomena. Concepts can be very abstract or they can be fairly concrete. In the abstract, concepts can be labels used to identify properties or they can be symbolic representations of reality that are difficult to describe. For example, poverty and prejudice are fairly abstract concepts. You know immediately what each of these means, but it would probably be difficult for you to write a concise description of what they mean and even more difficult to get a consensus among the class members of what they mean. The use of concepts allows researchers to break down questions and hypotheses but retain the flexibility, for now, of not having to describe specifically what is being studied. In the example above concerning the use of force by police, the concepts to be addressed are fairly straightforward: police officer, shift of duty, violent crime, educational level, and deadly force. Although these will require more definition, it is fairly easy to get a consensus about their meaning. In general, the more theoretical the research, the more abstract the concepts, the more policy oriented the research, the more concrete the concepts will be.

Operationalization

To be able to address concepts in terms of statistical analysis, they must be placed in a form that can be analyzed mathematically. This is accomplished through **operationalization**. This is a process of translating a concept, which is abstract and verbal, into a variable, which can be seen and tested, by describing how the concept can be measured. Operationalization was introduced by the physicist Percy Bridgman in 1927.

The process of converting the abstract to the concrete can best be seen in the example of poverty used in the preceding section. As discussed, poverty itself is an abstract term; it is difficult to come to a consensus about its meaning without defining it further. This process of operationalization is a specific form of defining concepts so they may be converted to data. For example, poverty can be operationalized by deciding that income will be used to measure a person's relative level of poverty. Income cut-offs can then be used to establish the income level that will be considered poverty status. In this case, poverty is a concept; it must be operationalized to reach a point where data can be gathered on it. Income, on the other hand, has data attached to it and does not need further measurement.

In the example of police use of deadly force, the distinction between the concepts and variables is much less clear. In cases like this, the difference between a concept and a variable may be determined by asking if the word is specific enough to be able to find data on it. *Police officer* is a fairly clear term, but for the purposes of research it is still a concept that needs to be operationalized. Questions to be answered include the following:

- Will all police officers be used, or just municipal officers?
- Will detectives or patrol officers be used?
- Will staff officers in patrol be included?

The answers to these questions and others will operationalize the somewhat vague concept of a police officer into someone who can be classified as an object of the research.

The process of transforming concepts into variables demonstrates a critical point in operationalization: Operationalized definitions used in research are the researcher's definitions and do not have to match the definitions others might use or definitions the same researcher might use in other research. For example, in this research project, the term *police officer* might be operationalized as only municipal patrol officers. Certainly, others could define the term *police officer* differently by including detectives and other police. Furthermore, a different research project might very well include detectives in an operationalized definition of police officer. For this particular research, however, *police officer* will be defined, or operationalized, specifically as a municipal patrol officer.

Gathering the Data

Gathering the data is the step in the research process where most people want to begin—and nearly the last place they should begin. Returning to the example of building a house, beginning by gathering data would be like deciding to build a house and, without developing any plans, ordering a truckload of 2 × 4s, 1000 pounds of nails, and 5 bags of concrete, and going to work. You might actually get the house built this way, especially if you are an expert, but it would be better to begin with carefully developed plans.

At this point, all decisions concerning the research should have been made. The researcher should have a well-thought-out theoretical model and a clear and complete research design detailing how the data will be collected and analyzed. The concepts should already be operationalized into variables that can be measured accurately. The only thing left to do is gather the data according to the research design, analyze it according to that design, and report the results.

The research process concerning police use of deadly force is shown in **Figure 1-2**. This exhibit shows each of the steps in a deductive process. The researcher established a research plan, working through the steps in the research process from theory to data and analysis to publication of conclusions. This scientific process is applicable whether conducting academic or practical research. Note that between the concepts and drawing the conclusions is the portion of the process that directly involves statistical analysis. This is the part of the process with which this book is most concerned.

Figure 1-3, an example taken from the strain theory of criminology, provides another illustration of the process of scientific inquiry. This is what might be expected in an academic research process. Although all of the steps are not included, the figure shows the flow of activity and the difference between theory and research. It also shows the types of work products that might be achieved at each step.

In this example, the observation is that poverty causes crime. The primary question or null hypothesis is the goal that drives the research: the relationship between poverty and crime. The research questions and hypotheses put the abstract primary question into testable statements. The concepts break the research questions and hypotheses down further into key elements that must be measured. The variables that are operationalized from the concepts are those elements from which the data will be drawn. Finally, the data represents the numbers and other information that can be examined in the process of statistical analysis. It is this statistical analysis to which the discussion now turns.

A minor but important point should be made here. Most statistics and research methods books approach a discussion of the research process as if original data were to be gathered (you were going to develop a questionnaire and survey people). It is very common, however, especially for students, to use existing data sets for research (called

Observation

- What makes police officers use deadly force?

Primary Question

- What factors most influence police use of deadly force?

Null Hypothesis

- There is no statistically significant difference between officers who use deadly force and those who do not.

Research Questions

- What is the relationship between an officer's shift and the likelihood that an officer will use deadly force?
- What is the relationship between the violent crime rate of an area and the likelihood that an officer will use deadly force?
- What is the relationship between an officer's level of education and the likelihood that an officer will use deadly force?

Research Hypotheses

- Officers on night shift have a greater likelihood of using deadly force than those on day shift.
- Officers are more likely to use deadly force in high-crime areas than in low-crime areas.
- College-educated officers are less likely to use deadly force than are non-college-educated officers.

Concepts

- Police officer, deadly force, shift of duty, violent crime rate, college education.

Variables

- Number of patrol officers, deadly force incidents, numer of officers by shift of duty, violent crime rate, college education of officers.

Data

- Data from the police department for each of the variables listed above would be obtained.

Statistical Analysis

- Since the variables here are interval level or can be dummy coded, correlation or regression may be appropriate for analysis.

Draw Conclusions

- Based on the analysis, there were no differences between officers using deadly force and those who did not.

Communicate Results

- A formal report on the result of the study might be presented to the police chief and/or city council.

Figure 1-2 Research Process: Use of Deadly Force

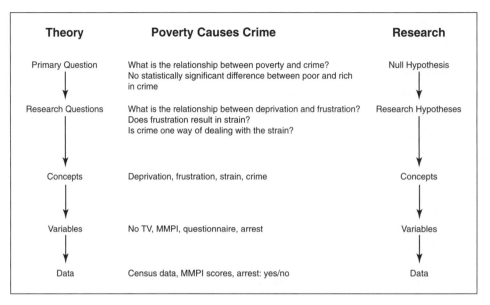

Figure 1-3 Process of Scientific Inquiry: Example from Theory to Data

secondary research), such as data taken from a police department or data from a repository such as the Interuniversity Consortium of Political and Social Sciences (ICPSR). When conducting original research, the typical methodology is to move from the primary question to the data, as discussed in this book. When conducting secondary research, however, it is more common to determine the primary question and perhaps a draft of research questions and then examine the data to see if data is available to support the concepts the researcher wants to use. If the data seems to support the required concepts, the researcher can return, establish the research questions, draw out the concepts, and operationalize them into variables supported by the data. If the data does not support one or more concepts, the researcher may be required to reconceptualize parts of the research questions or even drop certain portions of the research.

■ Statistical Analysis: The Art of Making Comparisons

A key element of research is statistical analysis. Statistical analysis is the nuts and bolts work that gives researchers the information needed to determine the success or failure of the research and the validity or fallacy of a theory. Statistical analysis deals specifically with the art of making valid comparisons, which is discussed in this chapter and is the focus of the rest of the book.

Foundations of Valid Comparisons

The art of making valid comparisons starts with the assumption that there is some relationship between the variables to be studied. This assumption is necessary because if there is no relationship between two variables, one cannot be the cause of the other, and the purpose of the research may be invalidated.

Several different types of comparisons are made in statistical analysis, depending on the goal of the research. Variables may be compared against themselves and what the researcher expects from them (**univariate statistics**); two variables may be compared to each other (**bivariate statistics**); or several variables may be compared to each other (**multivariate statistics**).

To make valid comparisons, there are some requirements that must be understood and followed. These are general requirements for making valid comparisons and are in addition to any requirements that individual methods of analysis may impose.

Comparing Appropriate Phenomena

It is easy to distort the presentation of findings, either consciously or unconsciously, when making comparisons. Take, for example, the following newspaper headline: "Murder rate jumps 100% in Polk County." This seems very dramatic until you discover that the number of murders only rose from two to four. Yes, this is technically a 100% increase, but a rise of two murders can be just an aberration that may last only one year. A more accurate comparison would have been to compare numbers or rates for Polk County to counties around it or to compare this year's increase to a ten-year average.

Although this headline is not necessarily the result of a research project gone astray, it is far too common that research produces results that can be manipulated to achieve a desired outcome. Thoughtful theory building and careful research design planning will usually prevent any unconscious distortion of comparisons. If a researcher truly works through the process of developing quality primary and research questions, it may become clear early in the research process that the comparisons to be made might be distorted. Even if it does not become clear during the theory-building process, work in developing concepts and operationalizing them should show any distortions.

Using Comparable Measures

Two levels of data may be collected: individual and aggregate. **Individual-level data** consists of one record that contains all the information to be used for each person or element. For each individual, the researcher could determine, from one data set or from several data sets in which each individual can be identified, such information as his or her age, income, number of arrests, and other data that is to be used in the research.

Aggregate-level data, on the other hand, consists of one or more data sets whose data cannot tie one person to all variables. Census data is an example of aggregate-level information. There is data on the number of persons living in houses, average income, age, and so on. A researcher could not, however, determine any information about any one person or family in the data set.

Comparisons must be based on the level of data available. For example, with individual-level data, a researcher can make comparisons based on individual-level analyses, and the same can be said for all aggregate-level analyses. If a data set contains some individual-level data and some aggregate-level data, however, the individual-level data should not be used at the individual level. The individual-level data should be combined to the same level as the aggregate data. For example, if some of the data is obtained on individual residents and some on a square-mile area, the residents' data should be aggregated to square-mile areas so the data is comparable. Although it is possible to aggregate individual-level data, you should not take the characteristics of aggregate-level data and attempt to apply them to an individual. This is called an *ecological fallacy* and should be avoided in any research.

Choosing Analysis Methods That Best Summarize the Data

Always strive to use graphs, tables, and statistical analyses that most clearly demonstrate the findings of the analysis. You have probably read articles from which you could not make sense of the statistical analysis, and it may not have been from an inability to understand the procedure. Often, results are presented such that they are almost uninterpretable—not necessarily because of their complexity but perhaps because the author did not use statistical analyses that best summarized the data. The goal of statistical analysis is to use methods that make sense of complex data and clearly support conclusions drawn.

It is a general rule of statistical analysis that it is best to use the highest-order statistics possible.[2] If the data will support it, analysis procedures designed for interval or ratio level data (covered in Chapter 2) generally provide the greatest explanatory power. Using higher-order analyses with lower levels of data can render the analysis unsound; alternatively, using lower-level analyses with higher levels of data is not taking full advantage of the data and information the data can reveal.

This is not as easy as it sounds, however. As discussed in Chapter 2 and beyond, it is generally necessary to use the statistical analysis method appropriate for the variable(s) with the lowest level of measurement. For example, to study the age of juvenile delinquents, with juvenile delinquency coded as yes/no/potentially delinquent, an analysis procedure appropriate for nominal-level data should be used because even though age could be analyzed using higher-order statistics, juvenile delinquency is nominal level; therefore, nominal-level statistical analyses should be used. There are, of course, exceptions to this rule that are discussed later, but the general rule is to use analyses that are appropriate for the lowest level of data available.

Drawing Conclusions

Mistakenly, many people believe that the process of statistical analysis, and even scientific inquiry, stops at the end of the analysis. Nothing could be further from the truth. Statisticians and researchers distinguish themselves in the interpretation of analyses and in the conclusions that can be drawn. This is also generally the most difficult part of statistical analysis. This step involves determining if the results of the statistical analysis support the hypotheses developed at the beginning of the research process.

At this step, the research process leaves statistical analysis and methodological issues and returns to theory. If the researcher is using a deductive process, this is the point at which the theory outlined in the first steps is compared to the results and a decision is made concerning whether the theory is supported or refuted. In an inductive process, this is the point at which the researcher makes initial conclusions about what he or she has seen. As evident from this discussion, theory never really leaves the research process, just as methodological issues and statistical issues are important in every step, from setting up the theory to drawing conclusions.

Moving from being a student to being a practitioner in terms of statistical analysis means sharpening your skills at interpreting analyses. It is one thing to be able to work problems or to coax a computer to provide an answer to a particular analysis; it is quite another to be able to take that analysis, return it to the process of scientific inquiry, and interpret what you have found in a way that brings new insights to the topic being studied.

Communicating the Results

The final step in the process of scientific inquiry is to communicate the results of the research. This step is also often overlooked. Many people believe that unless the findings support the hypotheses and are a monumental discovery, they are not worthy of communication. Although it is true that many of the more prestigious journals shy away from publishing negative results, it does not mean that they should not be communicated. It is important to people in the field that results, even negative results, be communicated. This will often prevent people from making the same mistakes or expending time on ground already covered. It can also aid in the process of scientific inquiry, as mentioned at the beginning of the chapter, by stimulating others to undertake research. This final, and vital, portion of the research process may be accomplished in a number of ways. Typically, the most desirable way is to publish the results in an academic journal or a book. The results can also be communicated in practitioner publications such as *Police Chief* or *Federal Probation*, in paper presentations at professional conferences, and through the release of technical bulletins.

■ Data and Purposes of This Book

Throughout this book, examples are taken from research conducted in criminal justice and criminology. This research does not always allow presentation of perfect tables or results of analysis. It is more important to be exposed to the reality of research and statistical analysis through actual works rather than to emphasize repetitious homework exercises and formulas using computational routines and shortcuts on small, error-free, fictitious data sets. This book features actual data in all of its noncooperative messiness and complexity that cannot be replicated outside the real world of research. **Figure 1-4** shows symbols and notations used in the book.

Most of the examples used in the book are drawn from data sets that were used in actual research projects undertaken by the authors. The first data set, LR_COP, is a survey of citizens in Little Rock, Arkansas, concerning their attitudes toward community policing. This data set contains a number of scale questions that represent lower-level data. This data was used in a number of technical reports and presentations to the Little Rock Police Department and the city council. The second data set, CENSUS, is taken from the US Census of 1990, and is combined with crime data from Little Rock, Arkansas. This data was used in a replication of Shaw and McKay's (1942) work examining social disorganization in Chicago and was published in *Varieties of Criminology* (Walker, 1993). The third data set, GANG, was a survey administered to juveniles appearing in the Little Rock Municipal Court. This data, published in the *Journal of Gang Research* (Walker, Watt, and White, 1994), addressed mundane, potentially delinquent, and delinquent activities of these juveniles.

The final two datasets are used for examples in the chapters dealing with multivariate statistics, primarily in relation to Chapters 12 and 13 on multiple regression techniques. The first dataset (AR_Sentencing) contains variables associated with sentencing outcomes in Arkansas. This dataset is concerned with exploring the number of months in prison an offender receives. The final dataset (gang_mem) comes from an ICPSR study on gang membership (Esbensen, 2003). This dataset is concerned with trying to predict what factors influence a youth's decision to join a gang in several metropolitan areas. While both of

Generally, formulas are presented and explained for each of the statistical procedures covered. These are presented primarily to demonstrate how each procedure works and to show how data behaves under certain circumstances. Some instructors will require students to work problems by hand to get a better feel for how the formulas work; others will not. Whatever the case, it is important that you understand the underlying concepts and mathematical derivatives behind each procedure, and you should strive to understand how the formulas work, even if you will never use them. In this book, computational formulas are generally replaced with formulas that, although sometimes more difficult, are more descriptive. This is done to enhance understanding the analysis rather than to make computations easier. The following symbols are used in the formulas.

N The sample size or total number of cases.

X_i A single score or case, where i is the score's location in the distribution. In grouped data this will be the midpoint of the class.

f_i The frequency of the ith category.

$\sum$ This is a summation operator. It requires you to add (or shows that something has been added) data, columns, or other material.

$\sum_{i=1}^{N}$ This tells exactly what to add. The starting point is shown on the bottom (where the lower limit is 1); numbers are added until the number on the top (N) is reached. If it is obvious that you are to add all of the numbers, cases, or other data, the upper and lower limits may be omitted.

$\sum f_i$ Add the frequencies in the column of i.

$\sum_{i=1}^{N} X_i$ Add all the raw scores, starting with one and ending when all numbers have been added.

Figure 1-4 Symbols and Notation Used in This Book

these datasets are used with higher level statistical analyses, the univariate and bivariate techniques discussed in Sections 2 and 3 of this book can be used on these data as well.

The primary advantage of these data sets is that they are small. This will make it easier for students to work problems by hand as well as to allow the use of student versions of statistical software, which are often limited in the number of variables and cases. The data sets have also been cleaned of errors and are provided in dBase, Excel, and SPSS (the statistical package used in examples in this book) formats, which will allow use of the data in practically any statistical analysis program used in courses. Also, since the examples provided are, with few exceptions, taken directly from the data sets, they can be replicated, modified, or extended using the data sets provided. Finally, the variables included give students and instructors the opportunity to use a variety of demographic, criminological, and other variables of interest.

Included in Appendix D is a list of the variables used in each of these data sets, along with a short description (the SPSS Value Label) of each. This will assist in understanding what variables are included in examples used in the book, and should help in deciding variables to be used in homework.

The focus of the book is on statistical reasoning and interpretation of analysis. To accomplish this goal, the data is presented both in formulas and in computer output. This allows you to focus on the proper identification of data (level of measurement, abnormalities, and the application of statistical analyses) and the interpretation of results. It is recognized that initially, the tables and output may be somewhat difficult to read, but this is the actual output from SPSS. It is certainly possible to place the output in cleaner and more understandable tables such as those that appear in academic journals, but this would not help you understand the output you would receive when conducting actual research. It is important, therefore, that you be able to read statistical output as it is presented in the real world. Once this is accomplished, reading other tables and output will be that much easier.

Also included for each chapter is a flowchart of the decision process included in the discussion of that chapter. Sometimes this flowchart will be replicated, either in part or entirely, in the text. This assists in understanding a particular part of the discussion. You are encouraged, however, to always look first at the flowchart for the chapter and then refer to it as you read, work through the examples, and complete the homework.

Many students approach a statistics course with great trepidation. Typical responses range from "I have never been good at math; I just barely made it through high school," to "I have a mental block with math," to "I haven't had a math course in 10 years." Learning statistics is not easy, primarily because you have to learn a new language, much like learning a foreign language. There is a review and practice test in Appendix A that may help determine how much math you remember. This review and practice test will also show that there is not a great deal of advanced math needed to study statistical analysis. If you can add, subtract, multiply, and divide, you will be able to work any formulas and undertake any statistical analyses in this book. After finishing the book, you should have the foundation needed to produce original research from statistical analysis and to be a consumer of statistical findings; that is, you should be able to read the tables and findings in journal articles and books rather than skipping over them. With the foundation of the process of scientific inquiry behind us, it is now time to move to the world of statistical analysis—hold on and have fun.

▩ Key Terms ▩

aggregate-level data
bivariate statistics
concept
deduction
descriptive statistics
discovery
individual-level data
induction
inferential statistics
multivariate statistics
null hypothesis
observation
operationalization

primary question
research
research hypothesis
research question
retroduction
scientific inquiry
social phenomena
statistical analysis
statistics
theory
univariate statistics
variable

■ Exercises

1. What is the difference between a null hypothesis and a research hypothesis?
2. What is the difference between a research hypothesis and a research question?
3. Using the list of variables for one of the data sets in Appendix D, develop a primary question, research questions, null hypothesis, and research hypotheses for a research project.
4. What is the difference between a concept and a variable? Provide examples of concepts and then operationalize them into variables. What data might match the variables chosen?
5. What are some of the concepts that would be associated with the variables you have chosen in Exercise 3?
6. Think about and write out a research design you think might be used to carry out your research.
7. Select a journal in criminal justice or criminology. See if you can determine the following from an article:
    a. The primary question
    b. The research questions
    c. The null hypothesis (you may have to develop this from the primary question)
    d. The research hypotheses
    e. Concepts
    f. Methods used to operationalize the concepts
    g. Variables

■ References

Bridgman, P. (1927). *The Logic of Modern Physics.* New York: Macmillan.

David, F. N. (1962). *Games, Gods, and Gambling.* New York: Hafner.

Dudycha, A. L., and L. W. Dudycha (1972). Behavioral statistics: an historical perspective. In R. E. Kirk (ed.), *Statistical Issues: A Reader for the Behavioral Sciences.* Pacific Grove, CA: Brooks/Cole.

Fisher, R. A. (1925). *Statistical Methods for Research Workers,* 11th ed. Edinburgh: Oliver & Boyd.

Galton, F. (1883). *Inquiries into Human Faculty and Its Development.* London: Macmillan.

Graunt, J. (1662). *Observations on the London Bills of Mortality.*

Pearson, K. (1895). Classification of asymmetrical frequency curves in general: types actually occurring. *Philosophical Transactions of the Royal Society of London,* Series A, Vol. 186. London: Cambridge University Press.

Robinson, W. S. (1950). Ecological correlates and the behavior of individuals. *American Sociological Review,* 15:351–357.

Shaw, C. R., and H. D. McKay. (1942). *Juvenile Delinquency and Urban Areas.* Chicago: University of Chicago Press.

Walker, J. T. (1993). Shaw and McKay revisit Little Rock. In G. Barak (ed.), *Varieties of Criminology.* New York: Pergamon Press.

Walker, J. T., B. Watt, and E. A. White (1994). Juvenile activities and gang involvement: the link between potentially delinquent activities and gang behavior. *Journal of Gang Research,* 2(2):39–50.

Yule, G. U. (1905). The introduction of the words "statistics," "statistical" into the English language. *Journal of the Royal Statistical Society*, 68:391–396.

Zimmerman, E. A. W., von (1787). *A Political Survey of the Present State of Europe.* London: Royal Statistical Society.

■ For Further Reading

Glaser, B. G., and A. L. Strauss (1967). *The Discovery of Grounded Theory.* Chicago: Aldine.

Kuhn, T. S. (1970). *The Structure of Scientific Revolutions*, 2nd ed. Chicago: University of Chicago Press.

Reynolds, P. D. (1971). *A Primer in Theory Construction.* Indianapolis, IN: Bobbs-Merrill.

Wallace, W. L. (1971). *The Logic of Science in Sociology.* New York: Aldine.

■ Notes

1. This example is used throughout the discussion to show the research process using a single example in each stage.
2. As discussed in Chapter 6 and beyond, higher-order statistics are those used with interval- and ratio-level data; lower-level statistics are generally used with nominal- and ordinal-level data.

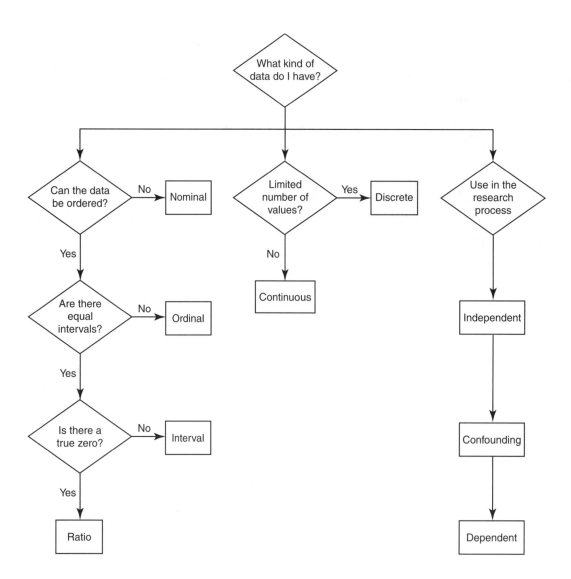

Variables and Measurement

Now begins the process of statistical analysis. To be able to discuss the characteristics (social phenomena) of our surroundings in terms of statistics, these characteristics must be in a form suitable for statistical analysis. Like most mathematical operations, this is undertaken by organizing the characteristics in terms of variables. In Chapter 1 you learned that this process is called operationalization. Operationalization is the process of measuring a variable. This is typically done by assigning numbers to the characteristics, thereby putting them in a form that can be analyzed mathematically, a process often called **measurement**. This chapter introduces variables as they are used in statistical analyses. The process of transforming characteristics into variables is also discussed. Finally, the ways in which data and variables can differ are explained.

■ The Variable Defined

At its most basic level, a *variable* is a social phenomenon, characteristic, or behavior that a researcher is trying to learn something about. These are called variables because they *vary* between cases. It would be silly to do research on people or things that did not vary; we could simply draw a picture and get the same results. It is how opinions, characteristics, and other factors differ among people that is the object of inquiry. For example, a topic of research might be to examine the difference in criminal behavior between males and females. It is the **variation** in criminal behavior that may be the result of variation between the sexes that is important; thus, the variables of crime and sex[1] would be the subject of research on this topic.

■ Transforming Characteristics into Data: The Process of Measurement

Measurement is an important part of the process of scientific inquiry and is a critical component of statistical analysis. The level of measurement of data that has been collected is a determining factor in the type of statistical analyses that can be used. Incorrectly identifying the level of measurement of a variable can be disastrous to research because statistical analyses that are not appropriate for the data could mistakenly be used, which would give erroneous results. Special attention, then, should be placed on understanding measurement.

Peter Caws (1959) offers one of the more complete definitions of measurement:

"Measurement is the assignment of particular mathematical characteristics to conceptual entities in such a way as to permit (1) an unambiguous mathematical description of every situation involving the entity and (2) the arrangement of all occurrences of it in a quasi-serial order."

Caws goes on to describe this quasi-serial order as establishing that for any two occurrences, either they are equivalent with respect to the characteristics being studied or one is greater than the other.

Measurement itself is a way of assigning numbers, or numbered symbols, to characteristics (people, social phenomena, etc.) using certain rules and in such a way that the relationships between the numbers reflect the relationships between the characteristics or variables being studied. For example, in early biological theory, a person's arms were often measured as one characteristic in determining criminality. A number could be assigned to each of 10 people (or each person's arms) by measuring an arm's length with a ruler (the number representing arm length in inches). If the number assigned to one person was greater than the number assigned to another person, the conclusion would be that the first person had longer arms than the second. A further inference from this relationship could be that the first person might be more likely to be criminal than the second. In this case, the relationship among the numbers used in measurement corresponds to the relationship among the persons or their characteristics.

In fields such as physics or engineering, transforming characteristics into variables and data is not a problem. A person conducting research on the space shuttle can operate in terms of the speed of the shuttle, height above the Earth, and so on, which are already in the form of numbers. Even in the example used above, there was a relatively easily measured characteristic: length of arms in inches. In social sciences such as criminal justice and criminology, getting data into a numerical format is not always this easy. In many instances, researchers must work to find ways to turn characteristics into data. This is sometimes accomplished by classifying the characteristics into categories of the variable defined by the researcher. For instance, a researcher might want to know whether or not a person being studied has been convicted of a crime. When developing the research methodology for this project, the researcher could decide there are two categories of concern: convicted and not convicted. To transform the characteristics of the people being studied into data, the categories to be used might be predefined as 1 to represent convicted and 0 to represent not convicted. Then, when the data is gathered, all that must be done is to code all convicted persons as 1 and all nonconvicted persons as 0; thus, the characteristics have been turned into data for the variable *criminal*. This is the process of measurement.[2]

A distinction should be made here between the measurement that will be undertaken in this book and the measurement required when conducting research. In conducting research, it is important to measure accurately the phenomena of study, such as whether the IQ test is a valid measure of intelligence. In fact, there is a field of study in measurement theory that addresses the validity of measurement. In this book the only requirement is to identify correctly the level of measurement of data. It will be assumed that the variables used are a valid measure of the phenomena being studied.

Sometimes there is also confusion about the difference between definition and measurement. *Definitions* usually replace a word with more words that simply clarify meaning. Measurement, on the other hand, assigns a numeric term that can be analyzed mathematically or statistically. A definition, then, helps clarify a concept but does not require great precision; a definition may be needed for part of the definition. Measurement, however, specifies how characteristics will be grouped on any particular social phenomenon in a manner that makes it clear what groups the characteristics fall into. There may be a few accepted definitions, or there may be only one (e.g., from Webster's dictionary). Measurement, however, is based on the research. It is how *you* are going to operationalize the concept (as discussed on Chapter 1).

As also stated in Chapter 1, it is important to remember that variables are only representations of the characteristics being studied and are not the characteristics themselves. As such, the value of a variable (especially for nominal- and some ordinal-level data) could be replaced with other values (1, 2 rather than 0, 1), the ordering could be changed (male, female rather than female, male), or the scale could be changed (XL, XXL, XXXL rather than S, M, L). It is good to attempt to categorize data accurately and unambiguously. Avoid the temptation, however, to place so much pride in a concept that you begin to view the reality that is to be measured by the concept (which may vary greatly) with the way you have defined it. As an example, not all people will define *gang behavior* the same way, and your measurement may be no better or worse than anyone else's.

■ How Variables Can Differ

In analyzing variables, it is important to know the ways in which they differ. Different types of variables have different characteristics and require different statistical methods for proper analysis. Proper classification of a variable is thus a very important step in research and in statistical analysis because everything that follows is dependent on this proper classification.

Variables can differ in three primary ways. First, they can differ in their level of measurement. The mathematical characteristics of the variable *crime*, discussed above, are different from the mathematical characteristics of a person's attitude toward crime. Variables can also differ in their scale continuity. Some variables, such as the number of children a person has, are whole numbers with no fractions in between. Other variables have many values in between any two values. For example, if asked how old you are, you would probably respond with a whole number such as 19 years old. If pressed, though, you would be able to describe your age in terms of years and months, or years, months, and days, and so on. Whole numbers represent **discrete variables;** while variables that can be further subdivided represent **continuous variables**. Finally, variables differ in the ways in which they are used in the research process. **Dependent variables** are the object of inquiry: for example, criminal behavior. These are the characteristics we want to know something about. **Independent variables** are used to describe characteristics of that object: for example, age and its effect on criminal behavior. A third use in the research process, **confounding variables**, has a bearing on how the first two types of variables influence each other. Each of the ways variables differ is discussed in this section.

Levels of Measurement

The **level of measurement** is a very important part of classifying variables for use in statistical analysis because the level of measurement will determine what analyses are appropriate. For example, in Chapter 4 you will learn that the mode is the most appropriate procedure for determining the measure of central tendency for nominal-level data, whereas the mean is more appropriate for interval- and ratio-level data. Misidentifying the level of measurement can be detrimental to a research project. Some statistical analyses are **robust** enough (i.e., they will withstand violations of their assumptions) that it is not a problem using ordinal-level data for procedures that generally require interval- or ratio-level data: for example, the t-test. Other statistical procedures are not robust enough to withstand misidentification of a level of measurement, which can result in seriously flawed analyses and conclusions. Similarly, it is best to use the highest-level statistical analysis available for the data. Even though it is not wrong to use nominal-level analyses with higher-level data, a great deal of detail and power that is available with the higher levels of measurement is lost by doing so.

Every variable can be put into one of four categories: nominal, ordinal, interval, and ratio. Each of these levels of measurement has different characteristics that must be considered when deciding on a statistical procedure.

Nominal Level

The lowest level of data in terms of levels of measurement is nominal level. **Nominal level data** is purely **qualitative** in nature, meaning that the variables are word oriented, as opposed to **quantitative** data, which is number oriented. Variables in this category might include race, occupation, and hair color. By assigning numbers to these characteristics, thereby making them useful for statistical analysis, we are actually doing nothing more than renaming them in a numeric format, and words or letters would serve just as well. For example, the categories of a variable could be classified as either *male* or *female*, or these categories could be abbreviated to M and F. Numbers could be substituted for these classifications, making them 1 and 2. The numbers are simply labels or names that can indicate how the groups differ; but they have no real numeric significance that can tell magnitudes of differences or how the numbers are ordered: Male/female, M/F, and 1/2 all mean the same thing here. Additionally, they could be switched around such that female/male, F/M, and 2/1 also mean the same as in the first three instances. The only reason to use numbers rather than letters is to make mathematical operations possible.

In operationalizing nominal level data, observations are simply placed into categories. The ordering of categories is usually arbitrary. For example, it makes no difference when operationalizing eye color whether the *brown* category is assigned 0, 1, or 2. The only requirement is that all like scores be coded the same.

For data to be considered nominal level, the categories of the variable should be distinct, mutually exclusive, and completely exhaustive. These requirements are the same for ordinal level data.

Distinct means that each value or characteristic of a variable can be easily separated from the other characteristics of that variable. For example, it is generally easy to separate male from female participants in a research project. These categories are distinct. This task is not always so easy, though. Take, for example, colors. Depending on whether you are working from a box of 8 crayons or from a box of 64, you may have a difficult time establishing the distinctiveness of, say, the color red. Also, when conducting attitudinal surveys, people are seldom totally *pro* or *con* on a subject such as the death penalty but represent a continuum of attitudes. Failure to maintain distinctiveness does not characterize the data as being on another level or as being less than nominal. It may, however, require that you reconsider or rework the operationalization of the variable to ensure it can meet this and other requirements.

Mutually exclusive means that values fit into only one of the categories that have been designated. In the example above, the distinctiveness of the categories should allow a person to fit into one, and only one, category. In a study, there should not be many subjects who can fit into both the male and female categories. If subjects in a study can be cross-classified only a few times (as often happens in any real research project), it can be dealt with on a case-by-case basis according to the methodological plan. Continually finding subjects who can fit into more than one category, however, may indicate that you need to reconsider the variable's operationalization.

Completely exhaustive means there is a category in which each characteristic can be placed. In a research project, there should not be subjects who do not fit into one of the categories that were developed. Exhaustiveness is achieved through careful planning and operationalization. Researchers must strive to develop categories that characterize the variable being studied accurately and completely. Although "other" categories are generally effective, they should be used sparingly. If a large number of subjects fall into the "other" category, you may need to reconsider your operationalization and create more categories that depict the variable more accurately. For example, if the variable *race* has the categories *white*, *black*, and *other* but there are a large number of Latino participants, an additional category may be necessary.

Ordering is the final characteristic of nominal level data. The primary difference between nominal level data and other levels of measurement is that nominal level data cannot be ordered. Since values are arbitrarily assigned, it cannot be argued that one value is greater than or less than another value. For example, coding blue eyes as 2 and brown eyes as 1 does not mean that blue eyes are better than brown eyes. The variable *eye color* cannot be ordered in terms of greater or lesser on the characteristic being studied.

Remember that the most important element of nominal level data is that the categories of the data cannot be ordered. The categories being distinct, mutually exclusive, and completely exhaustive is important, but violation of these requirements does not make the data something less than nominal; it is still nominal level data. It is very important, then, to establish the ordering (or lack thereof) of the categories. You should be very specific as to why the categories can be ordered or cannot be ordered. Failure to be very clear concerning the ability to order categories will almost always result in misclassification of the level of measurement. For example, you may have data such as patrol district number, which you obtained from the local police agency. These patrol district numbers may look as follows:

1

2

3

4

5

These are certainly ordered, right? Wrong. If you examined them closely, you would see that the *numbers* can be ordered but what they represent (patrol districts) is not ordered. You could just as easily have labeled them patrol districts A, B, C, D, E. This issue is addressed again later in our discussion of ordinal level data.

The key in operationalizing and assigning values to nominal level data is that all characteristics, sometimes called *attributes,* that are the same are assigned the same value, and characteristics that are different are assigned different values. To effectively utilize attributes measured at the nominal level, all values that are the same should be coded the same, and very few of the values should have the possibility of being coded into more than one category.

As discussed in Chapters 4, 5, and 6, most statistical analyses associated with nominal level data are simple counts and analyses based on those counts. In these analyses, some assumptions can be made about how many of the characteristics are in each category but little else. Since we are not able to ascribe true numbers to nominal level data, any procedures requiring a quantitative measure generally will not work with those data.

Ordinal Level

Ordinal level data is similar to nominal level data in that it must also be distinct, mutually exclusive, and completely exhaustive. Remember, though, that if an ordinal level variable is not, for example, distinct, it does not make it nominal. It is still ordinal level, but it has some methodological issues of distinctness. The difference between nominal and ordinal level data is that ordinal level data can be ordered.

In ordinal level data, phenomena are assigned numbers; the order of the numbers reflects the order of the relationship between the characteristics being studied. This order in the categories is such that one may be said to be less than or greater than another, but it cannot be said by how much. This ordering is similar to having a foot race without a stopwatch: It is possible to determine who finished first but it is difficult to determine exactly by how much that runner won. An example of an ordinal level variable might be income, where the categories are as follows:

$50,000 and more

$40,000–$49,999

$30,000–$39,999

$20,000–$29,999

$10,000–$19,999

Less than $10,000

Values assigned to the categories of ordinal level data may be expressed either in words—low, medium, high—or in numbers—1, 2, 3. Either set may be used as long as it conveys the ordering of the categories.

There are two different classifications of ordinal level data: partially ordered and fully ordered. The differences between them are the number of data points in the scale and the degree to which approximate intervals can be determined.

Partially ordered variables are often expressed in a small number of categories (typically, less then 5), such as small, medium, and large. Although it is easy to tell which has a greater value relative to the other, the magnitude of differences is impossible to determine. For example, there may be little difference between a small and a medium item and a great difference between a medium and a large item. Here, there are no equal intervals between these categories, precluding them from rising to interval level data.

Fully ordered ordinal level data provides more options upon which to place ranks. The typical data here is on a scale of 1 to 10. Variables that have 10 data points instead of three allow refined measurement and more closely approximate equal intervals. A scale of 1 to 100 would allow even greater precision and would come much closer to approximating interval level data.

A controversy often arises concerning where partially ordered data ends and fully ordered data begins. The issue here is that many people will attempt to use interval level analyses with fully ordered ordinal level variables. This is particularly the case with Likert scales (discussed later). Although there are no strict guidelines, general wisdom on the break between partially ordered and fully ordered data is that five or more categories is generally considered fully ordered. Thus, a scale with five or more items might be considered fully ordered, whereas a scale with four items would be considered partially ordered. Whether it is appropriate to use interval level analyses with fully ordered ordinal level data is a more complicated issue that is addressed later.

Statistical analyses associated with ordinal level data operate either from frequency counts, as with nominal level data, or are based on the rank of the characteristic. It should be emphasized that statistical procedures for ordinal level data still do not evaluate the number associated with the characteristic. What is being analyzed is the ordering of the characteristics and the rank of a particular score. It is this ordering that separates ordinal level data from nominal level data, but the intervals between the ordered variables are not such that they rise to interval level.

Interval Level

The move to interval level data represents a move from a qualitative to a truly quantitative scale. **Interval level data** builds on ordinal level data because the data may be ranked but there are also equal intervals between categories. Whereas with ordinal level data it is not possible to know exactly what the interval is between, for example, shirt sizes of small, medium, and large, with interval level data it is easy to establish that there are equal intervals between, for example, miles per gallon in fuel economy.

Another difference between ordinal and interval level data is the **scale continuity**. As discussed later in the section on scale continuity, interval level data may be **continuous**, where there are intermediate values in the scale that can be divided into subclasses, or *discrete*, in which there are no values falling between adjacent values on the same scale. Ordinal level data is rarely continuous in nature because, by its definition, ordinal level data is based on ranks that are generally represented by whole numbers (or perhaps gross decimal divisions such as 5.5).

Based on these characteristics, interval level data differs from ordinal level data primarily in having equal intervals between values, such as seconds in a measure of time, whereas ordinal level data does not, as with a ranking of fastest to slowest time. Interval level data is separated from ratio level data, though, because interval level data does not have a **true zero**, which is required for ratio level data.

Researchers generally want to make variables interval level because they can then use interval level statistical analyses, which are more powerful than ordinal level statistical analyses. Also, math can be used with interval level data to draw conclusions not obvious from the raw numbers.

Special Kinds of Interval Level Data. Since it is desirable to use interval level analyses when conducting research, special attempts are often made to make data interval level. Two such instances are using dichotomized data as interval level and attempting to make ordinal level scales interval level.

Dichotomized variables are variables that are divided into two categories, such as male/female and yes/no. Dichotomized variables are sometimes treated as interval level. Actually, if the variable represents the presence and absence of something, it may be considered ratio. The reason for this is that the two points can be treated, mathematically, the same as other interval level data.[3] One of the identifying characteristics of interval level data is that it can form a straight line between points (see **Figure 2-1**). The same is true for dichotomized data, with the straight line running between the two points. Additionally, since there are only two data points, there is only one interval; therefore, there can be no unequal intervals.

As discussed in the chapters concerning multivariate analysis, this is the reason dichotomized variables are often allowed in multivariate analysis even though these procedures are normally reserved for interval and ratio level data. It should be stressed, however, that even though the interval is equal, dichotomized variables are not always normally distributed (see Chapter 4). When sample sizes are small, therefore, interval level statistical analyses may not be appropriate even though the variable may be considered as being interval level. Additionally, if there is a high proportion of values concentrated in one category (almost all 1's or 0's), the data may be too skewed (see Chapter 6) for interval level analyses.

Although not necessarily a mathematical question for dichotomized data, there is the additional issue of whether the data is actually dichotomized. As discussed above, people are rarely either totally pro or totally con on a particular issue, such as the death penalty. People typically have a range of emotions, or their approval or disapproval depends on the situation. It is possible to force the data into two categories by making people choose one or the other, but the reality is that this attitude is actually repre-

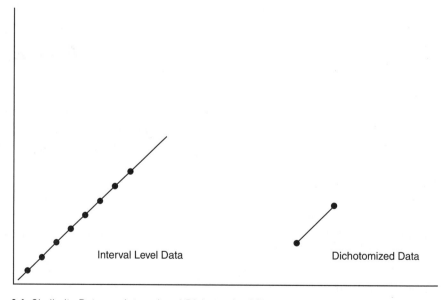

Figure 2-1 Similarity Between Interval and Dichotomized Data

sented as a continuum from pro to con, depending on the situation. This represents a special issue for an analysis with dichotomized data. If the purpose of the research is not dependent on a true dichotomy, for example, the research is examining how people choose rather than what they choose, there is no real problem in using a dichotomized variable with an underlying continuum. If the purpose of the research is to predict categories, however, the underlying continuum between pro and con represents a significant theoretical problem for the research. Another example can be drawn from medicine. If the purpose of the outcome is to study the success of a surgery based on whether the person lived or not, the use of dichotomized data is perfect: 1= the person lived; 0 = the person died. If the success is measured in the quality of the surgery, however, there is a continuum of success that must be accounted for; for example, the person lived but was permanently maimed.

As is apparent from this discussion, using interval level analyses with dichotomized data is appropriate, but it has its drawbacks. Researchers attempting to use this type of data should be aware of these problems and be prepared to consider them in the theoretical, methodological, and statistical plans of the research.

The debate surrounding the issue of ordinal level data versus interval level data intensifies when addressing the use of scales such as Likert[4] scales (1 to 5: *strongly disagree, disagree, neutral, agree, strongly agree*) that are used so frequently in opinion research. There are actually two issues here. The first is whether a scale of perhaps five items or fewer should be considered interval level; the second is whether any scale should be considered interval level.

The first issue can be settled more quickly than the second. Most researchers agree that any scale based on fewer than five items is difficult if not impossible to consider as interval level. Even if there was some argument that the intervals between the three or four items were equal, it is unlikely that the data would be sufficiently normally distributed to allow use of **parametric** (interval level) analyses.

The key element in the larger debate is whether a scale can be considered to have equal intervals between its items. For example, is the distance in attitude between *strongly disagree* and *disagree* the same as the distance in attitude between *disagree* and *neutral*? Although there are equal intervals between 1 and 2 and 4 and 5, it is much more difficult to state that there is the same interval between attitudes on this scale; people simply do not feel the same about an issue such that this conclusion can be drawn. There is even weaker support for this conclusion when the scale includes a *neutral* category. Can it be stated with certainty that the difference between *agree* and *neutral* is the same as between *disagree* and *neutral*, and is that distance the same as between *strongly agree* and *agree*? It is unlikely.

The reason this debate is important is because the nature of interval level analyses is also somewhat in debate. For example, in the strictest sense, means and standard deviations, and the statistics based on them, should not be used with ordinal level data (see Chapter 4 and beyond). As Stevens (1968) noted, however, researchers can often gain a better understanding of the data by using such procedures, and the procedures are often robust enough to allow using ordinal level data. Furthermore, as Abelson and Tukey (1963) and others have pointed out, if ordinal data can be verified as normally distributed (see Chapter 4 and Chapters 14–17), parametric (interval level) statistics may be appropriate. If, however, the data is not normally distributed, **nonparametric** (ordinal level) statistics are more appropriate.

This debate will not be settled here, but a grounded piece of advice is offered. Generally, when attempting to determine whether data is ordinal or interval, it is best to return to the theory and the research methodology. If the theory and methodology of a study is such that you can honestly state that the difference between the values of 1 and 2, representing, for example, strongly disagree and agree, is the same as between the values of 4 and 5, representing disagree and strongly disagree, use interval level analyses. If there are doubts, ordinal level analyses are more appropriate.

Ratio Level

Ratio level data is considered the highest order of data. It is interval level data with a true zero, where there is the possibility of a true absence of the characteristic in the variable. For example, income is generally considered a ratio level variable. You can certainly have zero income! This true zero may be implied, however. For example, scientists have addressed the issue of a true, or absolute, zero in temperature even though they have never been able to achieve it.

The difference between interval and ratio level data is that interval level data has no true zero. For example, ignoring for a moment any discussions of science or conception, it is generally accepted that something cannot have zero age. If something exists, by definition it should have age, based on how long it has existed. If this is true, age has no true zero and is considered interval level instead of ratio level data.

Another important characteristic of ratio data is that it generally includes things that can be counted: dollars, eggs, people, and so on. Under its definition, if something can be counted, it should represent whole numbers of the item, and it is therefore easy to envision an absence of that item (especially dollars!). This should not be confused, however, with the whole numbers that are a part of the ranks in ordinal level data. There, no true zero is possible because everything must have a **rank** (see **Box 2-1**). Even though the ranks are whole numbers, they are not ratio level.

Box 2-1

A Note About Precision

In this section, you will gain experience in classifying variables into their levels of measurement. Often, it will be difficult to obtain a consensus of whether a variable is interval or ratio. For example, is the air pressure in a tire interval or ratio? On the one hand, it is possible to have no air in your tire (and usually, a flat spare tire when it happens). This would make the variable ratio, right? Well, science tells us that there actually is air in that tire and that we are simply using too crude a measurement to gauge it accurately. Science also tells us that there is no true absence of air pressure, making it an interval level variable. In practice, the difference between an interval and a ratio level variable is minimal, and statistical analyses work equally well for either.

The same can be said for determining scale continuity. In this book, characteristics such as age are considered continuous variables. In the hard sciences, this variable would be discrete because a finite number of data points *are* available, even though there are many.

The moral of this story is that you should measure data as accurately as possible, but it is generally not necessary to ponder the infinity of the universe or to use quantum physics when deciding between interval and ratio levels of measurement.

The presence or absence of a true zero affects the mathematical procedures that can be used to analyze the data. Without a true zero, ratios cannot be employed. If we cannot use ratios, a particular score cannot be said to be twice (or three times, or four times, etc.) as much as another. The presence of a true zero allows researchers to work with the knowledge that the ratios will remain constant even though the numbers may change (2 is half of 4 and is the same ratio as 12 to 24). In real-world research, though, interval and ratio are treated the same. There have been no statistical analyses created that work with ratio level data to the exclusion of interval level data, although it is theoretically possible. Therefore, data that can be measured at the interval or ratio level is suitable for higher order statistical analyses.

The Process for Determining Level of Measurement

As discussed above, properly identifying data and classifying it into its correct level of measurement is a vital part of the research process. A practical process for determining the level of measurement is shown in **Figure 2-2**. Using this process will enable you to examine data (or questions from surveys) and classify them into the proper level of measurement.

The first step in this process is to determine if the data can be ordered. If the data cannot be ordered, or if the categories can be switched around such that the ordering does not matter, the data is considered nominal level. If the data can be ordered, it is at least ordinal level and requires further examination.

The next step in this process is to determine if there are equal intervals between data points. It should be noted here that equal intervals need not be present, just the possibility of equal intervals. For example, when surveying people about their age, there are definite gaps in the data. This is because there are usually not enough people

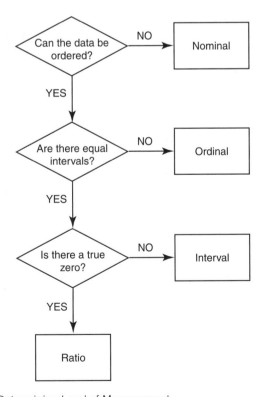

Figure 2-2 Process for Determining Level of Measurement

surveyed to have someone at every possible age. This does not mean the data is not interval, though, simply because there are not perfect intervals in the data. If there is no possibility of equal intervals but the data is ordered, the data is ordinal. If there is the possibility of equal intervals between the data points, such as age in years or months (where the interval between each data point is 1 year or 1 month, respectively), the data is at least interval level and requires further examination.

The final step in this process is to determine if there is a true zero, whether clearly identifiable or implied. If there is no true zero, the data is interval; otherwise, the data is ratio level.

There are a few important points to remember when using this process to determine the level of measurement. These will not make the process of determining the level of measurement easier, but it will make it more accurate.

First, it is important to be specific about the data. It is not enough to think abstractly about the variable. You have to ask how the data is arranged *in this case*. This is particularly important in establishing whether or not the data is ordered. It is not sufficient simply to state that the data can be ordered. How can it be ordered? What specifically supports an argument that it can be ordered?

Relatedly, look at how the data is being used in the research. Age, for example, can be used in a number of ways in the research process—ways that can change the level of measurement. If age is categorized into groups (0–10, 11–20, etc.), it is probably ordinal level, whereas someone's age in years is probably interval level. Do not simply assume that all variables are measured at the same level all the time. Check the specifics of the research before jumping to conclusions about the level of measurement.

It is also imperative that you look beyond the numbers in a frequency distribution and look at the underlying dimension to understand whether or not there are equal intervals. For example, if you are examining how long in months prisoners have been incarcerated, the underlying dimension is time in prison in months. In this case, the interval is always months, regardless of what actually is displayed in a frequency distribution. In measuring the taste of doughnuts on a scale of 1 to 10, though, the underlying dimension is taste. The interval is how one doughnut tastes compared to another. Here, the interval will probably be different. For example, the difference between a 5 and a 6 may be quite different than between a 1 and a 2, because when you move from a 5 to a 6, you are saying that the doughnut tastes good rather than just okay, so the interval that may take a doughnut from a 5 to a 6 is probably larger than to get from a 1 to a 2. The jump to get from a 9 to a perfect 10 may be the largest of all because it takes a great doughnut to be perfect. The bottom line here is that the intervals between different categories of taste are not the same, even though the difference in numbers is 1.

Finally, it is imperative to work through the process outlined in Figure 2-2 until a "no" answer is reached. Ordering must be shown for a variable before equal intervals matter. There are many variables that have a true zero but cannot be ordered. Having a true zero does not matter if the data cannot be ordered; it is still nominal level data. To state that a variable is measured at the ratio level, it is necessary to show all of the following:

1. The data can be ordered.
2. There are equal intervals.
3. There is a true zero.

Also, do not succumb to the temptation to find that a variable can be ordered and then quit, stating that it is ordinal. Find out first if there are equal intervals. If there are no equal intervals, the variable is ordinal; if there are equal intervals, the process continues.

Changing Levels

The level of data relies heavily on conceptual grounds, or how the data is used. For example, temperature could be measured by those living in the far north as either freezing or not freezing, which would be nominal level data. Others may talk of temperature as either warmer, cooler, or the same, which would be ordinal level. Generally, temperature is discussed in terms of degrees Fahrenheit, which is interval level; but in science a Kelvin scale is often used, which is ratio level. In criminological research, age is generally considered an interval level variable. If age is being used simply in its chronological context, it certainly is an interval level variable. If it is being used as a measure of maturity, however, not only may it not be an interval level variable, but there may be problems supporting age as a valid measure of maturity (see the later section on the use of variables in the research process).

As discussed in Chapter 1, statistical analysis generally requires the use of analysis procedures appropriate for the lowest level of data to be analyzed. For example, if only one variable in a study is nominal and the rest are interval, nominal level analyses would be required. When this occurs, it is sometimes advantageous to change the level of measurement for some of the variables. Most often, this is a process of converting higher levels of data to lower levels. For example, age, normally interval level, could be converted to ordinal level by categorizing it into age groups (0–13 years old, 14–21, 22–30, etc).

Other times, levels of measurement are changed for theoretical or methodological reasons. For example, asking people their exact income does not always result in an exact measure because people will guess, and asking the question in this way may cause them not to answer the question at all. Allowing respondents of a survey to choose an income category rather than write in their income may obtain a more accurate measure of income, and it may increase the response rate.

Although there are reasons to recode[5] variables to lower levels of measurement, it should be undertaken with caution and only for proper reasons. The reason to use interval and ratio levels of measurement is that it allows increased precision in the data. For example, there is less variation between data points in interval level data, and the equal intervals allow more precise estimations than do the unequal intervals of ordinal level data. On the other hand, it is sometimes necessary to recode a variable to get enough data to conduct any analysis at all. As discussed in Chapter 6, there are certain requirements for some statistical analyses that deal with the size of the groups being studied. Meeting this size requirement sometimes obligates a researcher to recode a variable to obtain a sufficiently large group. Typically, this involves combining higher order (interval level) data into a smaller number of categories.

Combining different levels of data raises another issue: When interval level data is categorized, is it still interval level or is it reduced to ordinal level? A general rule is that it is considered ordinal level because it is grouped into broader categories, often without equal intervals. In the example above, age probably would be considered ordinal because of the categorization and the unequal intervals. Even if there were equal intervals, the data would probably still be considered ordinal level because it is grouped into a few categories where it is more difficult to examine the true interval level relationship between the categories.

It is sometimes possible to change the data from lower to higher levels of analysis, but this should be approached cautiously and with a full understanding of the process. The researcher must be clear that such changes are not precise but only approximations. For example, the most common method of changing to a higher level of measurement is through correspondence analysis. Using this statistical procedure, data that is at the nominal or ordinal level can be analyzed at the interval level with certain assumptions and limitations. Dichotomizing a variable is another method of using nominal or ordinal level data for interval level analyses. This use has even been extended to the point that researchers will dichotomize a variable that has several categories and use them in separate analyses. For example, a researcher might take the categories of high, medium, and low and dichotomize them into high/medium, high/low, and medium/low. Although this procedure has gained widespread approval, it should be used very carefully because the overall procedure still violates the mathematical assumptions of interval level data.

The bottom line is that all data collection and research is an exercise in making trade-offs, and it is sometimes necessary to trade off the level of measurement to meet other methodological requirements. The best approach to take is to gather the data at the highest level. If the analysis or methodology then requires categorization that reduces the level of measurement, this can be done while retaining the data at a higher level for other analyses.

Scale Continuity

In addition to their level of measurement, variables can also be divided into continuous and discrete scales of measurement. Although not as vital to the selection and interpretation of statistical analyses as level of measurement, scale continuity is another method of classifying data, and is used in later discussions. At this point, only a brief discussion of the difference between discrete and continuous data is necessary.

Data is *discrete* if there are a limited number of values for a variable. For discrete variables, only integers (not fractions or decimals) are needed to label the categories. For example, the number of crimes committed is generally represented by integers: 1, 2, 3, and so on.

Continuous variables can have an infinite number of fractions between them. Age is a good example of a continuous variable. Depending on how fine the scale is set, for example, down to the minutes rather than years, it is not unlikely that 1000 different ages could be obtained from 1000 people in a survey. If one person is 20 years old and another person is 30, there are many ages that can fit between these two. Even if there are two people who are both 20 years old, one born in January and one born in December, there are still many people who have different ages between them. Even with two people born on the same day, there is the opportunity for others to be born between them by measuring the hours or minutes of the time of birth. As shown in this example, continuous data can be represented by a large number of values for any given variable. Some statistical procedures do not work well or are difficult to perform with continuous-level data. For example, the tables discussed in Chapters 7 through 10 get very complicated with continuous data because of the number of cells (boxes) involved in such an analysis.

As an example of determining scale continuity, if you collected 100 responses to a survey and then one more questionnaire was turned in, you could determine if there could be room for that response without duplicating a response already obtained. If not, if all possible categories had been taken, it is likely that the data is discrete. If, on the other hand, there were categories that had not been taken, the data might be continuous.

One final note about scale continuity. Be careful about identifying scale continuity. It is identified as the way the variable is *currently* being measured, not its *possible* measurement. For example, age can be discrete if it is categorized either in the data collection or in the analysis. Again, not that it will detract from the analysis, but proper recognition of the scale continuity is important in some decisions of which statistical analysis to use.

Use in the Research Process

The final way variables may be classified is by their use in the research process. Variables may operate in one of three ways in the research process: as a dependent variable, as an independent variable, or as a confounding variable.

Dependent Variable

In the research process, a dependent variable is one that is the focus of the research. Crime is the archetypical dependent variable in criminology and criminal justice research, although many others are possible. Dependent variables are thought to be influenced by other variables to behave in a certain way. For example, it is believed by many that the behavior of a person can be influenced through learning the behavior from his or her peers. In this type of research, the scores on the variable criminal behavior should vary as scores change on the independent variable(s), which might include frequency of contact and duration of contact.

It is sometimes difficult to determine the dependent variable, especially when someone else's research is being examined. It is sometimes helpful to see the dependent variable as your response if someone asked the topic of your research (or a term paper). You would probably respond to the question with something like "search and seizure" or "religion in prison." These, or the variables associated with these responses, are the dependent variables for that research.

Independent Variable

An independent variable is one that the researcher thinks is causing a reaction in the dependent variable, or it may explain why the dependent variable fluctuates from person to person. It is assumed that changes in the independent variable will usually precede any change in the dependent variable. In experiments, the independent variable is manipulated to determine the result on the dependent variable. For example, a researcher might make a rat's cage hotter to see if it reduces the time the rat spends on the running wheel. It is almost impossible to manipulate independent variables in this way in social science research because the researchers do not have control of the variables. As such, criminology researchers typically use variables from people's lives. Unemployment, education, and social disorganization are examples of independent variables used in criminological research.

An important issue that often gets overlooked in discussing independent variables is what actually constitutes an independent variable. It is common to see variables, often demographic variables such as age, sex, race, and others, included in research. Sometimes these are identified specifically as independent variables and actually are used that way in the research process. Sometimes these variables are called *control variables*.[6] Used thoughtfully and properly, these demographic variables are appropriate for research. Often, though, these variables are just "dumped" into the research with little thought, planning, or discussion. Take the variable *age*, for example. Many criminological studies include age as a variable. Indeed, there is a strong and sustained relationship between age and crime that has been supported by numerous studies. But is age really an independent variable? The answer is no. Why would researchers be interested in the elapsed time since birth? The reason is that age is actually a proxy for or an indicator of a more theoretically supportable variable, maturity. At least theoretically, an independent variable should be the cause of something in the dependent variable. The question to ask, then, is: Can *age* itself cause anything at all? The answer is, typically, no; age is more likely to be a proxy, and usually an imperfect proxy (some people never grow up), for something else, such as maturity, that should be included in the research. Using age as a proxy for a theoretically supportable variable often leads to imperfect measures; people will lie about their age and round off the years differently. This typically leads to an underestimation of the real variable. The point here is that many of the demographic variables included in research have no real place as a variable. If the researcher is careful to determine why a variable such as age is important to the research, he or she may be able to determine a more appropriate independent variable. That is the variable that should be used.

Confounding Variable

A *confounding variable* is not an independent or a dependent variable but a variable that influences the relationship between these two. It may represent one of several different relationships with the other variables. Two of those are discussed here: intervening variables and spurious variables.

An **intervening variable** is a variable that may be between the independent and dependent variable and is actually causing the change in the dependent variable. When intervening variables are accounted for in research, they may assist in drawing conclusions about the dependent variable (in effect, they become independent variables). Intervening variables become a problem, however, when they are not accounted for in the research. This often results in mistaken credit being given to the wrong independent variable. For example, a research project may examine the relationship between education and police use of deadly force (Fyfe, 1988). It may be hypothesized that the independent variable, *college education*, is causing reductions in the use of deadly force. This may not be the case, however; it may be that most of the college-educated officers participating in the study are a part of administrative units such as research and planning and therefore do not have the same chance to use deadly force as patrol officers. Here, the variable *duty position* is intervening between education and use of deadly force. The relationship between these variables is shown in **Figure 2-3**.

Spurious variables are similar to intervening variables in that they detract from the true relationship between independent and dependent variables, but spurious variables operate a bit differently. Spurious variables show a relationship because of a similar trend in both variables over time. They influence both the dependent and independent variables such that the relationship between the independent and dependent variables is

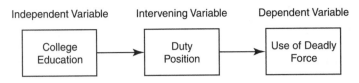

Figure 2-3 Intervening Variables

inflated. For example, in a research project examining poverty and crime, there are often very high correlations between income level and criminality. It is plausible, though, that it is not the mere lack of money that is causing the criminality. It is more likely that the poverty status of a person is one of many factors, such as where the person may live, opportunities, or discrimination, that may be causing the variation in the dependent variable. Each of these other factors are spurious variables. They are highly correlated with poverty status and highly correlated with crime; therefore, they inflate the relationship between the two. This relationship is shown in **Figure 2-4**.

Another difference between intervening and spurious variables is their inclusion in the research process, particularly the theoretical model. As stated above, intervening variables are sometimes included in the theoretical model. For example, in Figure 2-3, duty position is an intervening variable, but it could have been included in a theoretical model shown. Spurious variables, however, are almost never included in the theoretical model, or are included erroneously. For example, it can be argued that there is a relationship between the salaries of university professors and the price of Cuban cigars. Do university professors smoke so many Cuban cigars that their economic well-being drives that market? Of course not. Both of these variables are influenced by a third, spurious, variable: general increases in prices and earnings over time. Since both variables follow the same trends, they appear to be related, but they are not.

A final note about variable types: A particular variable can be a dependent variable in one study, an independent variable in another, an intervening variable in another, and a spurious variable in still another. For example, the education level of a correctional officer may be a dependent variable in studying the influence of management attitudes on increased education; an independent variable, such as education effects on inmate abuse; or as a confounding variable such as that between age and employment

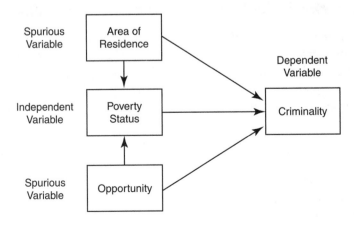

Figure 2-4 Spurious Variables

performance. You must be careful, therefore, about how you look at variables and make sure of the true role of a variable in a given study.

■ Conclusions

Variables are important to statistical analysis because they are the object of examination. The different roles that variables play determine how they will be examined. For example, in a regression (see Chapters 12 and 13), the dependent variable is placed in the equation first, typically followed by any controlling variables and then the independent variables. Confounding variables play a role in the research process in that the researcher must attempt to identify and control them. Even whether a variable is discrete or continuous will be important at later points in learning about and conducting statistical analyses.

Apart from the types of variables in the research process, measurement and the levels of measurement play a key role in statistical analysis. The importance of establishing the correct level of measurement has been stressed in this discussion. The level of measurement of variables determines what types of analyses will be conducted. Understanding how to determine levels of measurement will be essential in subsequent chapters.

■ Key Terms

completely exhaustive
confounding variable
continuous variable
dependent variable
dichotomized variable
discrete variable
distinct
fully ordered variable
independent variable
interval level data
intervening variable
level of measurement
measurement
mutually exclusive
nominal level data

nonparametric
ordering
ordinal level data
parametric
partially ordered variable
qualitative
quantitative
rank
ratio level data
robust
scale continuity
spurious variable
true zero
variation

■ Exercises

For each of the variables below: (a) discuss why it is distinct, mutually exclusive, and completely exhaustive or why it is not; (b) discuss whether the variable is discrete or continuous; (c) work through the process of determining the level of measurement, making sure to explain each step; (d) write the level of measurement for that variable based on your arguments; and (e) state how the variable would likely be used in the research process.

Variables in everyday life

1. paint color (red, blue, etc.)
2. miles per hour
3. weight (in pounds)
4. weight (underweight, appropriate, overweight)

5. shirt/blouse size (small, medium, large)
6. number of crimes committed
7. player's uniform numbers (00, 15, etc.)
8. signs of the zodiac (Aries, etc.)
9. undergraduate major (sociology, engineering, etc.)
10. political party preference (Democratic, Republican, etc.)
11. brands of light beer (Coors, Bud, Miller, etc.)
12. movie ratings (G, PG, R, X)
13. education (illiterate or literate)
14. occupation (type of industry: agriculture, construction, etc.)
15. education (years of schooling completed)
16. religion (Protestant, Catholic, Jewish, Muslim, etc.)
17. schooling (private or public)
18. grade in class (A, B, C, D, F)
19. schooling by type of diploma (elementary, high school, college)
20. street address numbers
21. size of car (compact, luxury, SUV, etc.)
22. number of people on a committee
23. type of workers (line or staff)
24. marital status (married or not married)
25. make of car (Chevrolet, Ford, etc.)

Variables from surveys

26. What is your marital status?
    (1) _____ never married
    (2) _____ married
    (3) _____ divorced
    (4) _____ separated
    (5) _____ widowed
27. What is the size of your community?
    (1) _____ rural (unincorporated areas)
    (2) _____ rural community (under 10,000)
    (3) _____ small city (10,000–49,999)
    (4) _____ medium-sized city (50,000–249,000)
    (5) _____ large city (over 250,000)
28. What is your social security number?

    _____ ___ _____
    Do you receive any of the following forms of financial aid?
29. Loan (1) _____ yes (2) _____ no
30. Grant (1) _____ yes (2) _____ no
31. How many hours did you transfer to State U. from another college or university?
    _____ hours
32. What is your overall grade-point average?
    (1) ___ under 2.00
    (2) ___ 2.00–2.49
    (3) ___ 2.50–2.99
    (4) ___ 3.00–3.49
    (5) ___ 3.50–4.00

33. What is your major? _____
34. Where do you live?
    (1) ___ residence hall
    (2) ___ fraternity/sorority house
    (3) ___ off-campus
35. What were your scores on the SAT/ACT test?
    _____
36. Do you consider yourself:
    _____ lower class
    _____ lower middle class
    _____ middle class
    _____ upper middle class
    _____ upper class
37. What is your age (in years)?
    _____

Variables from statistical output

38. JOB: What is your occupation?

Value Label	Value	Frequency	Percent	Valid Percent	Cumulative Percent
Professional	1	105	30.3	34.9	34.9
Clerical/technical	2	43	12.4	14.3	49.2
Blue collar	3	28	8.1	9.3	58.5
Retired	4	68	19.6	22.6	81.1
Housewife	5	16	4.6	5.3	86.4
Other (part-time)	6	28	8.1	9.3	95.7
Unemployed	7	13	3.7	4.3	100.0
Missing		46	13.3	Missing	
	Total	347	100.0	100.0	

39. HOME: Do you own or rent your home?

Value Label	Value	Frequency	Percent	Valid Percent	Cumulative Percent
Own	1	240	69.2	72.3	72.3
Rent	2	92	26.5	27.7	100.0
Missing		15	4.3	Missing	
	Total	347	100.0	100.0	

40. KIDS: How many children do you have?

Value Label	Value	Frequency	Percent	Valid Percent	Cumulative Percent
	0	107	30.8	33.1	33.1
	1	44	12.7	13.6	46.7
	2	75	21.6	23.2	70.0

Value Label	Value	Frequency	Percent	Valid Percent	Cumulative Percent
	3	39	11.2	12.1	82.0
	4	21	6.1	6.5	88.5
	5	15	4.3	4.6	93.2
	6	11	3.2	3.4	96.6
	7	3	0.9	0.9	97.5
	8	2	0.6	0.6	98.1
	10	1	0.3	0.3	98.5
	11	1	0.3	0.3	98.8
	12	2	0.6	0.6	99.4
	15	1	0.3	0.3	99.7
	16	1	0.3	0.3	100.0
Missing		24	6.9	Missing	
	Total	347	100.0	100.0	

41. ADDRESS: How long have you lived at this address?

Value Label	Value	Frequency	Percent	Valid Percent	Cumulative Percent
	0	20	5.8	5.9	5.9
	1	42	12.1	12.5	18.4
	2	18	5.2	5.3	23.7
	3	20	5.8	5.9	29.7
	4	14	4.0	4.2	33.8
	5	17	4.9	5.0	38.9
	6	21	6.1	6.2	45.1
	7	8	2.3	2.4	47.5
	8	10	2.9	3.0	50.4
	9	6	1.7	1.8	52.2
	10	13	3.7	3.9	56.1
	11	9	2.6	2.7	58.8
	12	10	2.9	3.0	61.7
	13	8	2.3	2.4	64.1
	14	5	1.4	1.5	65.6
	15	8	2.3	2.4	68.0
	16	8	2.3	2.4	70.3
	17	5	1.4	1.5	71.8
	18	2	0.6	0.6	72.4
	19	5	1.4	1.5	73.9
	20	9	2.6	2.7	76.6
	21	1	0.3	0.3	76.9
	22	6	1.7	1.8	78.6
	23	1	0.3	0.3	78.9
	24	3	0.9	0.9	79.8
	25	6	1.7	1.8	81.6
	26	1	0.3	0.3	81.9
	27	3	0.9	0.9	82.8
	28	3	0.9	0.9	83.7

Value Label	Value	Frequency	Percent	Valid Percent	Cumulative Percent
	30	7	2.0	2.1	85.8
	31	3	0.9	0.9	86.6
	32	4	1.2	1.2	87.8
	33	3	0.9	0.9	88.7
	34	1	0.3	0.3	89.0
	35	7	2.0	2.1	91.1
	37	1	0.3	0.3	91.4
	40	8	2.3	2.4	93.8
	41	3	0.9	0.9	94.7
	42	2	0.6	0.6	95.3
	43	3	0.9	0.9	96.1
	45	5	1.4	1.5	97.6
	47	2	0.6	0.6	98.2
	50	4	1.2	1.2	99.4
	51	1	0.3	0.3	99.7
	79	1	0.3	0.3	100.0
Missing		10	2.9	Missing	
	Total	347	100.0	100.0	

Other Variables to Ponder

Here are some variables whose level of measurement is often debated. See if you can determine the level of measurement and present arguments for why it should be a different level of measurement.

42. Calendar year
43. Hours in a day

Practical Exercise

44. Select three articles from a research journal in criminal justice or criminology. Identify and write down (a) the variables used in the research; (b) the level of measurement for each variable; (c) why you chose that level of measurement; (d) the scale continuity for each variable; and (e) use of the variable in the research process.

■ References

Abelson, R. P., and J. W. Tukey (1963). Efficient conversion of non-metric information into metric information. *Annals of Mathematics and Statistics*, 34:1347.

Caws, P. (1959). Measurement. In C. W. Churchman (ed.), *Measurement: Definitions and Theories*. Hoboken, NJ: Wiley.

Fyfe, J. J. (1988). Police use of deadly force: research and reform. *Justice Quarterly*, 5(2):165–205.

Stevens, S. S. (1968). Measurement, statistics and the schemapiric view. *Science*, 161(3844):849.

For Further Reading

Behan, F. L., and R. A. Behan (1954). Football numbers. *American Psychologist*, 9:262.

Duncan, O. D. (1984). *Notes on Measurement: Historical and Critical*. New York: Russell Sage Foundation.

Stevens, S. S. (1946). On the theory of scales of measurement. *Science*, 103(2684):677.

Stevens, S. S. (1951). Mathematics, measurement and psychophysics. In S. S. Stevens (ed.), *Handbook of Experimental Psychology*. Hoboken, NJ: Wiley.

Notes

1. In the current sense of political correctness, the term *gender* is frequently used as a description of whether a person is male or female. Biologically, however, the term *sex* is more accurate in this respect; whereas gender is more appropriate in examining masculinity/femininity. Throughout this text, the term *sex* will be used to distinguish whether a research subject is male or female.

2. Although there is no bright line rule about labeling dichotomized variables (which gets 0 and which gets 1, or whether to use 0, 1 or 1, 2), there are some advantages to a standardized system. For example, when using yes/no questions, it is often advantageous to set yes as 1 and no as 0. This way, the mean of the variable will be equal to the proportion (p) of yes answers in the data and the variance is $p(1 - p)$. This is particularly useful when the characteristic under study is set as 1. For example, if you are looking at the differences between male and female sentences in courts, you are really interested in how females are treated in relation to males. In this case, females would be given a score of 1.

3. For example, monotone and affine transformations are identical for dichotomized data and interval-level data.

4. To resolve another debate, people who knew Rensis Likert say he pronounced his name like licking a lollipop (Lick-ert).

5. *Recoding* is a procedure used in database management programs, statistical packages, and other software programs that allows the researcher to manipulate the data in some way, including categorizing it to a lower level.

6. A control variable is a variable "held constant" in an attempt to clarify a relationship. For example, we know there is a difference between males and females in their criminality. After preliminary analyses, then, the variable *sex* might be used to control for this difference, so the research would only compare males to males and females to females.

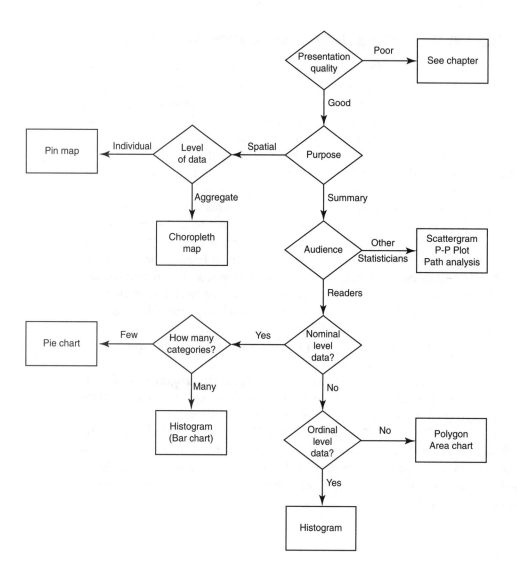

Understanding Data Through Organization

Summarizing Data in Tables, Frequency Distributions, and Graphical Representation

An important goal for statistical analysis is providing for others a clear understanding of the characteristics of a variable or the relationship between two or more variables. The majority of statistical analyses involve the manipulation and interpretation of numerical data. Statistical analysis is an effective method of summarizing variables and their relationships with other variables—effective for those who understand statistical analysis. For many, however, $r^2 = 0.79$, $p < 0.0001$ means nothing. To make the summarization of data more understandable to a wider audience, graphical representation of data is often used. This is especially effective in technical reports and other documents that may be a part of an evaluation or where the analysis will be provided to people who may not have a background in statistics. For example, some police chiefs are notorious for reading only the executive summary of a report and looking at the graphs before sending the report to an assistant chief or whoever does research and planning. If you are trying to get the chief to support a particular idea, you must put your most convincing information in the executive summary and in graphs.

There are two primary ways graphical representation is used in conjunction with statistical analysis: graphs (also known as charts) and maps. Although technically not the same as graphs and maps, tables can also be used effectively for graphical representation of data. Graphical representation of data is most often used in univariate analysis. As shown in **Figure 3-1**, this means that one variable is represented at a time: in this case, showing the difference between males and females in how they responded to a survey. As you will see, however, graphical representation can also be used effectively in bivariate (two variables) and multivariate (more than two variables) analyses.

The goal of graphical representation is to choose a graph that best summarizes the data without losing anything in the process. Some decisions concerning which method of graphical representation to use are relatively easy. For example, graphical methods typically associated with regression analysis include normal probability (P-P) plots and

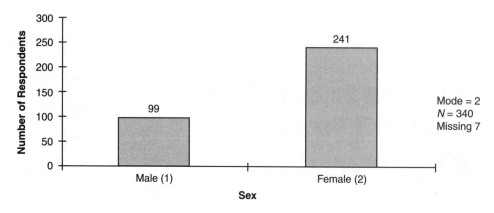

Figure 3-1 Sex of People Responding to a Survey

scattergrams. Other decisions are more difficult. For example, is it better to use a pie chart or a bar chart with nominal level data?

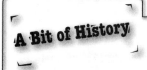

A Bit of History

Schmid (1954) traced the use of graphical representation back to 1786 and the publication of *The Commercial and Political Atlas* by William Playfair (1801). In this work and others, Playfair introduced and developed the basics of graphical representation, such as the pie chart and the bar chart.

Until relatively recently, charts and graphs were not created by researchers themselves, but were the work of draftsmen and architects. This was necessary because researchers and statisticians often did not have the drawing ability required for clear and informative graphs. The relationship between researchers and those skilled in creating graphs is illustrated by the fact that the first formalized standards for graphical representation were put forth jointly by the American Statistical Association and the American Society of Mechanical Engineers (Schmid, 1954). The necessity of others creating charts and graphs set a heavy price for the early use of graphical representation in presenting the findings of research. Nonetheless, the ease of interpretation inherent in graphical representation has caused researchers to turn to these methods for more than 200 years. The development of computer graphing programs and the graphing capabilities of spreadsheet and statistical analysis programs makes modern graphical representation so easy that no one should avoid the inclusion of graphical representation of analyses when appropriate.

Schmid and Schmid (1979) proposed that selecting appropriate graphical representation should be based on four criteria. The first criterion is the medium of presentation. Different methods of graphical representation work better than others, depending on how they will be displayed. For example, pin maps may not be appropriate where poor-quality copies will be used for presentation. The second criterion is the purpose of presenting the data in a graphical format. If the purpose is to show the spatial relationship of data, maps are a better method of graphical representation than charts or tables. The third criterion is the audience. If the primary audience is other researchers, charts that aid analysis may be more appropriate. If the audience has little or no statistical background, basic pie charts and bar charts probably are preferred. The type of data being represented is the final criterion in choosing a method of graphical representation. Nominal level data is best represented in graphs different from those suited for ordinal level data.

The purpose of this chapter is to present a foundation for creating and understanding graphical representation of data. After finishing this chapter, you should be able to understand which graphs or maps work best with a particular kind of data, and you should be able to read and interpret graphs that have been created by others, including understanding where their graphs are flawed. To that end, in this chapter we introduce the theory behind graphical representation and present generally accepted guidelines for determining which methods of graphical representation to use with particular data. The elements of graphs and procedures necessary to create effective graphs are also discussed.

■ Frequency Distributions: A Chart of a Different Color

Although technically a tabular rather than a graphic representation, **frequency distributions** represent the basic level of presenting statistical analyses in a form other than mathematical script or text. As such, frequency distributions are included in this chapter to show their usefulness in presenting findings, especially in preliminary univariate analysis.

Frequency distributions are helpful in summarizing data (scores) because they provide a visualization of how scores are spread over categories or scales. Frequency distributions are also an excellent method for checking data integrity. For example, if the variable **Arrested** is coded 1 for yes and 2 for no, and the frequency distribution displays a category labeled 21, there are problems with the data; either that particular case was coded wrong or there was a typographical error. At this point, you can return to the data and correct that mistake prior to conducting further analyses. Because of the ability of frequency distributions to summarize data for quick examination, they should almost always be completed as a first step in any statistical analysis.

There are many different types of distributions. The most common of these are frequency distributions, percentage distributions, and distributions that are a combination of each. A **combination distribution** from SPSS is shown in **Table 3-1**. This distribution displays frequencies, percentages (both total and valid), and cumulative percent.

Conventions for Building Distributions

Within statistical analysis, there are conventions and standards for presenting tables and distributions. You will soon come to realize that many people do not follow these conventions; however, if you adhere to these guidelines when building your own tables and distributions, they will be more clear and the results will be easier to interpret.

First, the title of the table or distribution should clearly explain the contents. It should contain the variables included in the table and any variable label or explanation of the variable that makes it clear what data the distribution contains. In the example

TABLE 3-1   Combination Table for Education from the 1993 Little Rock Community Policing Survey

What is your highest level of education?

Value	Value Label	Frequency	Percent	Valid Percent	Cumulative Percent
1	Less than high school	16	4.6	4.8	4.8
2	GED	59	17.0	17.6	22.3
3	High school graduate	8	2.3	2.4	24.7
4	Some college	117	33.7	34.8	59.5
5	College graduate	72	20.7	21.4	81.0
6	Postgraduate	64	18.4	19.0	100.0
	Missing	11	3.2	Missing	
Total		347	100.0	100.0	

in Table 3-1, the variable is *Education* and the explanation is the wording from the survey from which the variable was drawn: "What is your highest level of education?" If the table or distribution is a frequency or percentage distribution only, the title should include which measure is being used, frequency or percentage. In combination distributions, this requirement is often dropped because it would take too much space to identify all of the types of data being presented and because most readers can determine the type of distribution from the headings of the columns. If not expressed clearly in the discussion, the title should also contain the origin of the data. This may include the year the data was collected, the geographic region of collection, and any specific information about how the data was gathered.

Possibly more important than the title is proper labeling of the columns and categories. The categories of the data should be clearly identified in the left column, called the **stub** of the table or distribution. These categories should be labeled even if the values are included. In the example shown in Table 3-1, simply including the values 1 to 6 does not help the reader interpret the table unless he or she has the categories (less than high school, GED, high school, etc.) that correspond to the values. The words that represent the numbers in the stub are called the **value labels**.

Each of the columns should also be clearly labeled with **column headings**. This is especially important in combination tables or distributions where several columns contain different information. The headings included in Table 3-1 are the frequency, percent, valid percent, and cumulative percent. In this distribution, the column headings *frequency*, *percent*, *valid percent*, and *cumulative percent* are actually separate distributions. Each of these distributions is discussed below.

All relevant columns should be totaled. These totals are called **marginals**. This is important information for summary analysis, and prevents the reader from having to make independent calculations. It is also important to provide the reader with information on the way the table or distribution was developed. In Table 3-1, for example, the frequency and percentage totals are based on all 347 responses. Valid percentage and cumulative percentage, however, are based on the total minus the missing responses (336). Missing values are places where, for example, a person did respond to the survey but left one or more of the questions blank. In most research, especially survey research, respondents frequently do not answer all of the questions. When that occurs, there is missing data for a particular variable. There are two ways of dealing with missing data for a particular variable. Some researchers use a procedure whereby other values, such as the average, are substituted for the missing values. This results in no missing values for the variable, but some of the data values are not real responses. Other researchers simply exclude any missing cases from the analysis. In frequency tables, SPSS displays the number of data values that would be excluded for a variable. In Table 3-1, 11 data values would be excluded.

How do you do that?

Beginning in Chapter 4, there will be descriptions of how to perform the analyses discussed in this text in SPSS. This is meant to assist you in case you want to replicate or expand on what is in the book with analysis of your own. This procedure is begun here by listing the steps in SPSS, version 11.5, to obtain frequency distributions. Different versions of SPSS may have slightly different procedures, but these should be close enough for you to be able to work in just about any version.

1. Open a data set.
   a. Start SPSS.
   b. Select *File*, then *Open*, then *Data*.
   c. Select the file you want to open, then select *Open*.
2. Once the data is visible, select *Analyze*, then *Descriptive Statistics*, then *Frequencies*.
3. Make sure the "Display Frequency Tables" is checked.
4. Select the variables you wish to include in your distribution and press the ▶ between the two windows.
5. For now, do not worry about the options boxes at the bottom of the window and just press *OK*.
6. An output window should appear containing a distribution similar in format to Table 3-1.

Finally, footnotes should be added to tables and distributions when more information is needed, when clarification is necessary, or when there is something unusual about the table or distribution. If not included in the title, the source of information should be included in a footnote. This is particularly important when the data in the table or distribution is not collected by the presenter. For example, if information is drawn from the *National Crime Victimization Survey* (NCVS) and presented in a table, a footnote should be included that reports the source of the information. The method of computing results other than frequencies and percentages is also typically included in a footnote.

Frequency Distributions

Although nearly all statistical analysis programs will create frequency distributions, it is helpful to know how they are created and formatted. In their most basic form, frequency distributions are created simply by listing the category and tallying the number of observations fitting that category. In Table 3-1 the categories represent the operationalized variable *Education* and the concept of the educational level of the respondents. In statistical notation, these categories would be the X in a formula.

Notice that each of these categories is distinct from every other; it is easy to tell the difference in diplomas between a person with less than a high school degree and a person with a postgraduate degree. They are also mutually exclusive; a person with a highest degree at high school would not also have the highest degree at college. Finally, they are completely exhaustive; everyone should be able to select one category. The categories can also be ordered from less than a high school degree to a postgraduate degree. There are no equal intervals; however; the interval between a GED and a high school graduate is not the same as the interval between having some college credits and being a college graduate. Taken together, it means this variable is considered ordinal level data.

In the next column of Table 3-1 is the frequency associated with that category. This column is simply the tally associated with that category; that is, there were 16 respondents with less than a high school degree, 8 with a high school degree, and so on. This column of numbers represents the f (frequency) in statistical notation. For any given category, the frequency is called the *cell frequency* for that category, and is represented by the notation f_i, where i represents the category for that frequency. Take, for example, the cell frequency for f_1 ($f_{\text{less than high school}}$). The total of this column is the Σf or N for that variable.

Constructing a table like this is simple. With a limited number of values, a table could be created simply by tallying the values and placing them in a table. Take, for example, the following values[1]:

F, S, J, G, F, F, G, G, G, S, J, J, G, G, G, G, F, F, S, S, S, S, F, J, J, F, G, S, F, F, G, J, J, J

If these were tallied, they would represent a table that looked like the following:

Freshman	✚ ||||
Sophomore	✚ ||
Junior	✚ |||
Senior	✚ ✚|

These tallies could then be put into a frequency distribution, as shown in **Table 3-2**.

Frequency Distributions for Grouped Data

For nominal and ordinal level data, particularly partially ordered data, setting out the categories is easy: Simply make a category for each value. With continuous-scale, interval and ratio level data, however, the distinction is not always so easy. If a survey

TABLE 3-2 Frequency Distribution of Student's Class Standing*

Value Label	Value	Frequency
Freshman	1	9
Sophomore	2	7
Junior	3	8
Senior	4	10
Missing		0
Total		34

asks someone her age and she responds, "20," this can easily be put in a table as 20. What if the person was asked how old she was to the nearest month? How should that person's age be tallied? You can easily see the problem of fractions and when to count a person a certain age based on whether he or she is 19½ or 20½. This creates a particular problem in grouping the data. How old does a person have to be to fit into the 20- to 25-year-old group? What if a person is 19½, 19¾, or 19 years and 364 days? The answer lies in the rules for grouping data in frequency distributions.

When grouping data in frequency distributions, or when using frequency distributions with continuous data, the **real limits** of the scores are typically used. The real limits of whole numbers lie from one-half digit below the value to one-half digit above. For example, the real limits for a 20-year-old person would be from 19.5 to 20.5. The 19.5 would be the **lower limit** of the category (symbolized as L_i), while the 20.5 would be the **upper limit** (symbolized as U_i). The category would be referred to by its midpoint (20), also symbolized by X_i. This will become important when discussing univariate statistics such as the median, where the formulas are different for individual and grouped data. The final piece of information important for grouped, continuous data is the *class width* or **interval width**. The class width is symbolized by W_i and represents the interval associated with that category. Class width is calculated as follows:

$$W_i = U_i - L_i$$

For age, used above, the class width is 1 year (19.5 to 20.5). For grouped data, the class width might be larger. For example, the class width for the class 20 to 25 would be 6 (25.5 − 19.5). As an example, look at **Table 3-3**. This table contains data on a group

TABLE 3-3 Frequency Distribution for Grouped Data

Value	Real Limits	Midpoint, X	Frequency, f
31–35	30.5–35.5	33	2
26–30	25.5–30.5	28	3
21–25	20.5–25.5	23	4
16–20	15.5–20.5	18	5
11–15	10.5–15.5	13	4
6–10	5.5–10.5	8	3
1–5	0.5–5.5	3	2
			N 23

of people ranging in age from 1 to 35. The real limits of the age run from one-half digit below the lower value to one-half digit above the upper value, that is, 6 months before the actual birth date of the lower value to 6 months after the actual birth date of the upper value. For example, in the class 26–30, people who are 25 years and 6 months old up to people who are 30 years and 6 months old would be counted. The lower limit (L_{26-30}) would be 25.5, and the upper limit (U_{26-30}) would be 30.5. The class width (W_{26-30}) in this case would be 30.5 minus 25.5, or 5. The midpoint of each class is calculated by adding the upper limit and lower limit together and dividing by 2. The midpoints will represent the X in further calculations.

Percentage Distributions

Percentage distributions are created by dividing each **cell frequency** (f_i) by the total number of cases (N) and then multiplying by 100. Percent distributions are an important addition to frequency distributions because they provide the relative weight of a particular category to the variable. For example, in **Table 3-4** it is difficult to tell if eight respondents in the "junior" category is a large or small portion of the total. The percentage distribution, however, shows that this category contains 23.53% of all respondents. This value was obtained by taking the value of the cell (8) and dividing it by the total of respondents (34). Although this is just a little less than 25% (which is what you would expect if the school was evenly divided among the grade levels), it is the second-lowest value in the distribution (there are more seniors and freshmen than juniors).

Percentage distributions also allow researchers to make comparisons between similar data. For example, if data is drawn from the *Uniform Crime Reports* (UCR) for juvenile crime and aggregated (grouped) into Part 1 and Part 2 crimes (see **Table 3-5**), it would be difficult to determine the relative contribution of each to overall crime in a state based solely on the raw numbers. If percentages are developed, however, it would be possible to compare the two groups. As shown in Table 3-5, Part 1 offenses represented about 40% of total juvenile crime (high of 44%, low of 39%), while Part 2 offenses represented about 60% of total juvenile crime (high of 61%, low of 56%).

Percentage distributions may also include **valid percentages** created by dividing the cell frequency by N minus the number of any missing cases (data points). Using the example in Table 3-1, valid percentages can be calculated by dividing each cell frequency by N (347) minus the missing values (11). Each cell, then, would be divided by 336. For example, the *less than high school* category had 4.6% of the values in the distribution. If missing values are not counted, however, the category contains 4.8% of the values in the distribution, as calculated in the equation $16/336 = 0.048$.

TABLE 3-4 Frequency Distribution of Student's Class Standing

Value Label	Value	Frequency	Percent
Freshman	1	9	26.47
Sophomore	2	7	20.59
Junior	3	8	23.53
Senior	4	10	29.41
Missing	.	0	Missing
Total		34	100

TABLE 3-5 Trends in Juvenile Crime in Arkansas, 1980–1990

Year	Part 1	Percent	Part 2	Percent	Total
1980	5265	0.39	8190	0.61	13,455
1981	4810	0.40	7350	0.60	12,160
1982	4937	0.41	7200	0.59	12,137
1983	4680	0.41	6790	0.59	11,470
1984	4940	0.43	6650	0.57	11,590
1985	4600	0.40	6912	0.60	11,512
1986	5256	0.43	6930	0.57	12,186
1987	5200	0.44	6650	0.56	11,850
1988	5600	0.44	7029	0.56	12,629
1989	5460	0.44	7020	0.56	12,480
1990	5610	0.44	7260	0.56	12,870

Source: Federal Bureau of Investigation (1980–1990).

The final type of percentage distribution is the **cumulative percent** distribution. This distribution calculates the percent contribution to the total N for a particular category and all those below it. This assists researchers by making it easy to tell where certain critical points are in the distribution. As shown in **Table 3-6**, the cumulative frequency can be calculated for each category. In the first category, <u>Freshmen</u>, the cumulative percent is the same as the percent because there are no values below *Freshmen*. In the second category, <u>Sophomores</u>, the cumulative percent is 26.5 + 20.6, which equals 47.1%. This process continues until the final category, <u>Seniors</u>, where the percentage should equal 100%.

A word of caution about percentage distributions. Always make sure you know on what the percentages are based. In **Table 3-7**, the results of a survey seem to indicate strong support for increasing the number of law enforcement officers. Closer examination of the table, however, reveals that the majority of respondents did not answer the question and that the findings are only based on the responses of five people.

TABLE 3-6 Frequency Distribution of Student's Class Standing

Value Label	Value	Frequency	Percent	Cumulative Percent
Freshman	1	9	26.5	26.5
Sophomore	2	7	20.6	47.1
Junior	3	8	23.5	70.6
Senior	4	10	29.4	100
Missing	.	0	Missing	
Total		34	100	

TABLE 3-7 Misleading Percentage Distribution

Do you favor more law enforcement?

Response	Frequency	Percent
Yes	4	80
No	1	20
Missing	147	Missing
Total	152	100

Combination Distributions

The most common type of distribution for current statistical packages is a *combination distribution.* For example, SPSS includes frequency, percentage, valid percentage, and cumulative percentage distributions in the standard distribution. Combination distributions provide the researcher with the most extensive data on a particular variable because he or she can examine the frequency, percentages, or cumulative percentages in one distribution. This is the type of distribution shown in Table 3-1.

■ Graphical Representation of Frequencies

Although frequency tables and distributions are effective methods of presenting data in summary form, large frequency tables are often too complex for easy examination of data. Furthermore, the mathematical format of frequency tables decreases the effectiveness of data presented in this way if the reader is truly a statistical novice or has a fear of numbers. The simplest and often clearest method of presenting data, especially univariate data, in a graphical format is through the use of graphs.

Pie Charts

The most basic type of graph is a **pie chart**. A pie chart is best used for nominal level data where there are a small number of categories. The pie chart is good for this purpose because it is so easy to interpret. The pie depicts N, or 100% of the frequencies, so the relationship between the categories of the variable can be represented visually by the size of each slice of the pie.

Pie charts should be used only with discrete, nominal or ordinal level data with fewer than five categories. Nominal and ordinal level variables are best represented with a pie chart because of the obvious breaks between categories, whereas with continuous (interval and ratio level) data, the breaks would have to be forced, possibly misleading the reader concerning the nature of the data. Furthermore, variables with only a few categories should be used because pie charts with more than five categories begin to look cluttered.

In the example in **Figure 3-2**, respondents to the Little Rock Community Police Survey were asked if they had been a victim of a crime in the previous 12 months. To prevent respondents from having to estimate exact times, they were allowed to choose the approximate time of the crime from categories matching the patrol shifts of the police department. From this pie chart it is easy to see that most criminal victimizations occurred between 3 and 11 p.m., the next most victimizations occurred from 11

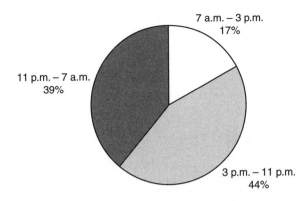

7 a.m. – 3 p.m.
17%

11 p.m. – 7 a.m.
39%

3 p.m. – 11 p.m.
44%

Figure 3-2 Pie Chart of Time Crime Occurred for Respondent Victims of Crime

p.m. to 7 a.m., and the fewest victimizations occurred between 7 a.m. and 3 p.m. This pie chart allowed for quick interpretation of the data with little mathematical skill required.

Be careful when using pie charts to show comparisons between different variables or when several pie charts will be used in close proximity to each other. If pie charts are used where the N is not the same, although the pie pieces and percentages will appear similar, they will not be based on directly comparable data. This may misrepresent the information to readers or cause them to draw erroneous conclusions. To someone not accustomed to analyzing information from graphs, pie charts that are close together give the impression they are based on similar information such as equal N's. To avoid confusion, only use pie charts to compare variables when they have equal N's. If graphing nominal level variables in close proximity to each other is unavoidable, or if the goal is to show comparisons on similar variables, the differences in N should be clearly noted either in the title or in a footnote to the chart. Otherwise, it is advisable to use a different chart, such as a histogram, for some of the data, to show that they are not directly comparable.

Histograms and Bar Charts

The next most basic chart is a **histogram**, a term that is actually a name for a family of similar charts that includes histograms, bar charts of several different varieties, Burchard charts, and others. This family of histograms is the graphical representation of a frequency distribution where categories of a variable are plotted along one axis, and responses are plotted as bars on the other axis. The bars in a histogram are plotted so the length of the bar represents the frequency of that category.

Histograms are generally used with ordinal level data, with nominal level data (in the form of a bar chart) with several categories, and when including more than one variable in the same chart. Histograms may also be used with interval or ratio level data when there are only a few categories. Histograms are not typically used with continuous, interval or ratio level variables because there are probably too many bars to present the data clearly. For continuous, interval level variables, polygons or area charts are more appropriate.

Constructing a histogram is essentially the same as stacking blocks (or beer cans, depending on your orientation). Think of having one block for each respondent, with

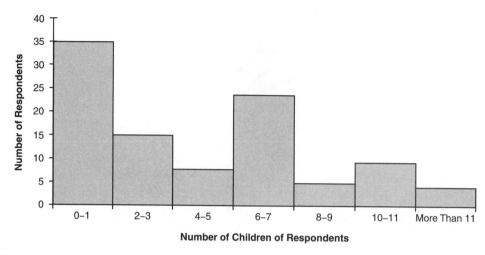

Figure 3-3 Histogram of Distribution of Children in Little Rock Community Survey

each block labeled for the score of the respondent. If you stack together all of the blocks labeled 0–1, all the blocks labeled 2–3, and so on, the result would look much like the graph in **Figure 3-3**. The Y axis, which is the vertical axis of the histogram, would be like setting down a scale beside the blocks to measure how many blocks were in each stack.

In Figure 3-3, the number of children for each respondent in the community survey is graphed. As shown in the graph, the majority of respondents had either one child or no children. Here, the bars represent 0 or 1 child, 2 or 3 children, and so on. The bar for the 0–1 category shows that 35 responses were in this category—more than any other category. Each of the other categories has bars that correspond to the number, or frequency, of responses in that category. Notice in this figure that the bars are not separated. Not separating the bars is typical for ordinal and categorized interval level variables to show the continuity of the data. With nominal level variables, the bars should be separated, creating a special type of histogram called a bar chart (see Figure 3-4 below).

There are conventions and standards that should be followed when constructing a histogram. First, the responses are generally placed on the **X axis**, which is the horizontal axis. The exception is a horizontal bar chart (discussed below). The second convention concerns placing the responses. If the responses are nominal, they should be placed in the most logical format; if they are ordinal or categorized interval, they should be placed in ascending order from left to right. Also, if the variable is nominal, there is typically a separation between the bars to denote that the responses are not linked or ordered. Another convention is to begin numbering the **Y axis** with zero and to use equal intervals throughout the scale. The X axis also generally begins at zero, but may begin at any number or text label as long as the unit of measurement is clear. As with a pie chart, labeling is always necessary. In addition to the title, always label the X and Y axes so the reader understands what is being represented. Also, if there is more than one set of bars, a **legend** should be added to show what each set of bars means (see Figure 3-4). Histograms should be built based on equal class intervals if used for grouped data. When class intervals are different, it is difficult to make comparisons between groups. For example, it would be difficult to interpret a graph where the ages

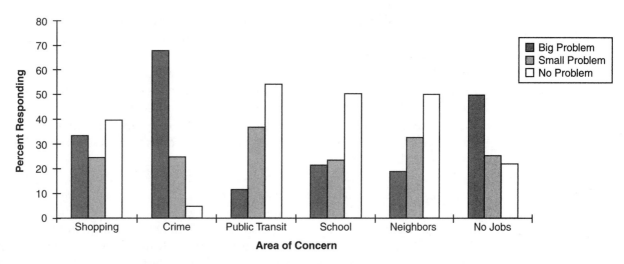

Figure 3-4 Multiple Bar Chart of Responses to Question About Greatest Concerns of Respondents

of responses were in the form 1–10, 20–50, and 50 and older. Either the categories were created artificially because the *N*'s were similar (which does nothing for interpreting the data), or the *N*'s will be so different that interpretation will be difficult. Only if there is a valid methodological or theoretical reason to use categories with unequal intervals should they be employed. Next, the total frequency represented by the bars should equal *N*. Finally, if there is anything out of the ordinary, make sure to include it in a footnote.

If a variable is nominal level, it is sometimes advantageous to use a variation of a histogram called a **bar chart**. In a bar chart, the bars of the histogram are separated to denote the division between the categories or variables, and as discussed above, a separator line is used to denote the lack of continuity between responses. Bar charts may be oriented either horizontally or vertically (see **Figures 3-4** and **3-5**).

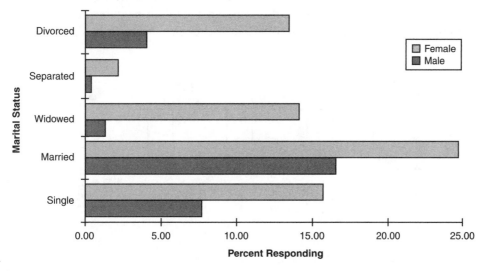

Figure 3-5 Horizontal Bar Chart of Marital Status Differences Between Male and Female Respondents

Figure 3-4 shows six different variables on one histogram (vertical bar chart). You can tell this is probably nominal-level data because this is a bar chart—the bars are separated and there is a separator line between the bars, denoting that the variables are not ordered. For each of these variables, the bars show how many respondents thought the issue was a big problem, small problem, or no problem. As expected, crime showed the greatest number of respondents stating it was a big problem and the fewest stating it was no problem. For the variable *crime*, almost 70% of respondents thought it was a serious problem, about 30% thought it was a small problem, and less than 5% thought it was no problem at all.

Nominal level variables are often placed in a horizontal bar chart, as in Figure 3-5, to further emphasize the nominal nature of the variable. The variable *marital status* in Figure 3-5 is nominal because the various categories of marital status generally cannot be ordered. As such, a horizontal bar chart was used to represent the nominal nature of the data. The representation of the data is the same as with a vertical bar chart. In this case, there was a higher percentage of respondents, both male (16%) and female (25%), who were married than in any other category. The percentage of females in all other categories was about the same, while males varied quite a bit between those who were single and those who were separated.

A variation of the bar chart is an **offset Burchard chart**. In this method of graphical representation, responses are dichotomized, with one response on one side of the axis and a second response on the other. The Burchard chart can be oriented either horizontally or vertically. This is often used for positive and negative responses or where there are some negative numbers. In **Figure 3-6**, a Burchard chart is used to show the dichotomized attitudes of residents concerning the safety of their neighborhood, the city, and the United States. The percentage of respondents who stated that the neighborhood, city, or United States was safer is shown above the *X* axis, and the percentage of those who felt safety had diminished is shown below the *X* axis. It is easy to see the dichotomy of responses in a Burchard chart, but the magnitude of differences is more difficult to determine than with a pie chart or histogram. Here, almost 80% of respondents felt safety in the United States is becoming worse. Contrary to popular belief, the greatest number of respondents felt their neighborhood safety was becoming better (almost 20%), with only a little over 40% feeling it had gotten worse.

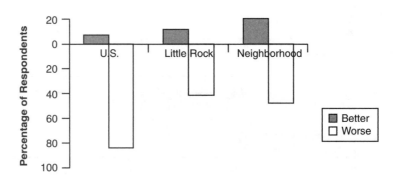

Figure 3-6 Burchard Chart of Attitudes Toward Neighborhood Safety

● Polygons and Area Charts

As stated above, although it is possible (and acceptable) to use histograms (not bar charts) with interval and ratio level data, these variables are generally best represented graphically by using a polygon or area chart. A *polygon* or **area chart** is used for continuous data (or possibly discrete data with many categories) instead of a histogram or pie chart for two reasons. First, there are often too many categories to display effectively on a histogram or pie chart. Also, the flow of the line of a polygon or area chart across the midpoints gives the visual impression of continuity and gradual shifts between data points that is characteristic of continuous and interval or ratio level variables rather than abrupt shifts between categories for nominal and ordinal level data.

A **polygon** is constructed essentially the same way as a histogram, with two exceptions. First, the points for a polygon are plotted above the midpoint of a category corresponding to the frequency of the category, whereas in a histogram, the box covers the interval of the category completely. Also, a polygon is a closed figure in that the line either touches the X axis at both ends or touches the Y axis on the left side and the X axis on the right side.

Polygons are helpful in two types of analysis. First, the line of the polygon can be used like any of the previous methods of graphical representation to examine the distribution of scores. A polygon is particularly useful, however, because it can be used to examine the extent to which the data conforms or deviates from a normal curve. As you will learn in Chapter 4 and beyond, linearity of data is important to proper statistical analysis. One way of visually establishing the normality of data is by comparing it to a normal curve, which can be approximated with a frequency polygon.

Figure 3-7 is a frequency polygon. Although this data appears to be ordinal since it appears to have larger categories, it is actually continuous interval level data where the graph has been drawn with larger intervals between the data points on the X axis. Here, each value has its own section of the X axis (even though the points between the categories shown are not displayed). Although it does not show on the chart, the line of the polygon runs through the midpoint of each category at a place representing the frequency of occurrences in that category.

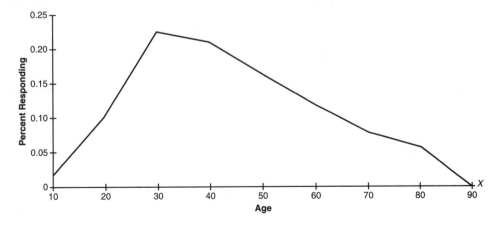

Figure 3-7 Frequency Polygon of Respondent Ages

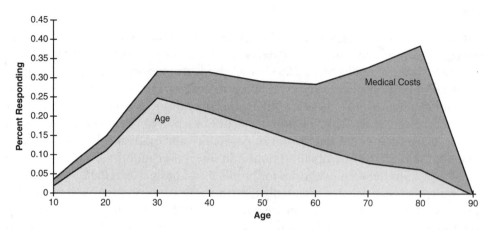

Figure 3-8 Area Chart of Age and Medical Costs

As shown in Figure 3-7, there were only a few people less than 20 years old who responded to the survey. The majority of people were in their 30s, 40s, or 50s, and there were even a few people 80 years old who responded. The portion of the polygon that extends beyond the over 80 category is created by including a zero-frequency category. This is used to enclose the polygon. We can also learn from this polygon that the data is generally normal in nature but that it has a slight positive skew (see Chapter 6). Although univariate analyses would have to be conducted prior to any analyses requiring normality, the appearance of this polygon would support an initial conclusion that the data is relatively normally distributed.

An *area chart* is a variation of the polygon in which the polygons of two variables are plotted such that the differences and overlap between the two can be shown visually. The only real difference between an area chart and a polygon, in addition to the fact that an area chart typically deals with two variables and a polygon deals with one variable, is that an area chart generally is filled in whereas a polygon is not.

One of the original reasons to use an area chart was the ability to illustrate and calculate the overlap of the variables. In **Figure 3-8**, for example, if the area covered by medical cost and the area covered by age were calculated, it would show that there is about a 55% overlap between the two variables. This can translate into a correlation between the two variables that represents a preliminary bivariate analysis. When all statistical analyses were done by hand, it was sometimes easier to calculate a difference in area than to do the analyses. The speed and ease of computerized statistical analysis programs have made the necessity of area charts for bivariate analysis obsolete, but area charts remain effective as a visual demonstration of the relationship between two variables.

■ Analyzing Univariate Statistics

The basic charts and graphs discussed above essentially focus on graphically displaying frequency tables of variables. The variables are represented graphically by the frequency of categories or by the percentage of contribution a category made to the variable. This information is appropriate for getting a message across to a reader, but statistical analysis requires more information than simple frequencies. One of the first ways that graphical representation can be used to assist in statistical analysis is by

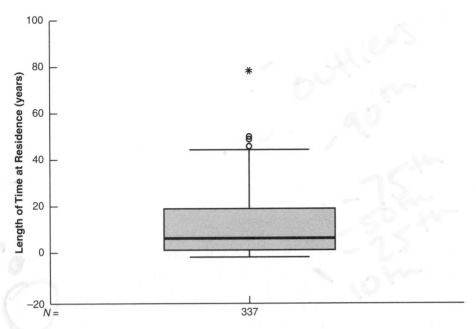

Figure 3-9 Box and Whisker Graph of the Length of Time Respondents Lived at Their Present Address

graphing univariate information about a variable. This can typically be completed using a box and whisker graph (**Figure 3-9**).

Box and whisker graphs are a method of showing the measure of central tendency and measure of dispersions of a distribution. Box and whisker graphs to show univariate measures of central tendency and dispersion were first suggested by John Tukey in *Exploratory Data Analysis* (1977). More recent computerized box and whisker graphs will also display the location of *outliers* in the distribution, represented by the circles and star in Figure 3-9.

In most box and whisker graphs, the box displays the **interquartile range**, which consists of the data values between the 25th and 75th percentiles of the distribution (some statistical packages use standard deviations to represent the box). Fifty percent of all the values for the variable fall within the boundaries of this box. Within the box, the heavy line generally represents the 50th percentile or the midpoint of the distribution. The whiskers of the graph typically extend from the 10th to the 90th percentile of the distribution, although newer versions of SPSS are using the more ambiguous 1.5 box lengths from the 75th percentile.

In the box and whisker graph in Figure 3-9, the midpoint of the distribution is at the 8-year mark. The box covers those having lived at that particular address for from 3 to 20 years. The whiskers of the distribution are at 1 and 45 years. Also notice there are several **outliers** in the distribution. These represent the seven respondents who had lived at their address for between 47 and 51 years (between 1.5 and 3 box lengths from the 75th percentile) and the one person who had lived at her address for 79 years (more than 3 box lengths from the 75th percentile). Also notice that the heavy line representing the median value is not in the center of the box. This means the data for this variable is positively skewed (see Chapter 6).

■ Analyzing Change

The majority of graphs are used to display variables graphically at one point in time. There are ways, however, to graph a variable over time. Two charts that are useful for this purpose are line charts and ogives.

Line Charts

Line charts are very common in the media, especially newspaper stories, because they can be used to show change or trends over time, such as changes in the stock market. Line charts are used less often in graphical representation of statistical analysis and academic research. Line charts are similar to polygons except they are not enclosed at the margins of the graph. Line charts may be used to show trends in one variable or in plotting two or more variables on one graph.

The line chart shown in **Figure 3-10** is typical of line charts used in criminology and criminal justice, showing a variable distributed by years. This type of chart, sometimes called a *trend line,* is used to show the changes that have taken place in a variable over the course of time. The line chart in Figure 3-10 shows trends of juvenile crime over a 10-year period. What can be seen from this chart is that juvenile crime has not made the tremendous jumps that some cities or states have experienced and that has been reported in the media. The bad news this graph shows is that while Part 2 juvenile crimes dropped in the early 1980s and then remained relatively stable, Part 1 crimes rose over this 10-year period. The line chart displays these changes clearly and in one graph.

Ogives

Ogives show the cumulative frequency or cumulative percentage of a variable. In its basic form, an ogive is much like a line chart, showing the change in frequency over categories of a variable. Ogives represent cumulative frequency well but not much better than can easily be shown in a frequency or percentage distribution or simply stated in a description of the variable. A better use of an ogive is to show the cumulative frequencies or cumulative percentages of a variable as they extend over time (see **Figure**

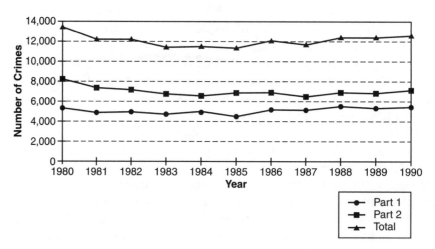

Figure 3-10 Line Chart of Juvenile Part 1 and Part 2 Crimes in Arkansas, 1980–1990
Source: Federal Bureau of Investigation (1980–1990).

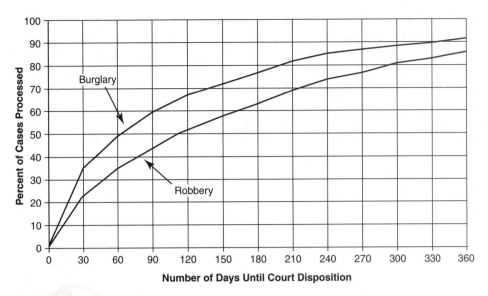

Figure 3-11 Ogive Chart of Time from Arrest to Conviction for Burglary and Robbery

3-11). An ogive is interpreted by looking at where each data point is for each time period. Each of the data points in an ogive moves up to represent change in cumulative frequency and moves to the right to denote a change in time.

In Figure 3-11, the number of days from arrest to court disposition is plotted for robbery and burglary. Although a trend line or bar chart could be used to represent the data, an ogive shows clearly how long it takes for each type of case to make it through the court system. For example, the greatest number of convictions within 30 days is for burglary (probably because of plea bargaining). It can also be said that a higher percentage of burglaries than robberies are cleared within 360 days.

■ Analyzing Bivariate and Multivariate Data

Graphical representation does not end at the display of a single variable or variables. Graphs can also be useful in analyzing bivariate and multivariate relationships, even beyond what can be displayed in an area chart or line chart. With the exception of path diagrams, most of the methods of graphical representation discussed in this section are used by researchers primarily to aid statistical analysis, and rarely are they included in any reports or publications.

Scatter Plots

Scatter plots (also called *scattergrams* or *scatter graphs*) are used most often in correlation and regression analyses. Scatter plots assist in determining visually if the two variables are associated. Scatter plots may also show the nature of the relationship. For example, a scatter plot may assist in determining if there is a curvilinear relationship that may be adversely influencing the correlation. Scatter plots may also assist in determining if there are any outliers that may be disrupting the distribution.

There are actually two different types of scatter plots, each with totally different characteristics and with completely different methods of interpretation. In a normal bivariate scattergram it is best to have the points in a relatively uniform distribution; a linear distribution of points is best. Regression analysis also uses scatter plots, also called **residuals plots**. In a residuals plot, it is best if there appears to be no pattern of the points. You must be careful, then, when examining scattergrams to determine the type of scatter plot and not to confuse the scatter plots discussed here with the residuals plot from regression analysis. To construct a scatter plot, each point, or corresponding set of scores, is plotted on the graph based on the scores for that point. For example, if a person who is 14 years old has committed 10 crimes, this person would be plotted at the 10 on the *X* axis and the 14 on the *Y* axis, as in **Figure 3-12**.

In research, scatter plots may have from a few points to thousands of points. Most statistical packages have even created methods of showing where many points may overlap. It is the distribution of these points that makes scatter plots an effective method of preliminary analysis; and it is the location of points standing alone that helps in analyzing outliers in a data set. An example of an outlier would be the lone point between 14 and 16 in **Figure 3-13**. As shown in the figure, there is a definite pattern to the relationship between age and number of children. This relationship is not linear, however. The vast majority of respondents at all ages have no children, and the number of children decreases in an almost triangular pattern, seemingly independent of age. This is to be expected, however, in that a person with 16 children cannot be expected to be less than about 30 years old.

Scatter plots that researchers most want to see are those showing a linear relationship between the variables. These relationships may either be positive (as one variable increases in frequency, the other variable increases with it) or negative (as one variable increases, the other variable decreases). These two instances are shown in **Figure 3-14**. It is important to note that these are not perfect associations (all points on or very near the line). Perfect associations rarely occur in social science research, so you should become familiar with scatter plots that are less than perfect, such as those shown here. Also note the line in these charts. This is a regression line and is an important part of statistical analysis, as discussed later in the chapters related to regression.

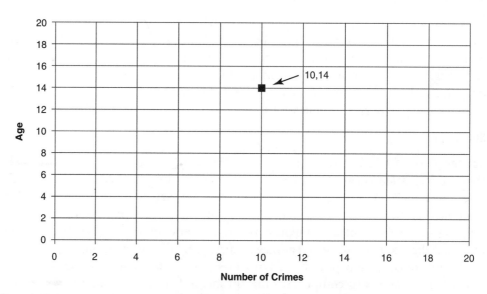

Figure 3-12 Sample Scatter Plot of a Single Point from a Frequency Distribution

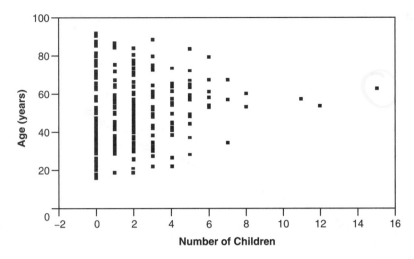

Figure 3-13 Scatter Plot of Relationship Between Age and Number of Children

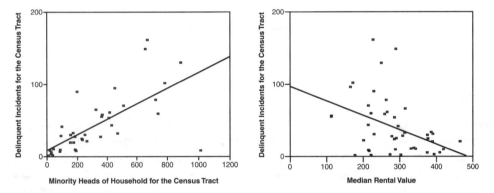

Figure 3-14 Scatter Plots of Positive and Negative Associations

Normal Probability Plots

Normal probability plots, also known as *P-P plots*, are used in higher-order, parametric analyses such as multiple regression. Normal probability plots are another method of determining if a distribution is normally distributed. A normal probability plot is created by making a scatter plot of the scores of a variable and the expected value of that score. If the data is normally distributed, the expected values of a score should match fairly well the observed values of that score. If that is true, the scatter plot should be represented by a fairly straight line. In **Figure 3-15**, the data is in a fairly straight line. In social science research, it is almost impossible to find data that are exactly normally distributed, so you should not expect to see a straight line in any analyses you perform.

The normal probability plot is most effective at showing when data substantially deviates from a normal distribution and must be transformed prior to any further analyses. Obtaining a plot such as the one in Figure 3-15 tells the researcher that the data is fairly normally distributed and that transformation is not yet called for. It is still necessary, however, to conduct exploratory analyses on the data to determine the actual deviation from normality and whether transformation is necessary.

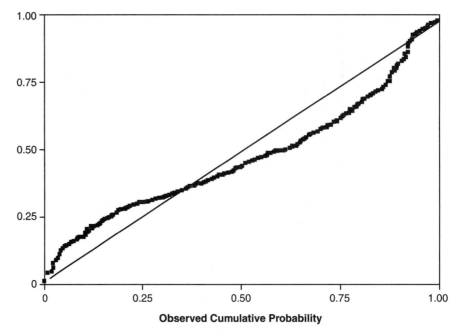

Figure 3-15 Normal Probability Plot of Juvenile Gang Crimes

Path Diagrams

Path diagrams are very effective for visualizing a theoretical model (relationships between variables) and the results of analyses. Developed by Wright in the early 20th century (1921, 1934), path diagrams come in several forms. Currently, the most popular uses of path diagrams show (1) a theoretical model, (2) the results of a path analysis regression model, or (3) the results of a structural equation modeling (SEM) analysis.

A path diagram is created by drawing the variables (or concepts) with arrows to represent relationships between the variables. The arrows are then labeled using the results of statistical analyses to show the strength or significance of the relationship. In **Figure 3-16**, concepts are used in the path diagram rather than variables. Each of these concepts was developed by a factor analysis. Each of the concepts (or factors) contains

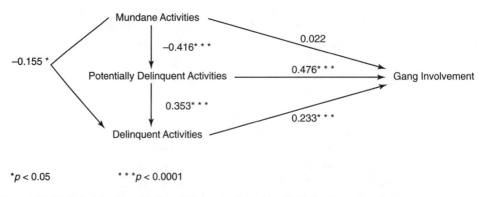

Figure 3-16 Model of the Progression of Behavior from Normative to Gang Involved
Source: Walker, Watt, and White (1994).

several variables that have been shown theoretically and statistically to be measures of the same social phenomenon. The factors were then placed in a regression model, and a path analysis was conducted to determine the relationship between the factors. The numbers shown in the path model represent the partial regression coefficients (Beta) taken from the regression model for that particular relationship.

It can be seen from this diagram that mundane activities are negatively and significantly associated with potentially delinquent activity. Mundane activities are also negatively and significantly associated with delinquent activities, although to a lesser degree. Finally, mundane activities are negatively but not significantly associated with gang involvement. This shows that mundane activities have the greatest influence on potentially delinquent activity, less on actual delinquency, and almost no influence on gang involvement. On the other hand, potentially delinquent activities have a significant and positive influence on both delinquent activities and gang involvement. In fact, potentially delinquent activities show a stronger relationship with gang involvement than does actual delinquent behavior (which was also positively and significantly associated with gang involvement). What can be concluded from this path diagram is that certain mundane activities have some influence on whether a person will become involved in potentially delinquent activities. Involvement in potentially delinquent activities, in turn, has some influence on whether a person will become involved in delinquent activities and involved in a gang. These relationships can easily be seen in a path diagram, even without a full understanding of statistical analysis.

■ Analyzing Geographic Distributions

Graphing and analyzing geographic or spatial data present unique problems and opportunities for graphical representation. It is possible, of course, to display geographic data in either tabular or chart form. The spatial character of the data is lost or obscured, however, when presented in this manner. For example, the distribution of crime across a city can easily be aggregated and displayed in a table or represented graphically in a bar chart or other graphical method; however, the trouble spots of the city could not be determined without prior knowledge of the area or moving between the graph and a map. Even a person not familiar with the city, however, could easily see the distribution of crime from a map where each crime was plotted with a point or where areas were shaded to show different rates of crime. The essence of mapping, then, is to show the distribution of a variable over a geographical area. The two most frequently used methods of showing geographic or spatial distribution are pin maps and choropleth maps.

Pin, Spot, or Point Maps

Pin, spot, or point maps have long been a natural part of criminological and criminal justice work. Pin maps and choropleth maps were used extensively in some of the very first criminological research (Shaw and McKay, 1942), and pin maps have been used by law enforcement agencies for many years to show the distribution of crimes, calls for services, and other data. When pin maps were first used, they were difficult and time consuming to construct because each point had to be placed by hand. Hand placement of pins or points also led to inaccuracies in maps because even if a map was developed at the city block level, the placement of the pins often included a lot of interpolation and guesswork. The development of graphical mapping programs has revolutionized the use of spot maps. Most of these programs will read data directly from a statistical program, database, or other computerized listing of addresses. This makes it much easier, for example, to display census data on a map or to read data from a researcher's data set and have it displayed by street address. Many current mapping programs will even place the points on the proper side of the street. An example of a pin map generated by one mapping program is shown in **Figure 3-17**.

Pin maps display the actual data points or scores of the data. The arrangement and number of points on the map represent the frequency of values. This arrangement of points makes it easy to determine whether one area has more or fewer points than another. The ability to analyze pin maps visually becomes difficult as the number of points increases, however. Once points begin to stack on each other, their relative contribution to the visual presentation of the map begins to diminish. For example, in Figure 3-17, one point has 52 values on it, but it still appears as one point. One way of handling this problem is to change symbols or change to numbers in places where values will overlap. This method must be balanced, however, by the loss of clarity that comes from mixing symbols. Different symbols are more useful for those maps where a variety of points will be displayed. For example, in a map showing the pickup points and the body dump sites of serial murder victims, two different symbols are desirable. For maps displaying the spatial distribution of gas stations, banks, fast-food marts, and other frequent targets of crime, several different symbols may be used.

Choropleth Maps

Pin maps are generally used to display individual-level data. If aggregate-level data is used, however, choropleth maps are a better choice. **Choropleth maps** display frequencies, rates, or other summary data for a variable by using shaded areas, crosshatching, or other methods of demarcating one area from another. The scale and unit of analysis for choropleth maps depends on the area to be mapped. A map of a city is often broken up into census tracts or other census boundaries, although some cities have other natural spatial limits, such as identified neighborhoods or the "natural areas" in Chicago. A map of a state could easily be broken up by counties, and a map of the United States could be divided by state or region.

Typical choropleth maps are shown in **Figures 3-18** and **3-19**. These maps show the distribution of juvenile crime and the distribution of vacant and boarded-up homes from the research project on social disorganization used in this book. Two observations can be made from these maps. First, the distribution of juvenile crime no longer follows the concentric ring pattern (Shaw and McKay, 1942) that dominated social disorganization research for a number of years. The current distribution of juvenile crime follows hot spots of high levels of crime distributed across the center of the city and interspersed with areas of low juvenile crime.

Comparing these two maps leads to a second observation, that there is similarity between areas of high juvenile crime and areas with many vacant and boarded-up homes. Although a full analysis of the relationship between juvenile crime and boarded-up homes cannot be derived from examining these maps visually, they can serve as a starting assumption that some relationship may exist. In fact, this is the way in which much of the early research from the Chicago School began. Robert Burgess began to notice similarities in maps of Chicago turned in by students as part of their graduate coursework. These similarities lead to the work of Shaw and McKay (1942)

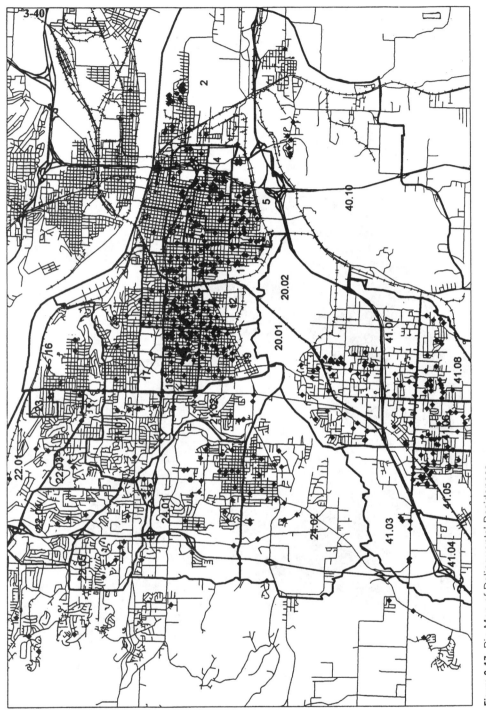

Figure 3-17 Pin Map of Delinquents' Residences.
Source: Walker, Jeffery T. (1992). Ecology and Delinquency in 1990: A Partial Replication and Update of Shaw and McKay's Study in Little Rock, Arkansas. Unpublished dissertation, Sam Houston State University, Huntsville, Texas.

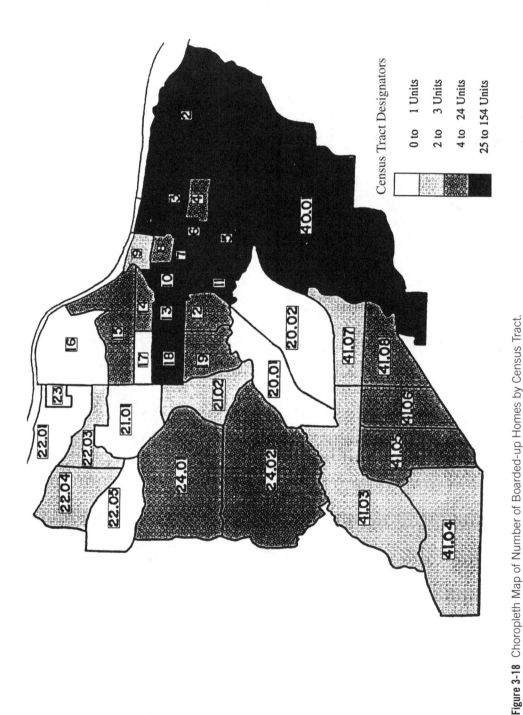

Figure 3-18 Choropleth Map of Number of Boarded-up Homes by Census Tract.
Source: Walker, Jeffery T. 1992. Ecology and Delinquency in 1990: A Partial Replication and Update of Shaw and McKay's Study in Little Rock, Arkansas. Unpublished dissertation, Sam Houston State University, Huntsville, Texas.

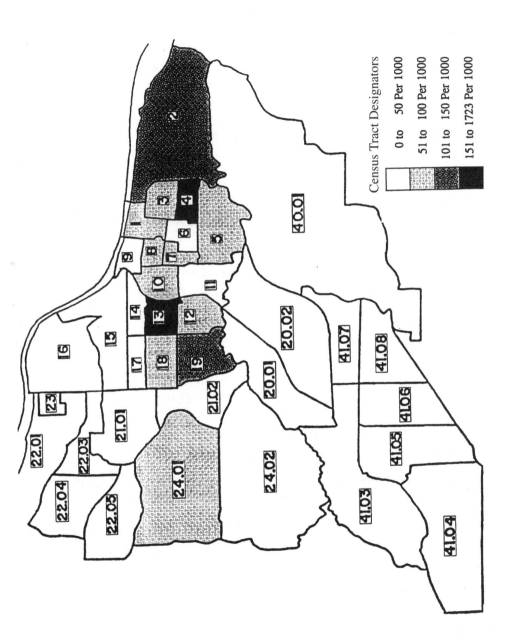

Figure 3-19 Choropleth Map of Delinquency Rate per Thousand Juveniles by Census Tract.
Source: Walker, Jeffery T. 1992. Ecology and Delinquency in 1990: A Partial Replication and Update of Shaw and McKay's Study in Little Rock, Arkansas. Unpublished dissertation, Sam Houston State University, Huntsville, Texas.

and others who substantiated the relationships between many of the variables found by Burgess.

Constructing Effective Choropleth Maps

Most of the conventions in creating choropleth maps concern how they will be shaded. This is because the shading or crosshatching will generally determine the interpretability and clarity of the map. When you are using shading, it is generally accepted that light (lower density of lines, dots, etc.) will mean less frequency or value, and dark will mean greater frequency. Typically, the least values will be white, especially if that category contains a zero, or very lightly shaded. As the values increase, so do the number of lines or dots, thickness of lines, or some other means of portraying increased intensity, concluding with black shading to represent the highest value. In this scheme, each type of crosshatching will represent a different interval or value of the data (as shown in Figures 3-18 and 3-19). Although this method of shading allows determination of differing values, a legend should be used to show the values associated with each level of shading. This is especially important when several maps will be used with different scales for each map. A legend should be used even if the values are discussed in the narrative.

Unlike other forms of graphical representation, it may be better to use unequal intervals when creating statistical maps. Whereas histograms and other charts rely on relatively precise measures of frequency or other intervals, maps are more robust in interval shading. It may be more advantageous, then, to divide intervals based on theory or natural breakpoints in the data rather than imposing divisions artificially. The warning remains, however, that class intervals should not be too disparate because of the possibility of skewed representation. Even though there is more latitude in the categories of choropleth maps, it is best to keep the number of categories small. Research (Jenks and Knos, 1961) has shown that maps with more than eight categories of shading are difficult to interpret because the human eye has difficulty distinguishing between the various levels of shading.

Problems with Choropleth Maps

Although the use of choropleth maps is an effective method of presenting spatial information, it is not without its problems (Schmid and Schmid, 1979). First, the entire geographic unit is shaded as if the variable was distributed equally throughout. In fact, most areas contain high and low spots of any given variable, and very few are even minimally consistent in any characteristic. Also, choropleth maps have distinct boundaries that promote the image of abrupt changes in the characteristics being mapped. In reality, though, most characteristics change gradually and not on boundary lines.

■ Conclusion

As you have learned in this chapter, graphical representation of data can be an effective method of examining, interpreting, and communicating the nature and characteristics of data. Frequency distributions aid in showing how the data is arranged; charts and graphs ease the examination and interpretation of frequency distributions;

and maps are especially effective for displaying data that is geographically or spatially oriented.

One word of caution, however: Graphical representation should not be used as a cosmetic appendage or in place of interpretation and analysis. The charts must serve a useful purpose and they should be explained thoroughly in the narrative. Charts presented without supporting discussion give the appearance of hiding something or not providing all necessary information. If used wisely and in proper form, however, graphical representation of data and analyses can serve a vital role in the presentation of research findings.

■ Key Terms

area chart
bar chart
box and whisker graph
cell frequency
choropleth map
column headings
combination distribution
cumulative percentage
frequency distribution
histogram
interquartile range
interval width
legend
line chart
lower limit
marginals
normal probability plot

offset Burchard chart
ogive
outliers
path diagram
pie chart
pin, spot, or point map
polygon
real limit
residuals plot
scatter plot
stub
upper limit
valid percentage
value labels
X axis, Y axis

■ Exercises

1. For the following data set for the number of previous crimes committed by people appearing in court on one day, create a frequency distribution, a percentage distribution, and a cumulative percent distribution. Make sure you label your distribution and all of the parts correctly.

   Number of Previous Crimes Committed by People Appearing in Court on One Day

   2, 4, 4, 6, 2, 3, 2, 0, 2, 2, 0, 4, 8, 1, 2, 6, 1, 6, 9, 8

2. For the following data set, create a grouped frequency distribution. Use the categories 90–99, 80–89, 70–79, and so on. Make sure you label your distribution and all the parts correctly.

   72, 38, 43, 81, 79, 71, 65, 59, 90, 83, 39, 42, 58, 56, 72, 63, 49, 81, 56, 60,
   83, 89, 60, 52, 62, 32, 28, 39, 49, 48, 65, 92, 81, 58, 95, 82, 73, 73, 89, 95

3. For the following: (a) identify and discuss the level of measurement of each variable; (b) explain what type of graph would be most appropriate for this type of data and why; and (c) draw the graph (artistic ability is not evaluated).

AGE: Age group of respondents.

Value Label	Value	Frequency	Percent	Valid Percent	Cumulative Percent
0–10	1	0	0.00	0.00	
11–20	2	7	2.15	2.15	2.15
21–30	3	31	9.51	9.51	11.66
31–40	4	76	23.31	23.31	34.97
41–50	5	70	21.47	21.47	56.44
51–60	6	54	16.56	16.56	73.01
61–70	7	40	12.27	12.27	85.28
71–80	8	27	8.28	8.28	93.56
81–92	9	21	6.44	6.44	100.00
	Missing	0	0	Missing	
Total		326	100.00	100.00	

KIDS: How many children do you have?

Value Label	Value	Frequency	Percent	Valid Percent	Cumulative Percent
	0	107	30.8	33.1	33.1
	1	44	12.7	13.6	46.7
	2	75	21.6	23.2	70.0
	3	39	11.2	12.1	82.0
	4	21	6.1	6.5	88.5
	5	15	4.3	4.6	93.2
	6	11	3.2	3.4	96.6
	7	3	0.9	0.9	97.5
	8	2	0.6	0.6	98.1
	10	1	0.3	0.3	98.5
	11	1	0.3	0.3	98.8
	12	2	0.6	0.6	99.4
	15	1	0.3	0.3	99.7
	16	1	0.3	0.3	100.0
	Missing	24	6.9		
Total		347	100.0	100.0	

JOBS: What is your occupation?

Value Label	Value	Frequency	Percent	Valid Percent	Cumulative Percent
Professional	1	105	30.3	34.9	34.9
Clerical/technical	2	43	12.4	14.3	49.2
Blue collar	3	28	8.1	9.3	58.5
Retired	4	68	19.6	22.6	81.1
Housewife	5	16	4.6	5.3	86.4
Other (part-time)	6	28	8.1	9.3	95.7
Unemployed	7	13	3.7	4.3	100.0
	Total	301	86.7	100.0	
	Missing	46	13.3	Missing	
Total		347	100.0		

4. From current journals, select three articles that contain graphs or maps like those discussed in this chapter. For each graph or map, (a) discuss why it was appropriate for that data (or why not); (b) discuss how it compares with the conventions for design discussed in this chapter; and (c) explain what it is attempting to show.

■ References ■

Federal Bureau of Investigation (1980–1990). *Uniform Crime Reports.* Washington, DC: US Government Printing Office.

Jenks, G. F., and D. S. Knos (1961). The use of shading patterns in graded series, *Annals, Association of American Geographers,* 51:316–334.

Playfair, W. (1801). *The Commercial and Political Atlas,* 3rd ed. London: Wallis.

Schmid, C. F. (1954). *Handbook of Graphical Representation.* Hoboken, NJ: Wiley.

Schmid, C. F., and S. E. Schmid (1979). *Handbook of Graphical Representation,* 2nd ed. Hoboken, NJ: Wiley.

Shaw, C. R., and H. D. McKay. (1942). *Juvenile Delinquency and Urban Areas.* Chicago: University of Chicago Press.

Tukey, J. W. (1977). *Exploratory Data Analysis.* Reading, MA: Addison-Wesley.

Walker, J. T., B. Watt, and E. A. White (1994). Juvenile activities and gang involvement: the link between potentially delinquent activities and gang behavior. *Journal of Gang Research,* 2(2):39–50.

Wright, S. 1921. Correlation and causation. *Journal of Agricultural Research,* 20:557–585.

Wright, S. 1934. The method of path coefficients. *Annals of Mathematical Statistics,* 5:161–215.

■ Notes ■

1. F = freshman; S = sophomore; J = junior; and G = senior.

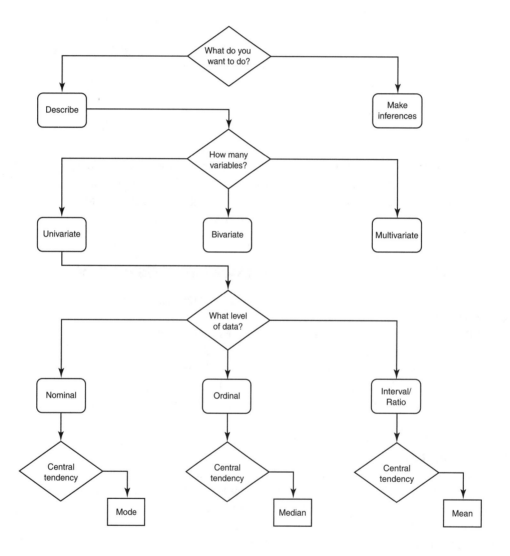

Measures of Central Tendency

■ Univariate Descriptive Statistics

Using frequency distributions and graphical representation, as in Chapter 3, helps researchers determine how the data is arranged and to summarize it. Frequency distributions and graphs, however, cannot always tell the entire story. It is usually necessary to summarize the data further. Instead of summarizing entire distributions, it is more often efficient to compare only certain characteristics of the data. To conduct this comparison, it is helpful to know certain information, such as the form of the distribution, the average of the values, and how spread out they are within the distribution.

This is where univariate descriptive statistics come into play. Univariate descriptive statistics are used to describe and interpret the meaning of a distribution. They are called univariate because they pertain to only one variable at a time and do not attempt to measure relationships between variables. Univariate descriptive statistics make compact characterizations of distributions in terms of three properties of the data. First is the **central tendency**, which translates to the average, middle point, or most common value of the distribution. The second property is the **dispersion** of the data. This relates to how spread out the values are around the central measure. Finally, there is the **form** of the distribution. The form of a distribution relates to what the distribution would look like if displayed graphically. Included in the form of a distribution is the number of peaks, skewness, and kurtosis. In this chapter we address the first univariate descriptive procedure: measures of central tendency. Measures of dispersion and measures of the form of a distribution are covered in Chapters 5 and 6, respectively.

■ Measures of Central Tendency

Measures of central tendency examine where the central value is in a distribution or the distribution's most typical value. There are three common measures of central tendency, one for each level of measurement (interval and ratio are combined). These are the mode for nominal level data, the median for ordinal level data, and the mean for interval and ratio level data.

Mode

At the lowest level of sophistication is the mode (symbolized by Mo). The **mode** is used primarily for nominal data to identify the category with the greatest number of cases. The mode is the most frequently occurring value, or case, in a distribution. It is the tallest column on a histogram or the peak on a polygon or line chart. The mode has the advantage of being spotted easily in a distribution, and is often used as a first indicator of the central tendency of a distribution.

How do you do that?
Obtaining Univariate Statistics in SPSS

This chapter and the two that follow address univariate statistics: measures of central tendency, measures of dispersion, and form. You can obtain all of these statistics in the same procedure in SPSS, and it is just an extension of the same procedure you used in Chapter 3 to obtain a frequency distribution. The steps to follow are:

1. Open a data set.
   a. Start SPSS.
   b. Select *File*, then *Open*, then *Data*.
   c. Select the file you want to open, then select *Open*.
2. Once the data is visible, select *Analyze*, then *Descriptive Statistics*, then *Frequencies*.
3. Make sure the "Display Frequency Tables" is checked.
4. Select the variables you wish to include in your distribution and press the ▶ between the two windows.
5. Select the *Statistics* button at the bottom of the window.
6. Check the boxes of any of the univariate measures you want to include in your research.
   a. For measures of central tendency (this chapter), check the boxes in the frame "Central Tendency," typically the mode, median, and mean.
   b. For measures of dispersion (Chapter 5), check the boxes in the frame "Dispersion," typically the standard deviation, variance, and range.
   c. For measures of form (Chapter 6), check the boxes in the frame "Distribution," specifically skewness and kurtosis
7. Select "Continue," then "ok."
8. An output window should appear containing a distribution similar in format to Table 4-3.

The mode is the only measure of central tendency appropriate for nominal variables because it is simply a count of the values. Unlike other measures of central tendency, the mode explains nothing about the ordering of variables or variation within

TABLE 4-1 Ungrouped Data

7	5	4	3	2
7	5	4	3	1
6	5	3	3	1

the variables. In fact, the mode ignores information about ordering and interval size even if it is available. So it is generally not advised to use the mode for ordinal or interval level data (unless it is used in addition to the median or mean) because too much information is lost.

There is no formula or calculation for the mode for either grouped or ungrouped data. The procedure is just to count the scores and determine the most frequently occurring value. Consider the data set in **Table 4-1**, which is the number of prisoner escapes from 15 prisons over a 10-year period. Here, there are 15 total escapes. There are two 7's, one 6, three 5's, two 4's, four 3's, one 2, and two 1's. The mode in this data set would be 3 escapes, since there are more 3's than any other value.

For grouped data, determining the mode is often even easier because the numbers are already counted. The data from Table 4-1 has been grouped in **Table 4-2**. What is the mode of this data set? Here you simply determine the category that has the highest value. In this case it would be the 3–4 category because it has a frequency of 6. If the data were plotted on a bar chart or polygon, the distribution would look like that in **Figure 4-1**. Here, you can see that the category 3–4 has the highest bar on the bar chart and it forms a hump in the polygon. This highest bar or hump indicates the mode for that variable.

One caution when discussing the mode. The mode is not the frequency of the number that occurs most often but rather, the category (or class) itself. It is easy to want to state that the mode in Table 4-2 is 6 because that is the frequency that is highest. This is not the mode, however; the mode is the category of the value that has the highest frequency: in this case, 3–4.

Data that is in a frequency table also makes calculating the mode easy. What is the mode in the frequency table in **Table 4-3**? The mode in this case is *4*, or *some college*. Note that in this case, the mode can be written as either *4* or *some college*. When using nominal or ordinal data where value labels are assigned to the values, the mode can be expressed as either the value (number) or the value label.

TABLE 4-2
Modal Value for Grouped Data

X	f
7–8	2
5–6	4
3–4	6
1–2	3

Figure 4-1 Bar Chart and Polygon of Grouped Data of Table 4-2

TABLE 4-3 Combination Table for Education from the 1993 Little Rock Community Policing Survey

What is your highest level of education?

Value Label	Value	Frequency	Percent	Valid Percent	Cumulative Percent
Less than High School	1	16	4.6	4.8	4.8
GED	2	59	17.0	17.6	22.3
High School Graduate	3	8	2.3	2.4	24.7
Some College	4	117	33.7	34.8	59.5
College Graduate	5	72	20.7	21.4	81.0
Post Graduate	6	64	18.4	19.0	100.0
Missing		11	3.2	Missing	
Total		347	100.0	100.00	

N	Valid	336
	Missing	11
Mean		4.08
Median		4.00
Mode		4
Std. Deviation		1.460
Variance		2.131
Skewness		−.477
Std. Error of Skewness		.133
Kurtosis		−.705
Std. Error of Kurtosis		.265
Range		5

The histogram with a polygon overlay for the data in Table 4-3 is shown in **Figure 4-2**. As shown in the figure, the highest bar on the histogram or the hump in the polygon is at the *4* or *some college* level. The mode as calculated here is what is obtained from SPSS. In the output in Table 4-3, the mode is identified as *some college* (*4*), with a frequency of 117. Notice also that the median, mean, and other measures are also included in this table. This is typical univariate output from SPSS. It provides most of the univariate descriptive statistics discussed in this chapter and the two that follow. Table 4-3 may look somewhat daunting right now, but by the time you finish Chapter 6, a frequency table and univariate output such as this should be shorthand for everything you need to know about a distribution.

A distribution is not confined to having only one mode. There are often situations where a distribution will have several categories that have the same or similar frequen-

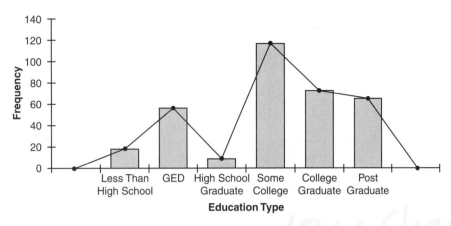

Figure 4-2 Histogram and Polygon of Education Responses

cies. In these cases, the distribution can be said to be bimodal or even multimodal. It is also possible for a distribution to have no mode if the frequencies are the same for each category. If the data in Table 4-1 is modified, a bimodal, multimodal, and a data set with no mode can be created, as shown in **Figures 4-3** to **4-5**. In Figure 4-3, categories 3 and 4 both have the same frequency: 3. In this case, both the 3 and the 4 would be the modes because each has the same (highest) value.

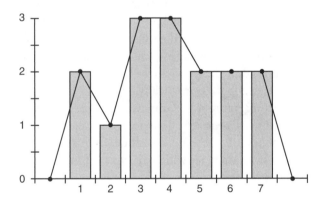

Figure 4-3 Bimodal Distribution

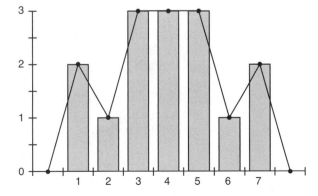

Figure 4-4 Multimodal Distribution

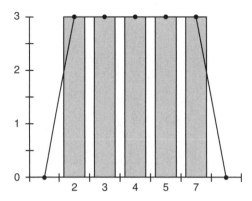

Figure 4-5 No-Modal Distribution

In Figure 4-4, the 3, 4, and 5 categories all have frequencies of 3. This means that all three categories would be the mode for this distribution. When almost half of the categories in the distribution represent the mode, its use as a measure of central tendency is reduced.

In Figure 4-5, all of the categories have the same frequency. This does not happen very often, but it is possible, especially in survey research or with other data that have a limited range of categories. The mode as a measure of central tendency in this case is practically useless, although it would be beneficial as a way of stating that all of the values have the same frequency. Although each of the modes in Figures 4-3 (3 and 4), 4-4 (3, 4, and 5), and 4-5 (2 through 7) have the same frequency, that does not always have to be the rule. There is some debate as to what constitutes a bimodal or multimodal distribution. Some propose that the frequencies have to be the same for a distribution to be multimodal. Others argue that practically any peaks in a distribution can represent modes. For example, in Figure 4-1, some would argue that both the 1–2 category and the 3–4 category represent a mode. These people argue that any peak in a polygon, or any spike in the frequency, may represent a mode. In this book, only the category or categories with the highest frequencies will be designated the mode.

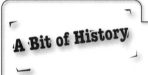

A Bit of History

Measures of central tendency are among the oldest of all descriptive statistics. The *mean*, for example, can be traced back to Pythagoras in the 6th century BC, although its development is surely much earlier. Galton (1883) coined the term *median* during his work on percentiles, but the procedure was used before this by Fechner for arriving at a value of the "middlemost ordinate." Finally, Karl Pearson reduced the concept of the "abscissa corresponding to the ordinate of maximum frequency" to the *mode* in his 1895 work.

As can be seen from these distributions, the mode quickly becomes ineffective when there are multiple modes and is worthless (except for an understanding of the nature of the distribution) when each category has a modal value. This is why the mode is not widely used as a measure of central tendency in statistics except for nominal level data.

Median

If the data is at least ordinal level, the median (symbolized by Me) may be a better choice for examining the central tendency of the distribution. The **median** is the point of the 50th percentile of the distribution. This means that the median is the exact midpoint of a distribution or the value that cuts the distribution into two equal parts. For the simple distribution 1, 2, 3, the median would be 2 because it cuts this distribution in half. Note that 2 is not the most frequently occurring or the product of some formula but simply the value in the middle. The median will always be the middle value, but sometimes it will be necessary to resort to math to determine the exact middle value.

The median is used with ordinal level data because it does not imply distance between intervals, only direction: above the median or below it. Recall from Chapter 2 that the nature of ordinal level data is that you can determine which category is greater than or less than another category, but there are not equal intervals so there is no way to determine how much greater or lesser the category is. The median also works on this principle, determining the midpoint of a distribution such that a category can be said to be less than or greater than the median, but there is no way to tell by how much. For example, take the following two distributions:

$$1, 2, 3, 3, 4, 4, 5 \qquad 1, 1, 1, 3, 10, 50, 100$$

Each has the same number of values, 7, although each has very different numbers. In this case, the modes would be different: 3 and 4 in the first; 1 in the second. Also, the means would be different: 3.14 in the first, 23.71 in the second. The median for both of these distributions, however, would be 3, the middle value in the distribution. In both distributions there are three values below the median and three values above the median.

The median may be used instead of the mean in a special circumstance where the distribution takes on the quality of being skewed (see the section on form in Chapter 6). The mean (discussed in the next section) is often highly influenced by extreme scores. For example, if you were to calculate the mean, or average, age of four people who are 2, 3, 4, and 50 years old, the mean would be 14.75 years old. Obviously, 14.75 is not a good measure of the central value in this distribution, but because of the way the mean is calculated, that is the value that would be obtained. The median of that distribution would be 3.5, which is much more like the central value. Even in the example above, the mean of the second distribution is 23.71, which is not really representative of the distribution. Note, however, that if the variable is interval and the distribution is not skewed, some of the explanatory power of the data is lost using the median as the measure of central tendency.

Median for Ungrouped Data

Calculation of the median for ungrouped data is relatively simple. All that is needed is the N for the distribution. If the N is not given, simply count the number of scores (remember—do not *add* the scores, *count* them). The N is then placed in the

EXHIBIT 4-1 Ungrouped Data

7	5	4	2	2	$\dfrac{N+1}{2} = \dfrac{23+1}{2}$
6	5	3	2	1	
6	5	3	2	1	$= \dfrac{24}{2}$
6	5	3	2		
6	4	2	2		$= 12$

formula $\dfrac{(N+1)}{2}$. If the data from Table 4-1 were expanded, the median can be calculated as shown in **Exhibit 4-1**.

There are 23 values here. Adding 1 to this number and then dividing by 2 obtains the exact middle of the distribution, in this case the 12th value. Once this value is calculated, if the numbers are not arranged in order, you should do so. This ensures that the middle value of the distribution is actually in the middle and that the numbers are arranged in order. Then, beginning with the lowest value, simply count up the ungrouped data until the value obtained in the formula is reached (the 12th value in Exhibit 4-1). This is the median. In this case, counting to the 12th value would produce a score of 3. So the median number of escapes in this distribution is 3.

There are several issues to note about the median. These are important to understanding when interpreting the median. First, be careful when calculating the median because two different numbers must be dealt with. The value that is obtained from the formula is *not* the median but simply the number of values to count up in the distribution to find the median (or median class for grouped data). The median is the score or class that contains the number from the formula. In the example above, the median is not 12. Twelve is only the number to count up from the beginning of the distribution to find the median, which is 3.

Also, if there is more than one of the same score in the median class (there are three 3's in Figure 4-5), the median is still that score even though it occurs more than once. The key value to look for is the middle value, regardless of how many there are of that particular category. This will be brought up again below in terms of calculating the median for grouped data where the class interval is greater than 1.

Finally, unlike the mode, the median does not have to be a value in the distribution. For an odd number of scores (as in Exhibit 4-1) the median will be one of the scores because it is the point that cuts the distribution in half. If there are an even number of scores, however, the median will fall in between two of the scores. For example, in the distribution 3, 4, 5, 6, 7, 8, 9, 10, the formula $\dfrac{(N+1)}{2}$ would give a value of 4.5.

This would put the median between the 6 and 7. When this occurs, the median is a value halfway between the two scores. In this case, the median would be 6.5. This holds true even if the two numbers do not have an interval of 1. For example, in the distribution 5, 6, 8, 10, 11, 12, the number from the formula puts the median between the score of 8 and the score of 10; therefore, the median would be 9.

Median for Grouped Data

For grouped data where the class interval is 1 or where the entire class can be used as the median, the process for finding the median is essentially the same as that for ungrouped data. The first step is to find the value to count up in the distribution using the formula $\frac{(N + 1)}{2}$. Then, simply count up the frequency of each class to find the median class. If the data from Exhibit 4-1 is grouped into a frequency distribution it would look like **Exhibit 4-2**.

The first step is to determine the midpoint using the formula. Since the data in Exhibit 4-2 has not changed, there are still 23 values (escapes). Plugging this value into the formula would, as before, result in a value of 12. Since the values are in a frequency distribution, they are probably already ordered. Although it is possible to find the median beginning from either the lowest or highest category, it is best for consistency to begin with the lowest value. In this case, you would begin with the class of 1 and count the frequencies until the 12th value is reached, which is 3. Note that it is possible here to count from 10 to 12 and still be in the 3 class. This is fine, as long as the value we are looking for, 12, is one of the numbers in that class. If the middle value from the calculation had been a 10, 11, or 12, the 3 class would still have been the median class.

The same procedure is used if the data is grouped with a class interval greater than 1 but where a median class is sufficient. The process for calculating the median where only the median class is desired is shown in **Exhibit 4-3**. This frequency distribution has the same N as the previous distributions, so the first step will be the same: calculating the value to count up to. Here again, the value is 12. Beginning with the lowest class and counting up to 12 will put you in the 16–20 class. This class contains between the 10th and 14th cases, but since it contains the value from the calculation, it is the median class.

Looking again at Table 4-3, the data in this distribution could be either nominal or ordinal. It could be argued, for example, that including a GED in the distribution disrupts the ordering of the categories such that the data should properly be called nominal. It could also be argued, however, that the categories are sufficiently ordered to be called ordinal. For that reason, and to ensure some consistency of examples, the same

EXHIBIT 4-2 Grouped Data with an Interval Class of 1

X	f
7	1
6	4
5	4
4	2
3	3
2	7
1	2
N	23

$$\frac{N + 1}{2} = \frac{23 + 1}{2}$$

$$= \frac{24}{2}$$

$$= 12$$

EXHIBIT 4-3 Calculating Median Class for Grouped Data

X	f
31–35	2
26–30	3
21–25	4
16–20	5
11–15	4
6–10	3
1–5	2
N	23

Step 1. Find the median interval $\dfrac{N+1}{2} = 12$.

Step 2. Count up in the frequency for that class.

Step 3. That is the median class for this distribution (16–20).

frequency distribution used to discuss the mode is used here to discuss output for the median.

The median in Table 4-3 is the same as the mode, *some college* (4). This was obtained in the same manner as described above:

$$\frac{347 + 1}{2} = 174$$

Counting up in the distribution (beginning at the 1 category, which is on the top in this example) puts the median in the 4th category (16 + 59 + 8 + 117). Since this category contains between the 83rd and 200th cases, it contains the 174th value. Also, since the category containing the median is sufficient in this instance, the median is said to be *some college*, or 4.

Calculating an exact median for grouped data with a class interval greater than 1 is somewhat more complicated. Using the data set in Exhibit 4-3, the procedure to calculate an exact median begins as all others, with the formula $\dfrac{(N + 1)}{2}$. This produces the value 12, the same as in previous examples. This means that the 16–20 class is the median class. As stated above, we could count to 14 in this class, beyond the 12 needed to establish a median value. The question then becomes: Where in this class does the median lie? To find out requires interpolation within the class. Assuming the scores are evenly distributed within the class,[1] the formula for calculating the exact median is

$$Me = L_m + \left(\frac{0.5N - cf_{bm}}{f_m} \right) i$$

where L_m is the lower limit of the median class, cf_{bm} the cumulative frequency of the interval below the median class, f_m the frequency of the median class, and i the width of the interval of the median class. Using this formula with the data from Exhibit 4-3, the median is calculated as follows:

$$Me = 15.5 + \left(\frac{0.5(23) - 9}{5} \right) 5$$

$$= 15.5 + \left(\frac{11.5 - 9}{5} \right) 5$$

$$= 15.5 + \left(\frac{2.5}{5} \right) 5$$

$$= 15.5 + (0.5)5$$

$$= 15.5 + 2.5$$

$$= 18$$

The value of 15.5 is the lower limit of the 16–20 class. N is 23, as in all other examples. The cumulative frequency is determined by adding up all the frequencies below the class containing the median. In this case, there are three classes below the median class: 1–5, 6–10, and 11–15. The frequencies of these classes (2, 3, and 4, respectively) equal 9 (cf_{bm}). The frequency of the class containing the median in this case (16–20 class) is 5 (f_m). Finally, the interval is calculated by subtracting the lower limit of the median class from the upper limit ($20.5 - 15.5 = 5$). In this case, the result of the calculations shows that the exact midpoint of the distribution is 18.[2]

Calculating the exact median in actual research is less often necessary. Most statistical programs report the exact median from the ungrouped data, or the researchers report only the median class or report the midpoint of the median category as the median. There are times, however, when it is necessary to determine the exact median from information in journal articles. For example, you may wish to know the exact median from **Table 4-4**. This table shows categories that are not only greater than 1 but

TABLE 4-4 Research on Racial or Ethnic Issues

What portion of your professional research focuses on racial or ethnic issues?

Portion	Number	Percentage
0–10%	6	15
11–25%	1	2
26–50%	12	30
51–75%	14	35
Over 75%	7	18
	40	

Source: Edwards, White, and Pezzella (1998).

are unequal. The procedure would be the same as discussed above, however. Here, the median category would be 51 to 75% $[(40 + 1)/2 = 20.5]$. Interpolating where in that class the exact median would be involves using the formula given above. Application of that formula for the data in Table 4-4 is shown below.

$$Me = L_m + \left(\frac{0.5N - cf_{bm}}{f_m} \right)i$$

$$= 50.5 + \left(\frac{0.5(40) - 19}{14} \right)25$$

$$= 50.5 + \left(\frac{20 - 19}{14} \right)25$$

$$= 50.5 + \left(\frac{1}{14} \right)25$$

$$= 50.5 + 0.07(25)$$

$$= 50.5 + 1.75$$

$$= 52.25$$

As you would expect, the median does not go very far into the median class in this example. This is evident because the frequency below the median class is 19, and the exact median is only 20.5.

This process is complicated somewhat when the median class is open-ended. For example, what is the midpoint in a distribution where the upper category for annual income is *$30,000 and greater?* There are several methods of dealing with this issue. Probably the best is to attempt to determine what a reasonable midpoint might be. This is also shown in the example in Table 4-4, which has two open-ended categories: *less than high school* and *postgraduate.* This would make it difficult to determine, for example, where the midpoint of a postgraduate degree would lie (some graduate work, master's degree, law degree, etc.). This would have to be a judgment call by the re-searcher based on theory and an understanding of the data.

Mean

A statistician is a person who stands in a bucket of ice water, sticks his head in an oven, and says, "On average, I feel fine."

—Unknown

The most popular measure of central tendency, both among statisticians and the gen-eral population, is the mean. The **mean** is used primarily for interval and ratio level data. Because it assumes equality of intervals, the mean is generally not used with nominal or ordinal level data. The mean is very important to statistical analysis because it is the basis, along with the variance (see the discussion of measures of dis-persion in Chapter 5), of many of the formulas for higher-order statistical procedures. The mean also serves as a check on the integrity of the data. As discussed above, the mean is often heavily influenced by extreme scores. So if a 17 has been mistyped as 177, the mean will be much larger than expected. Mean scores outside what would be expected for the data should be a signal to recheck the data.

There are actually several different versions of the mean. The mean discussed in this chapter is the *arithmetic mean* (from here on, called the mean). There are variations of the mean that are less utilized in social science research and are not discussed here. These include the *weighted mean, harmonic mean,* and *geometric mean.*

The symbolic notation for the mean is different than symbols that have been used to this point. The mean is symbolized either by μ or $\overline{X}$, depending on whether the data is a population or sample estimate (this distinction is used most often in Chapter 11 and beyond). It is interesting that descriptive statistics deals with a population, but it has become convention that the mean most commonly used in descriptive statistics is actually the symbol for the sample mean ($\overline{X}$). Since most texts use this notation for the mean in descriptive analyses, it will also be used here, even though the more proper notation would be the population mean (μ).

The mean is simply the average of all the values in a distribution. To obtain the mean, add up the scores in a distribution and divide by N (just as in calculating an average). In statistical terms, the mean is calculated as

$$\overline{X} = \frac{\Sigma fx}{N}$$

where *fx* is calculated by multiplying X times the frequency for each value. In the example used in Exhibit 4-2, the mean would be calculated as in **Exhibit 4-4**. Here, each X is multiplied by the frequency for that category ($7 \times 1, 6 \times 4$, etc.). That creates an *fx* column in the table, which is then summed to obtain Σfx (84). That value is then divided by the N for the distribution (23) to obtain the mean for the distribution. In this case, there were 23 prisons that had a total of 84 escapes, so the mean (average) number of escapes for these 23 prisons was 3.65 escapes.

The procedure for calculating the mean for grouped and ungrouped data is the same. The only difference is that for grouped data where the class interval is greater than 1, the midpoint of the class is used as X.[3] For example, in the frequency distribution in Exhibit 4-3, the midpoints of the classes would be 2.5 ($5.5 - 0.5 = 5$; $5/2 = 2.5$), 8.5, 13.5, and so on. These are the values that would be used for X in the formula for the mean.

The mean can be estimated from the example output that was used for the mode and median (as shown in Table 4-3). Note that this data is *not* interval or ratio level and

EXHIBIT 4-4 Calculating the Mean

X	f	fx
7	1	7
6	4	24
5	4	20
4	2	8
3	3	9
2	7	14
1	2	2
N	23	Σfx 84

$$\overline{X} = \frac{\Sigma fx}{N}$$
$$= \frac{84}{23}$$
$$= 3.65$$

is used here only to show the similarities and differences among the mean, median, and mode. Even though this is nominal/ordinal level data, SPSS treats it as interval level and uses the formula above for calculating the mean. In this example, each of the category values (1 through 6) are multiplied by the frequency for that category (1 × 16, 2 × 59, etc.). This *fx* value is summed to achieve a total of 1370. This is then divided by *N* minus the 11 missing values for a total of 336. The result is 4.077, which is what SPSS reported (rounded to 4.08).

In most cases in real research, the mean may not be accompanied by a frequency distribution, or the frequency distribution will be more for presentation than for analysis. In such cases, the mean may be reported alone, or it could be reported as part of a discussion or table of univariate statistics associated with the research.

The mean has several advantages over other measures of central tendency. From a practical standpoint, the mean is preferred because it is standardized. This means it can be compared across distributions. This is very beneficial when comparing similar data from different sources, such as the mean number of prisoners per institution in several states, because the two values can be directly compared. The mean is also important because the sum of the deviations of the scores from the mean is always zero. That is, if each value in a distribution were subtracted from the mean, the sum of those scores would be zero. This is discussed in detail in Chapter 5. A final important characteristic of the mean is that the sum of the squared deviations from the mean is the smallest value for summed deviations (smaller than if the same calculations were made for the mode or median). This principle of **sum of squares** will become very important in our discussions in Chapter 5 of the variance and sum of squares as they relate to regression lines.

As discussed above, the greatest problem with the mean is that it is greatly influenced by extreme scores in the distribution. The example in the section on the median, where a mean age of 15 was obtained when all but one of the values was less than 5, shows how much the mean can be influenced by extreme scores. That is why the median is used in cases where the data is skewed.

■ Selecting the Most Appropriate Measure of Central Tendency

The goal of many statistical analyses is to be able to develop summary statements, often about a large amount of data. Proper summarization depends on several factors, including the level of data, the nature of the data, the purpose of the summarization, and the interpretation.

The level of data has a substantial influence on which measure of central tendency should be used. As stated above, one measure is most appropriate for a particular level of data. The mode is most appropriate for nominal level data, and its use with ordinal and interval level data would result in a loss of power in terms of the information that could be gained from the data. For example, the median is most appropriate with ordinal level data. Although it can be used with interval level data (especially skewed distributions), it should not be used with nominal level data because the rankings assumed in the median cannot be achieved with nominal level data. Finally, the mean should be used only with interval or ratio level data because it assumes equal intervals of the data that cannot be achieved by nominal and partially ordered ordinal level data. The exception here is that the mean can be used with dichotomized nominal level data since this type of data approximates interval level characteristics.

Selection of the most appropriate measure of central tendency is also sometimes based on the nature of the distribution. As discussed above, if a distribution is highly skewed, or if it can be determined that there are some extreme values (outliers) in the distribution that would make the mean inaccurate as a measure of central tendency, the median should be used rather than the mean.

The second criterion for choosing a measure of central tendency is the purpose of summarization, typically in terms of what you are trying to predict. Imagine that you were asked to state one measure that would best capture the nature of a distribution. How would you go about that? To put it another way, you might bet $100 to guess a number drawn at random from a distribution. Which number would you choose? One way to address these questions would be to find the score that would be at the "heart" of the distribution: the most common score, the one that cut the distribution in half, or the average score. That is the goal and the role of measures of central tendency. There are several ways to go about this.

If you knew all the values in the distribution, you could calculate the mode easily and quickly. If you are interested in predicting an exact value, you should probably use the mode because it has the highest probability of occurring in any given distribution. Both the median and the mean may produce values that are not in the distribution, so if you must guess and be absolutely right as to the number, use the mode. For example, say you are taking a multiple-choice test and have no idea which answer to a certain question is correct. If you had the distribution of correct answers for that professor for that test, you would want to choose the modal answer rather than the median or mean. This is because you must get the answer correct or it does not count. As another example, consider a prediction based on driving a car around an obstruction placed in front of it. If tests occur over a number of drivers, the distribution would be bimodal: some steering to the left and some to the right. A suggested course of action would not be the median or mean, however, as that would have the vehicle crashing into the obstacle even though it minimized the error in steering.

If, on the other hand, you want to maximize your prediction by getting closest to the number over several tries, thereby minimizing your error, the median might be a better choice. Here, whether you miss high or low is irrelevant; what is important is the size of the error. In a popular game show, contestants are given $7 and required to guess the exact numbers included in the price of a car. For each number they are off, they lose $1. If they have money left over after making all the guesses, they win the car; if they run out of money, they lose. The probability of response plays a big part in the first two or three numbers. You would not want to guess 9 for the first number, for example. If contestants are at the fourth or fifth number, however, and still have money left, they may want to choose the median value (probably a 5) to minimize the error (loss of dollars). Being high or low does not matter here, only deviation from the number.

Finally, if you have the opportunity to average your misses over several guesses and the signs do matter (high guesses can offset low guesses), the mean is the best choice. The mean is good in that if you do not know a value, it is often best to choose the average. For example, if you had to guess the weight of a woman whom you had never seen, you should probably choose the mean weight for women because this would minimize the error. The mean is also practically the only choice when using estimates in higher-order analyses because the mathematical properties of both the mode and the median are such that they do not lend themselves to inclusion in other formulas. The mean is less efficient, however, with highly skewed distributions.

The final criterion for selecting a particular measure of central tendency is the interpretation. If you chose the wrong level of measurement and base your measure of central tendency on that choice, your interpretation may very well not make sense. For example, for a nominal level variable such as paint color, the mode makes sense (more people chose red than any other color). The median does not make much sense, however. For example, if you say half or fewer of the respondents chose red, what does that mean? There is no reference point because there is no order. The same holds true for the mean. How could you interpret an average of 1.8 on paint color; that the average color chosen was slightly different than red? It is easier to use lower measures of central tendency with higher levels of measurement, but you lose some of the power of your interpretation. For example, it is technically correct to say the modal age in a class is 20, but it is not as precise as saying the average age is 22.4.

■ Conclusion

In this chapter, we introduced univariate analyses by discussing the first of the univariate descriptive statistics, measures of central tendency. Measures of central tendency are one of the most used descriptive statistics and provide the most information. For example, if you were to ask someone about a group of people, you might provide an answer in terms of an average age or average income.

The measures of central tendency provide the information that their name implies: a measure of the central value. Think of a seesaw. For a seesaw to work properly, it must have a balance point in the middle so the weight is distributed generally equally on each side (as in **Figure 4-6**). Here, the measure of central tendency is at the balance point of the distribution. The picture of a seesaw, however, could easily be replaced with a histogram of a frequency distribution. If only the X axis were retained, the seesaw could look like the bar chart in **Figure 4-7**. This distribution is actually unique in that, mathematically, the mean equals 4. Since 4 is the most frequently occurring value, it is also the mode; and because 4 is the middlemost point in the distribution, it is also the median. If the values of the distribution were changed some, the

Figure 4-6 Balancing a Distribution on the Measure of Central Tendency

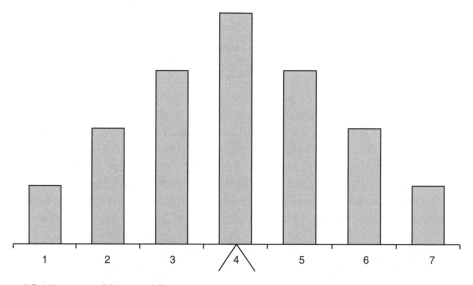

Figure 4-7 Histogram of Balanced Frequency Distribution

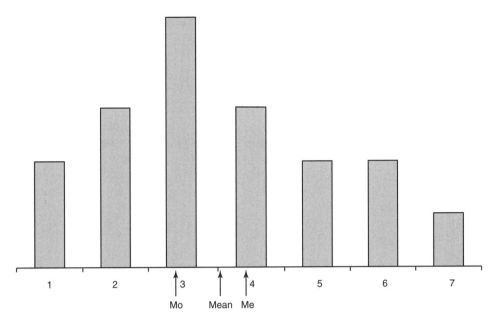

Figure 4-8 Histogram of Unbalanced Distribution

balance point would have to shift to keep the balance of the distribution. For example, in **Figure 4-8**, the mean, median, and mode are at different points. This is because of the spread and alignment of the values in the distribution.

You can see that just knowing the measure of central tendency is not always enough. Sometimes it is also important to know how spread out the values are or how they are arranged in the distribution. This is the reason that more than measures of central tendency are needed for a proper description of data. In Chapter 5, we address how spread out the values are in the distribution, and in Chapter 6 we discuss the arrangement of the data within the distribution. Together, these three pieces of information make up the complete analysis of a single variable (univariate analysis).

■ Key Terms

central tendency	median
dispersion	mode
form	sum of squares
mean	

■ Summary of Equations

Median (*Me*) for ungrouped data

$$\frac{N + 1}{2}$$

Median (*Me*) for grouped data

$$Me = L_m + \left(\frac{0.5N - cf_{bm}}{f_m} \right) i$$

Mean ($\overline{X}$)

$$\overline{X} = \frac{\sum fx}{N}$$

■ Exercises ■

The exercises for this chapter and Chapters 5 and 6 use the same examples. This will allow you to work through problems using all three types of univariate descriptive statistics.

1. For the set of data below, calculate:
   a. The mode
   b The median
   c. The mean

   6, 7, 8, 10, 10, 10, 12, 14

2. For the set of data below, calculate:
   a. The mode
   b. The median
   c. The mean

   7, 4, 2, 3, 4, 5, 8, 1, 9, 4

3. For the set of data below, calculate:
   a. The mode
   b. The median
   c. The mean

Interval	Midpoint	Frequency
90–99		6
80–89		8
70–79		4
60–69		3
50–59		2

4. For the set of data below, calculate:
   a. The mode
   b. The median
   c. The mean

Interval	f
90–100	5
80–90	7
70–80	9
60–70	4

5. For each of the variables in the frequency tables that follow (from the gang database), describe the level of measurement for each variable and how you determined your answer.

6. Using the frequency tables that follow (from the gang database), discuss the three measures of central tendency.

HOME: What type of house do you live in?

Value Label	Value	Frequency	Percent	Valid Percent	Cumulative Percent
House	1	280	81.6	82.4	82.4
Duplex	2	3	.9	.9	83.2
Trailer	3	34	9.9	10.0	93.2
Apartment	4	21	6.1	6.2	99.4
Other	5	2	.6	.6	100.0
Missing		3	.9	Missing	
Total		343	100.0	100.0	

N	Valid	340
	Missing	3
Mean		1.41
Std. Error of Mean		.051
Median		1.00
Mode		1
Std. Deviation		.945
Variance		.892
Skewness		2.001
Std. Error of Skewness		.132
Kurtosis		2.613
Std. Error of Kurtosis		.264
Range		5

ARREST: How many times have you been arrested?

	Value	Frequency	Percent	Valid Percent	Cumulative Percent
	0	243	70.8	86.2	86.2
	1	23	6.7	8.2	94.3
	2	10	2.9	3.5	97.9
	3	3	.9	1.1	98.9
	5	2	.6	.7	99.6
	24	1	.3	.4	100.0
Missing		61	17.8	Missing	
Total		343	100.0	100.0	

N	Valid	282
	Missing	61
Mean		.30
Std. Error of Mean		.093
Median		.00
Mode		0
Std. Deviation		1.567
Variance		2.455
Skewness		12.692
Std. Error of Skewness		.145
Kurtosis		187.898
Std. Error of Kurtosis		.289
Range		24

TENURE: How long have you lived at your current address (months)?

Value	Frequency	Percent	Valid Percent	Cumulative Percent
1	14	4.1	4.3	4.3
2	6	1.7	1.8	6.1
3	4	1.2	1.2	7.3
4	4	1.2	1.2	8.6
5	6	1.7	1.8	10.4
6	6	1.7	1.8	12.2
7	1	.3	.3	12.5
8	3	.9	.9	13.5
9	2	.6	.6	14.1
10	1	.3	.3	14.4
11	1	.3	.3	14.7
12	11	3.2	3.4	18.0
14	1	.3	.3	18.3
18	5	1.5	1.5	19.9
21	1	.3	.3	20.2
24	30	8.7	9.2	29.4
30	1	.3	.3	29.7
31	1	.3	.3	30.0
32	1	.3	.3	30.3
36	22	6.4	6.7	37.0
42	1	.3	.3	37.3
48	12	3.5	3.7	41.0
60	24	7.0	7.3	48.3
72	14	4.1	4.3	52.6
76	1	.3	.3	52.9
84	8	2.3	2.4	55.4
96	18	5.2	5.5	60.9
108	4	1.2	1.2	62.1
120	9	2.6	2.8	64.8
132	11	3.2	3.4	68.2

Value	Frequency	Percent	Valid Percent	Cumulative Percent
144	21	6.1	6.4	74.6
156	13	3.8	4.0	78.6
168	11	3.2	3.4	82.0
170	5	1.5	1.5	83.5
180	7	2.0	2.1	85.6
182	2	.6	.6	86.2
186	1	.3	.3	86.5
192	14	4.1	4.3	90.8
198	1	.3	.3	91.1
204	24	7.0	7.3	98.5
216	3	.9	.9	99.4
240	2	.6	.6	100.0
Missing	16	4.7	Missing	
Total	343	100.0	100.0	

N	Valid	327
	Missing	16
Mean		88.77
Std. Error of Mean		3.880
Median		72.00
Mode		24
Std. Deviation		70.164
Variance		4923.055
Skewness		.365
Std. Error of Skewness		.135
Kurtosis		−1.284
Std. Error of Kurtosis		.269
Range		239

SIBS: How many brothers and sisters do you have?

Value	Frequency	Percent	Valid Percent	Cumulative Percent
0	39	11.4	11.5	11.5
1	137	39.9	40.5	52.1
2	79	23.0	23.4	75.4
3	39	11.4	11.5	87.0
4	17	5.0	5.0	92.0
5	13	3.8	3.8	95.9
6	6	1.7	1.8	97.6
7	4	1.2	1.2	98.8
9	1	.3	.3	99.1
10	1	.3	.3	99.4

	Value	Frequency	Percent	Valid Percent	Cumulative Percent
	12	1	.3	.3	99.7
	15	1	.3	.3	100.0
Missing		5	1.5	Missing	
Total		343	100.0	Total	

N	Valid	338
	Missing	5
Mean		1.94
Std. Error of Mean		.098
Median		1.00
Mode		1
Std. Deviation		1.801
Variance		3.245
Skewness		2.664
Std. Error of Skewness		.133
Kurtosis		12.027
Std. Error of Kurtosis		.265
Range		15

References

Edwards, W. J., N. White, I. Bennett, and F. Pezzella (1998). Who has come out of the pipeline? African Americans in criminology and criminal justice. *Journal of Criminal Justice Education*, 9(2):249–266.

Galton, F. (1883). *Inquiries into Human Faculty and Its Development*. London: Macmillan.

Pearson K. (1895). Classification of asymmetrical frequency curves in general: Types actually occurring. *Philosophical Transactions of the Royal Society of London*, Series A, Vol. 186. London: Cambridge University Press.

Notes

1. This may not be a valid assumption, and it is possible for example that all the scores could be 14, but it would be impossible to calculate the median without deconstructing the values, so an assumption is made that all values in the median class are equally distributed.
2. For future reference, this formula is the same (except for the 0.5) as the one used for computing percentiles because the median is the 50th percentile of the distribution.
3. This procedure assumes closed intervals for each class. If you have a situation, say, where the oldest category of an age distribution is "6 and above," it is more difficult to determine the midpoint. It is sometimes necessary to make an estimate of where the central value of the class might be.

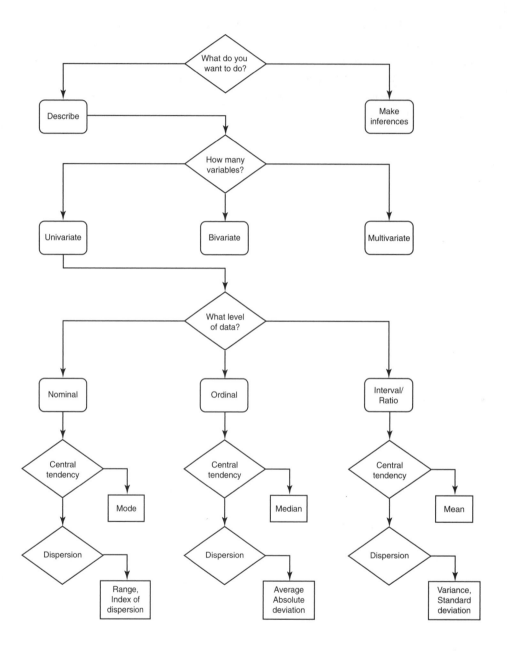

Measures of Dispersion

CHAPTER

5

It is important to know the central value of a distribution; but it is just one characteristic of the distribution. It is also important to know how spread out the values are, together with the central value, because it is possible to have two data sets with the same measure of central tendency but with very different distributions. For example, two tests were given, with the following results:

Test 1:   0   80   85   90   95   100

Test 2:   75   75   75   75   75   75

Each test has a mean of 75. The scores of test 1 range from 0 to 100, whereas all of the scores for test 2 are 75. These distributions have the same mean but very different spreads of scores. If you did not know your score and had to choose which group you were going to be in, you would want to know the spread of scores. If you are a weak student, it might be better to go with the all-75 group, whereas if you are a good student, you would definitely want to go with the spread group. This is the reason it is important to know the dispersion of the distribution together with the measure of central tendency. Measures of dispersion measure how narrow or how spread out the values are around the central value.

■ Deviation and Dispersion

Dispersion and deviation are largely synonymous. The only practical difference is that *deviation* typically refers to the difference between a single value or case and the measure of central tendency, whereas **dispersion** is used more to refer to the overall difference between all cases or values and the measure of central tendency.

Dispersion and deviation are important to research. All variables have dispersion; if not, they would not be variables (values that change between cases), they would be constants (a constant value no matter what the case). Take juvenile delinquency, for example. Some juveniles do not commit any delinquent acts during their teenage years (at least theoretically), many juveniles commit some delinquent acts, and some juveniles commit many delinquent acts.[1] This is what makes the variable delinquency a variable: The number of delinquent acts varies among juveniles. Delinquency is not a constant because not every juvenile commits the same number of delinquent acts. This

103

also means there is dispersion among the number of juvenile acts committed: Some juveniles commit a small number of acts, some a larger number, and some commit many delinquent acts. If a researcher asked how many delinquent acts a juvenile committed on average, the answer would represent the central value of juvenile delinquency. That does not tell the entire story, however. The researcher also needs to know how the juveniles differ from each other in their delinquency. This is the measure of dispersion.

Not only is dispersion an important addition to measures of central tendency but the two are closely related. As introduced in Chapter 4, two properties of the mean are that (1) the sum of the deviations from the mean always equals zero, and (2) the sum of the squared deviations from the mean is the smallest of any of the measures of central tendency. The first of these properties can be illustrated here. If we took the data (X) shown in **Table 5-1** and subtracted each value from the mean $(\overline{X} = 3.934)$, the sum of those scores would be zero. Here, deviations from the mean are also compared to deviations from the median and the mode. As can be seen, both of these numbers produced nonzero deviations, whereas the deviations from the mean are zero.

As also discussed in Chapter 4, the **sum of the squared deviations** from the mean is the smallest value for summed deviations, smaller than if the same calculations were made for the mode or median. This principle is shown in **Table 5-2**, where the second column shows the result of subtracting the mean (3.934) from each value and then squaring the result. When these values are summed, the result is 50.93. The third column shows the result of subtracting the median (4) from each value and then squaring the result. When these values are summed, the result is 51, larger than the sum of

TABLE 5-1 Sum of Deviations from the Mean, Median, and Mode

X	$X - \overline{X}$	$X - Me$	$X - Mo$
7	3.066	3	4
7	3.066	3	4
6	2.066	2	3
5	1.066	1	2
5	1.066	1	2
5	1.066	1	2
4	0.066	0	1
4	0.066	0	1
3	−0.934	−1	0
3	−0.934	−1	0
3	−0.934	−1	0
3	−0.934	−1	0
2	−1.934	−2	−1
1	−2.934	−3	−2
1	−2.934	−3	−2
Σ 59	0	−1	14

TABLE 5-2 Sum of Squared Deviations from the Measures of Central Tendency

X	$(X - \overline{X})^2$	$(X - Me)^2$	$(X - Mo)^2$
7	9.40	9	16
7	9.40	9	16
6	4.27	4	9
5	1.14	1	4
5	1.14	1	4
5	1.14	1	4
4	0.00	0	1
4	0.00	0	1
3	0.87	1	0
3	0.87	1	0
3	0.87	1	0
3	0.87	1	0
2	3.74	4	1
1	8.61	9	4
1	8.61	9	4
Σ 59	50.93	51	64

squared deviations from the mean. Finally, the fourth column shows the result of subtracting the mode (3) from each value and then squaring the result. When these values are summed, the result is 64, substantially larger than the sum of the squared deviations from the mean. This characteristic of the mean will become very important in the chapters on correlation and regression. The characteristics of deviations and dispersion are also important in discussing measures of dispersion as a univariate descriptive statistic.

■ Measures of Dispersion

Although there are measures of dispersion that are more appropriate for certain levels of measurement, it is not as clear cut as measures of central tendency. The dispersion of nominal level data typically is best analyzed with the range or an index of dispersion. These two, together with the mean absolute deviations from the median, can be used with ordinal level data. Interval and ratio level data can be examined with the variance or standard deviation. Each of these measures of dispersion is discussed in this chapter.

Range   *Typically Nominal ; partially ordered ordinal*

The simplest measure of the dispersion of a distribution is the **range**. The range is generally used with nominal and partially ordered ordinal variables, although it is sometimes included as an additional measure of dispersion for higher-level variables. The range is simply the difference between the highest and lowest values in a distribution. For example, if the spread of ages in a class was from 18 to 35, the range would be 17.

Although the range is generally stated in terms of a single figure, it is sometimes stated simply as the two extreme values. In the example above, the range of scores could be said to be 18–35. This is particularly true for nominal and ordinal level data. For example, in **Table 5-3** the range could be stated as 5(6 − 1). This does not make much sense, though. It is more easily interpreted if the range is stated as "less than high school to postgraduate."

The range is useful in that it is a quick and easily calculated measure of dispersion. Although most statistical packages include the range as a measure of dispersion (see Table 5-3), it is generally not necessary to use this feature except when examining a very large data set where the number of values might make it time consuming to determine the highest and lowest value. In the SPSS printout in Table 5-3, the range would be simple to calculate even without knowing the value from the descriptive statistics. Here, the range of values would be 5(6 − 1). It might be cumbersome, however, to determine the range simply by examining the raw data. Once the data is in a table, however, it is a simple matter.

TABLE 5-3 SPSS Output of Dispersion

What is your highest level of education?

Value Label	Value	Frequency	Percent	Valid Percent	Cumulative Percent
Less than High School	1	16	4.6	4.8	4.8
GED	2	59	17.0	17.6	22.3
High School Graduate	3	8	2.3	2.4	24.7
Some College	4	117	33.7	34.8	59.5
College Graduate	5	72	20.7	21.4	81.0
Post Graduate	6	64	18.4	19.0	100.0
Missing		11	3.2	Missing	
Total		347	100.0	100.0	

N	Valid	336	
	Missing	11	
Mean		4.08	
Median		4.00	
Mode		4	
Std. Deviation		1.460	
Variance		2.131	
Skewness		−.477	
Std. Error of Skewness		.133	
Kurtosis		−.705	
Std. Error of Kurtosis		.265	
Range		5	

The range does have limitations, however. First, it gives only a general indication of the variability of the values in a distribution. For example, if the range is reported as 5, you must still examine the data to understand what that means; it could mean that the data ranges from 1 to 6, or it could mean that the data ranges from 101 to 106. Also, the range does not account for intracategory variation, or the way the values are distributed between the extremes. Stating that the range is 5 provides little information about how the data is arranged within the range. There could be a relatively equal number in each category, as in the freshman class in **Table 5-4**, or there could be only a few values in all but one category, as in the senior class in Table 5-4. Here, there is a relatively even proportion of freshman in each number of arrests. Most of the seniors, however, have never been arrested. The range for these two classes is the same, however (5 or 0–5).

Finally, the range is heavily influenced by extremely high or low values that may be atypical of the distribution. Take, for example, the distribution of inmate deaths in two different prisons in a state, shown in **Table 5-5**. In this table, prisons A and B have similar records of inmate deaths. In fact, only one year is different between the two prisons; in 1990, prison B had a riot during which 26 prisoners were killed. This will heavily influence the range for this prison. The range for prison A is 3 (3 − 0) or 0–3, but the range for prison B is 26 (26 − 0) or 0–26. This demonstrates the problems with using the range as a measure of dispersion.

TABLE 5-4 Distribution of Arrests for Two Criminal Justice Classes

Number of Arrests	Freshman Class	Senior Class
0	25	99
1	20	0
2	18	0
3	15	0
4	12	0
5	10	1
Total	100	100

TABLE 5-5 Number of Prison Deaths, 1990–1995

Year	Prison A	Prison B
1990	0	26
1991	0	0
1992	1	1
1993	3	3
1994	1	1
1995	2	2

(Handwritten margin notes: "ranges from 0–1", "(where 1 = equal distribution across all possible cats)", "0 = minimum variability (all scores in 1 cat)")

(Handwritten note above heading: "D")

(Handwritten note top right: "Nominal part. ordered ordinal")

Index of Dispersion

The **index of dispersion**, D, is another measure of dispersion for nominal and partially ordered ordinal variables. The index of dispersion is not a commonly used measure of dispersion; but there are not many measures of association that are properly used with nominal and ordinal level data, resulting in many researchers using the variance or standard deviation when it is not appropriate for the data. Since there are not many measures of dispersion for nominal level data, the index of dispersion is discussed here as a possible measure.

As shown in the formula below, the index of dispersion is a ratio of the number of pairs of scores:

$$D = \frac{\text{actual numbers of pairs}}{\text{maximum number of pairs}}$$

In this equation, the denominator is the maximum number of pairs that could be created out of the scores if each set of numbers was in a different category; the numerator is the maximum number of unique pairs that can be made with the data at hand. Completing this formula requires two calculations. To calculate the maximum number of pairs, multiply the values of each combination of scores. The example in **Table 5-6** is a simplified example of the education level of probation officers. Since there are 15 total officers and 3 categories, maximum variability would occur when each category had 5 officers. Calculating the maximum number of pairs would then be accomplished by multiplying the frequencies of each of the categories as follows: 5 scores for college grad × 5 scores for some college = 25; 5 scores for college grad × 5 scores for no college = 25; and 5 scores for some college × 5 scores for no college = 25. Adding the 3 combinations achieves a score of 75. The actual number of pairs is then calculated by multiplying the actual frequency of the combinations of each category. Here, the actual number of pairs would be 66: 2 college grad × 8 some college, 2 college grad × 5 no college, and 8 some college × 5 no college. The index of dispersion for this data, then, is simply the ratio of the maximum number of possible pairs and the actual number of pairs:

$$D = \frac{\text{actual numbers of pairs}}{\text{maximum number of pairs}} = \frac{66}{75} = 0.88$$

The value of the index of dispersion runs from 0 to 1. If the cases were equally distributed across all of the possible categories, the actual number of pairs (numerator)

TABLE 5-6 Education of Probation Officers

Value	f
College grad	2
Some college	8
No college	5
N	15

would equal the maximum possible number of pairs, and the index of dispersion would equal 1 (see **Exhibit 5-1**). If all of the scores are in one category, there would be minimum variability and the index of dispersion would be zero (see **Exhibit 5-2**). In Table 5-6 and in the equation above, the value of 0.88 shows there is almost equal dispersion within the categories.

For more complicated data sets, the index of dispersion can be computed using the formula

$$D = \frac{k(N^2 - \Sigma f^2)}{N^2(k - 1)}$$

$k = \# \text{ cats}$

where k is the possible number of categories in the data set (in the example above, 3; even categories with 0 scores would be included); N is the number of cases (in the example, N would be 15); and Σf^2 is the sum of the squared frequencies.

Using the data from Table 5-6, use of this formula to calculate the index of dispersion can be demonstrated. In this example, there were 3 categories and 15 total cases. The only other calculation that needs to be made is Σf^2. In this simple case, f^2 for the

EXHIBIT 5-1 Index of Dispersion at Maximum Dispersion

Value	f
College grad	5
Some college	5
No college	5
N	15

In this set of scores, the frequencies are divided evenly among the three categories. This represents the maximum variability of the numerator. Here, the maximum number of scores that could be created is 75, as outlined above. Also, since the number of scores in each category is the same as the maximum number of pairs, the actual number of pairs is the same value (75).

$$D = \frac{\text{actual number of pairs}}{\text{maximum number of pairs}} = \frac{75}{75} = 1$$

EXHIBIT 5-2 Index of Dispersion at Minimum Dispersion

Value	f
College grad	0
Some college	0
No college	15
N	15

In this set of scores, the frequencies are grouped entirely into one category. This represents the minimum variability of the numerator. Here, the maximum number of scores that could be created is 75, as outlined above. However, since the number of scores in each category is 0 except for one category, all calculations for the actual number of pairs equal 0. Therefore, the calculation of the index of dispersion is

$$D = \frac{\text{actual number of pairs}}{\text{maximum number of pairs}} = \frac{0}{75} = 0$$

three categories would be $4(2^2)$, $64(8^2)$, and $25(5^2)$, for a total of 93. Entering these values into the formula above would result in the following calculation:

$$D = \frac{3(15^2 - 93)}{15^2(3 - 1)}$$

$$= \frac{3(225 - 93)}{225(2)}$$

$$= \frac{3(132)}{450}$$

$$= \frac{369}{450}$$

$$= 0.88$$

This is the same value that was obtained in the previous calculations using Table 5-6. There is no provision in SPSS to calculate the index of dispersion, so any use in actual research will require that this formula be calculated by hand.

Mean Absolute Deviation

The **mean** (*average*) **absolute deviation** is somewhat different from other measures of dispersion. This measure could be used with any measure of central tendency. For example, you could calculate the mean absolute deviation of scores from the mean, the mean absolute deviation of scores from the median, or even the mean absolute deviation of scores from the mode. Why an absolute deviation? It makes sense that if we wanted to examine deviation (dispersion) for a set of data, we could simply calculate the mean and then measure how far each value is from the mean. As discussed above and in Chapter 4, however, when the deviations from the mean are summed, the result is zero. That value does not tell anything about the dispersion. What is needed is a way to remove the signs of the values so they will sum to a positive value. That can be done by taking the absolute value of each deviation, that is, simply removing the sign.

The mean absolute deviation from the median has advantages over other measures because the mean absolute deviation from the median is a smaller number than for either the mean or the mode. The mean absolute deviation from the median is not often used as a measure of dispersion. There is no standard measure of dispersion for ordinal level data, however, and since the median is the most appropriate measure of central tendency for use with ordinal and skewed interval and ratio level data, it can be argued that the mean absolute deviation from the median is a valid measure of dispersion for this type of data.

The mean absolute deviation from the median is determined by calculating the median, subtracting that value from each score, taking the absolute value (removing the sign), and dividing by N. A simple example is shown in **Table 5-7**. Notice that this is the same prisoner escape data that was used in Chapter 4 and Table 5-1. Here, the median is 4 escapes, the mean number of escapes is 3.93, and the modal number of escapes is 3. Subtracting each score from the median is shown in column d, and the same calculations for the mean and mode are shown in columns b and f. Note here that

TABLE 5-7 Comparison of Absolute Values of Mean, Median, and Mode

a X	b $X - \bar{X}$	c $\|X - \bar{X}\|$	d $X - Me$	e $\|X - Me\|$	f $X - Mo$	g $\|X - Mo\|$
7	3.066	3.066	3	3	4	4
7	3.066	3.066	3	3	4	4
6	2.066	2.066	2	2	3	3
5	1.066	1.066	1	1	2	2
5	1.066	1.066	1	1	2	2
5	1.066	1.066	1	1	2	2
4	0.066	0.066	0	0	1	1
4	0.066	0.066	0	0	1	1
3	−0.934	0.934	−1	1	0	0
3	−0.934	0.934	−1	1	0	0
3	−0.934	0.934	−1	1	0	0
3	−0.934	0.934	−1	1	0	0
2	−1.934	1.934	−2	2	−1	1
1	−2.934	2.934	−3	3	−2	2
1	−2.934	2.934	−3	3	−2	2
Σ	0	23.066	−1	23	14	24

[handwritten: Smallest /N (15) = 1.53]

as discussed above, when the mean is subtracted from the values, the sum of these deviations is zero and is smaller than the deviations from the median (column d) or the mode (column f). When the absolute value of the deviations is calculated, however, a different distribution emerges. The absolute values of the scores from the median are shown in column e, and the absolute values of the mean and mode are shown in columns c and g. The total of the absolute deviations from the median is 23. This is the smallest value for the three measures, with the sum of absolute deviations from the mean being 23.066 and the sum of absolute deviations from the mode being 24. Dividing by N (15) gives a mean absolute deviation from the median of 1.53.

Using the mean absolute deviation from the median has an additional advantage. The result is a standardized measure, meaning it can be compared across groups. Like the index of dispersion, there is no direct mechanism in SPSS for calculating the mean absolute deviation from the median. It can be calculated with macros, however, and makes a good addition to the other measures of dispersion. *[handwritten margin: Standardized measure (comparable)]*

Variance

At the opposite end of the scale of sophistication are the variance and standard deviation. The standard deviation is simply the square root of the variance, so these measures are often spoken of together. These two measures of dispersion are most appropriate for interval and ratio level data, although they are often used with ordinal (especially fully ordered ordinal) level data because of their power over other measures of dispersion. The variance and standard deviation do not work well with nominal and *[handwritten margin: Intv. Ratio (or fully ordered ordinal sometimes)]* partially ordered ordinal level data because they measure the deviation about the mean; if the mean is not appropriate for nominal and ordinal level data, neither would be measures of dispersion based on the mean.

The variance was developed by Ronald A. Fisher in 1918 in connection with his work on the analysis of variance (see Chapter 18). Fisher used the variance in studying human traits and inheritibility because it was mathematically superior to the standard deviation, which was developed by J. F. Enke (1832).

The variance and standard deviation are based on the mean. In fact, the variance, and therefore the standard deviation, are nothing more than a special kind of mean—the average of the squared deviations from the mean. Squaring the deviations is desirable because, as discussed above, the sum of the differences of each value in a distribution from the mean is zero and because the sum of the squared differences of each value in a distribution from the mean yields the minimum summed value. These two properties make the variance and standard deviation ideal for mathematical calculations in the statistical procedures to come and play a big part in calculating the "best-fitting line," which is a goal when comparing two or more variables.

The variance (represented by s^2 for samples or σ^2 for populations) measures the average of squared deviations of scores around the mean. This was the process that was undertaken in Tables 5-1 and 5-7. The formula for calculating the variance is

$$\sigma^2 = \frac{\Sigma(X - \overline{X})^2}{N}$$

If the differences are squared and then summed, the result will be the smallest distance from any given value to the mean (as in Table 5-2). Dividing the figure by N results in a formula that is the mean of the squared deviations, just as when calculating the mean of any data set.

Recall that in this section of the book we are dealing with descriptive statistics. That means we are assuming that the data being analyzed represents a population. Again, $\overline{X}$ is used rather than μ because of convention, even though μ would be more appropriate because we are examining a population. If the data were considered a sample of a population, the formula for the variance would become

$$s^2 = \frac{\Sigma(X - \overline{X})^2}{N - 1}$$

Here the denominator of the formula is $N - 1$ rather than N. Dividing by $N - 1$ rather than N is a procedure used to account for bias in small samples. This is discussed more thoroughly in the section of the book on inferential analysies. In this portion of the book, the formula for dealing with populations is used because descriptive statistics assumes a population is being analyzed.

Calculating the variance requires calculating the mean, calculating the difference between the mean and each value (X or midpoint), squaring the differences, and dividing by N. In **Table 5-8**, the mean is 4(92/23). The third column represents calculations that subtract the mean from each value of X. For example, the first prison had 7 escapes in 10 years. Since the mean is 4, this prison had 3 more escapes than the mean. This can be repeated for each score, where each prison's score can be seen as higher

TABLE 5-8 Calculation of the Variance

X	f	$X - \bar{X}$	$(X - \bar{X})^2$
7	1	3	9
7	1	3	9
6	1	2	4
6	1	2	4
6	1	2	4
5	1	1	1
5	1	1	1
5	1	1	1
5	1	1	1
4	1	0	0
4	1	0	0
4	1	0	0
4	1	0	0
4	1	0	0
3	1	−1	1
3	1	−1	1
3	1	−1	1
3	1	−1	1
2	1	−2	4
2	1	−2	4
2	1	−2	4
1	1	−3	9
1	1	−3	9
Σ 92	23	0	68

$$\bar{X} = 4$$

$$\sigma^2 = \frac{\Sigma(X - \bar{X})^2}{N}$$

$$= \frac{68}{23}$$

$$= 2.96$$

than, lower than, or the same as the mean. Notice also that the sum of this column is zero (as shown in Table 5-1).

The fourth column takes each of these values and squares them (the 3 escapes higher that the first prison was above the mean is squared, resulting in a value of 9). The sum of these calculations equals 68. It is then simply a matter of dividing this value by N (23), which is the same as calculating the mean (average) of the squared deviations. The result is 2.96, so the mean of deviations is 2.96 (this is the variance).

To interpret the variance, if all the values in a distribution are the same (the value of the mean), the value of the variance will be zero. In this case, if all the values of X had been 4, the mean would have been 4 and the variance would have been zero. The variance will be at a maximum when all the scores are grouped in the extremes of the distribution (the highest and lowest values in the distribution). For example, if the data from Table 5-8 were altered so each value was in the extremes of the distribution (each

TABLE 5-9 Calculation of the Maximum Variance

X	f	$X - \overline{X}$	$(X - \overline{X})^2$
7	1	2.87	8.2369
7	1	2.87	8.2369
7	1	2.87	8.2369
7	1	2.87	8.2369
7	1	2.87	8.2369
7	1	2.87	8.2369
7	1	2.87	8.2369
7	1	2.87	8.2369
7	1	2.87	8.2369
7	1	2.87	8.2369
7	1	2.87	8.2369
7	1	2.87	8.2369
1	1	−3.13	9.7969
1	1	−3.13	9.7969
1	1	−3.13	9.7969
1	1	−3.13	9.7969
1	1	−3.13	9.7969
1	1	−3.13	9.7969
1	1	−3.13	9.7969
1	1	−3.13	9.7969
1	1	−3.13	9.7969
1	1	−3.13	9.7969
1	1	−3.13	9.7969
Σ 95	23	0.1	202.61

$$\overline{X} = 4.13$$

$$\sigma^2 = \frac{\Sigma(X - \overline{X})^2}{N}$$

$$= \frac{206.61}{23}$$

$$= 8.98$$

value was either a 7 or a 1), it would be represented by the distribution shown in **Table 5-9**. Here, although f remains 23, the mean will be different because the ΣX has changed to 95. Additionally, although the sum of differences from the mean remains 0 (actually, 0.1 because of rounding), the square of this value is much larger (206.61 rather than 68). This would be the first indicator that the variance is going to be much larger. Indeed, calculation of the variance for this data set produces a variance of 8.98, almost three times as large as the variance in Table 5-8.

Table 5-10 shows the output that has been used with all of the other measures of central tendency and with the range. Here the variance is 2.131. This is substantially lower than either of the values in the two earlier examples. Part of the reason for this is the nature of the data; the data in Tables 5-8 and 5-9 ranges from 1 to 7, while the data in Table 5-10 ranges from 1 to 6. It is also evident that there is less dispersion in the data in Table 5-10 than in Table 5-8 or 5-9.

TABLE 5-10 SPSS Output of Dispersion

What is your highest level of education?

Value Label	Value	Frequency	Percent	Valid Percent	Cumulative Percent
Less than High School	1	16	4.6	4.8	4.8
GED	2	59	17.0	17.6	22.3
High School Graduate	3	8	2.3	2.4	24.7
Some College	4	117	33.7	34.8	59.5
College Graduate	5	72	20.7	21.4	81.0
Post Graduate	6	64	18.4	19.0	100.0
Missing		11	3.2	Missing	
Total		347	100.0	100.0	

N	Valid	336
	Missing	11
Mean		4.08
Median		4.00
Mode		4
Std. Deviation		1.460
Variance		2.131
Skewness		−.477
Std. Error of Skewness		.133
Kurtosis		−.705
Std. Error of Kurtosis		.265
Range		5

The variance is important to statistical analysis for several reasons. First, as discussed above, the variance is the mean of the squared deviations, which makes it useful in higher-order statistical procedures such as regression (discussed later in the book). The squared deviations of the variance are also important because they give weight to extreme values. This represents an increased measure of spread as the deviation rises. For example, a deviation of 2 from the mean has more weight than two deviations of 1 each ($2^2 = 4 > 1^2 + 1^2 = 2$). Because of this characteristic, data sets with many small deviations will show smaller variance than data sets with fewer but larger deviations. Since we are attempting to measure deviation from the measure of central tendency, this characteristic is important. Finally, the expected value of the variance of a sample (using $N - 1$ as the denominator of the formula) is equal to the population variance, whereas the standard deviation is not. Differences in statistical procedures for samples and for populations are discussed in the section of the book on inferential analyses, but it can be stated here that statistical procedures that can cross over between samples and populations are important for some analyses.

The squaring of deviations presents a problem for the variance, however. Since all values are squared, the units of measurement no longer fit the original values, which makes the variance difficult to interpret. For example, it is difficult or impossible to interpret a value of 8.98 (from Table 5-9) when the values range only from 1 to 7. This is the reason the standard deviation has become important.

A final note about the variance (which also applies to the standard deviation) is that it is difficult to calculate a variance for grouped data. There are three reasons for this. First, the variance assumes the data is interval or ratio level. Even if a variable is interval level, when it is grouped, it generally becomes ordinal level, thereby violating the assumption of the variance. Further, when a variable is grouped, the midpoint is used rather than the numbers themselves (as in Table 3-3). The midpoint is an approximation of the values in a particular category and is thus not a true representation of the data. This means the variance will be slightly off (at best)—you will not get a value of zero if you add all the scores. Finally, even if you do want to use the variance with grouped data, you would have to ungroup the data. For example, if you had a midpoint of 18 and a frequency of 5, as in Table 3-3, you could not simply calculate $X - \bar{X}$ using the midpoint (18). You would have to list out $18 -$ the mean, five times and repeat this for each value.

Standard Deviation

As discussed above, the variance is often difficult to interpret. You can see from the example that when the range of values is from 1 to 7, a variance of 8.98 makes little intuitive sense. The **standard deviation** resolves this problem by putting the dispersion in the same units as the distribution. The standard deviation is calculated by taking the square root of the variance. Since the differences from the mean were squared to account for the sign (thus putting the values on a scale different from that of the data), it is a simple mathematical procedure to take the square root of the sum of those values and put the sum back on the same scale as that of the original data.

The standard deviation (represented by s or σ) was developed by J. F. Enke (who was a student of Gauss) in 1832. The term *standard deviation* was formally coined by Karl Pearson in 1894. Calculation of the standard deviation is both simple and complex. The simple part is that if the variance is known, you can simply take the square root of that value. If the variance is not known, however, that value must be determined prior to calculating the standard deviation. The formula for calculating the standard deviation without knowing the variance beforehand is

$$\sigma = \sqrt{\frac{\Sigma\left(X - \bar{X}\right)^2}{N}}$$

This is, of course, the square root of the variance. In the example from Table 5-9, the standard deviation would be

$$\sigma = \sqrt{\sigma^2} = \sqrt{8.98} = 2.99$$

This value is much easier to interpret in terms of the range of scores in this distribution. It is also important to note that the standard deviation can be interpreted in terms of the number of standard deviations from the mean (Z scores; Chapter 6). This is important in discussions of the normal curve and inferential statistics.

Uses for the Variance and Standard Deviation

Using the variance and standard deviation is not unlike using other commonly used standard measurements. This is because they can be used to compare two distributions. In the example given at the beginning of Chapter 4, you cannot tell much about the distributions from the measure of central tendency; you also need to know how spread out the values are. When comparing two distributions, both pieces of information are needed. For example, assume that you take tests in criminal law and in police administration and score 80 on one test and 75 on the other. You really need to know not only this information but also what other students scored. You need the average score *and* how spread out the scores were. An 85 when the dispersion is zero is probably not as important to your grades as a 75 where the dispersion is very large, because in the first instance, you know you are going to get an 85 on the test, whereas in the second example, you may get any score from an A to a C.

The variance is also important in other statistical procedures. Aside from the fact that the variance, along with the mean, is the foundation of many of the statistical formulas discussed later in the book, the variance is a fundamental component of some tests, such as analysis of variance (ANOVA).

Also, the most important use for the standard deviation is yet to come. When we begin to examine the normal curve and to apply the use of the normal curve in inferential analysis, the standard deviation will become very important because this is how the normal curve is divided and discussed, as well as how to determine where a particular value stands in a distribution.

■ Selecting the Most Appropriate Measure of Dispersion

It is more difficult to determine the appropriate measure of dispersion than to select a measure of central tendency because the measures of dispersion are not as tied to the level of measurement. The range can always be used regardless of the level of data or form of the distribution, although it is limited as to the information provided. For nominal level data and for some ordinal level data, the index of dispersion is a good choice. For interval/ratio level data and some ordinal level data, the variance and standard deviation are preferred. There is a caveat, however: If the median is the preferred measure of central tendency because the data is ordinal or the distribution is skewed, the mean absolute deviation may be preferred over the variance or standard deviation. The reason for this is simple but often overlooked. If you are using the median rather than the mean, it should be because the data is such that the mean is not an appropriate measure of central tendency, either because the data is ordinal level or because of skew. Why would you want to use a measure of dispersion based on the measure of central tendency that is not appropriate for the data? In this instance, it is more appropriate to use the mean absolute deviation because the mean absolute deviation about the median is a complementary measure of dispersion for the median.

■ Conclusion

This chapter added to your ability to describe data in terms of the central or most frequently occurring value by outlining procedures to assess how values differ from each other within a single variable. Although there are no widely accepted measures of variability for categorical (nominal and ordinal level) data, because this level of data is not expected to vary much or in a uniform fashion, some potentially useful measures of dispersion were offered. The discussion of the variance and standard deviation has set the stage for their use in later chapters as well as for their use in measuring deviation or dispersion within a variable.

The next chapter will conclude the description of one variable by describing the form or shape of the variable. Topics will include how spread out the overall distribution is (kurtosis), whether the data is grouped in one end of the values (skewness), and if there is more than one mode for the distribution.

■ Key Terms

dispersion
index of dispersion
mean absolute deviation
range

standard deviation
sum of squared deviations
variance

■ Summary of Equations

Index of Dispersion

$$D = \frac{\text{actual number of pairs}}{\text{maximum number of pairs}}$$

Index of Dispersion (complicated data sets)

$$D = \frac{k(N^2 - \Sigma f^2)}{N^2(k - 1)}$$

Variance (samples)

$$\sigma^2 = \frac{\Sigma(X - \overline{X})^2}{N}$$

Variance (population)

$$s^2 = \frac{\Sigma(X - \overline{X})^2}{N - 1}$$

Standard Deviation

$$\sigma = \sqrt{\frac{\Sigma(X - \overline{X})^2}{N}}$$

■ Exercises

1. For the following set of data, calculate:
   a. The range
   b. The variance
   c. The standard deviation

6, 7, 8, 10, 10, 10, 12, 14

2. For the following set of data, calculate:
   a. The range
   b. The variance
   c. The standard deviation

$$7, 4, 2, 3, 4, 5, 8, 1, 9, 4$$

3. For the following data set, calculate:
   a. The range
   b. The variance
   c. The standard deviation

Interval	Midpoint	Frequency
90–99		6
80–89		8
70–79		4
60–69		3
50–59		2

4. For the following set of data, calculate:
   a. The range
   b. The variance
   c. The standard deviation

Interval	f
90–100	5
80–90	7
70–80	9
60–70	4

5. Using the frequency tables that follow, discuss the measures of dispersion that would be appropriate for each.

   HOME: What type of house do you live in?

Value Label	Value	Frequency	Percent	Valid Percent	Cumulative Percent
House	1	280	81.6	82.4	82.4
Duplex	2	3	.9	.9	83.2
Trailer	3	34	9.9	10.0	93.2
Apartment	4	21	6.1	6.2	99.4
Other	5	2	.6	.6	100.0
Missing		3	.9	Missing	
Total		343	100.0	100.0	

N	Valid	340	
	Missing	3	
Mean		1.41	
Std. Error of Mean		.051	
Median		1.00	
Mode		1	
Std. Deviation		.945	
Variance		.892	
Skewness		2.001	
Std. Error of Skewness		.132	
Kurtosis		2.613	
Std. Error of Kurtosis		.264	
Range		5	

ARREST: How many times have you been arrested?

	Value	Frequency	Percent	Valid Percent	Cumulative Percent
	0	243	70.8	86.2	86.2
	1	23	6.7	8.2	94.3
	2	10	2.9	3.5	97.9
	3	3	.9	1.1	98.9
	5	2	.6	.7	99.6
	24	1	.3	.4	100.0
Missing		61	17.8	Missing	
Total		343	100.0	100.0	

N	Valid	282	
	Missing	61	
Mean		.30	
Std. Error of Mean		.093	
Median		.00	
Mode		0	
Std. Deviation		1.567	
Variance		2.455	
Skewness		12.692	
Std. Error of Skewness		.145	
Kurtosis		187.898	
Std. Error of Kurtosis		.289	
Range		24	

TENURE: How long have you lived at your current address (months)?

Value	Frequency	Percent	Valid Percent	Cumulative Percent
1	14	4.1	4.3	4.3
2	6	1.7	1.8	6.1
3	4	1.2	1.2	7.3
4	4	1.2	1.2	8.6
5	6	1.7	1.8	10.4
6	6	1.7	1.8	12.2
7	1	.3	.3	12.5
8	3	.9	.9	13.5
9	2	.6	.6	14.1
10	1	.3	.3	14.4
11	1	.3	.3	14.7
12	11	3.2	3.4	18.0
14	1	.3	.3	18.3
18	5	1.5	1.5	19.9
21	1	.3	.3	20.2
24	30	8.7	9.2	29.4
30	1	.3	.3	29.7
31	1	.3	.3	30.0
32	1	.3	.3	30.3
36	22	6.4	6.7	37.0
42	1	.3	.3	37.3
48	12	3.5	3.7	41.0
60	24	7.0	7.3	48.3
72	14	4.1	4.3	52.6
76	1	.3	.3	52.9
84	8	2.3	2.4	55.4
96	18	5.2	5.5	60.9
108	4	1.2	1.2	62.1
120	9	2.6	2.8	64.8
132	11	3.2	3.4	68.2
144	21	6.1	6.4	74.6
156	13	3.8	4.0	78.6
168	11	3.2	3.4	82.0
170	5	1.5	1.5	83.5
180	7	2.0	2.1	85.6
182	2	.6	.6	86.2
186	1	.3	.3	86.5
192	14	4.1	4.3	90.8
198	1	.3	.3	91.1
204	24	7.0	7.3	98.5
216	3	.9	.9	99.4
240	2	.6	.6	100.0
Missing	16	4.7	Missing	
Total	343	100.0	100.0	

N	Valid	327
	Missing	16
Mean		88.77
Std. Error of Mean		3.880
Median		72.00
Mode		24
Std. Deviation		70.164
Variance		4923.055
Skewness		.365
Std. Error of Skewness		.135
Kurtosis		−1.284
Std. Error of Kurtosis		.269
Range		239

SIBS: How many brothers and sisters do you have?

	Value	Frequency	Percent	Valid Percent	Cumulative Percent
	0	39	11.4	11.5	11.5
	1	137	39.9	40.5	52.1
	2	79	23.0	23.4	75.4
	3	39	11.4	11.5	87.0
	4	17	5.0	5.0	92.0
	5	13	3.8	3.8	95.9
	6	6	1.7	1.8	97.6
	7	4	1.2	1.2	98.8
	9	1	.3	.3	99.1
	10	1	.3	.3	99.4
	12	1	.3	.3	99.7
	15	1	.3	.3	100.0
Missing		5	1.5	Missing	
Total		343	100.0	100.0	

N	Valid	338
	Missing	5
Mean		1.94
Std. Error of Mean		.098
Median		1.00
Mode		1
Std. Deviation		1.801
Variance		3.245
Skewness		2.664
Std. Error of Skewness		.133
Kurtosis		12.027
Std. Error of Kurtosis		.265
Range		15

References

Enke, J. F. (1832). Über die methode der kleinsten quadrate. *Berliner Astronomisches Jahrbuch für 1834*, pp. 249–312. Translated 1841 in R. Taylor (ed.), *Scientific Memoirs*, Vol. 2, pp. 317–369.

Fisher, R. A. (1918). The correlation between relatives on the supposition of Mendelian inheritance. *Transcripts of the Royal Society of Edinburgh*, 52:399–433.

Pearson, K. (1894). On the dissection of asymmetrical frequency-curves: General theory. *Philosophical Transactions of the Royal Society of London*, Series A, Vol. 185. London: Cambridge University Press.

Zeller, R. A. (2000). On teaching about descriptive statistics in criminal justice. *Journal of Criminal Justice Education*, 10(2):349–360.

For Further Reading

Galton, F. (1883). *Inquiries into Human Faculty and Its Development*. London: Macmillan.

MacGillivray, H. L. (1981). The mean, median, mode inequality and skewness for a class of densities. *Australian Journal of Statistics*, 23:247.

Pearson, K. (1895). Classification of asymmetrical frequency curves in general: Types actually occurring. *Philosophical Transactions of the Royal Society of London*, Series A, Vol. 186. London: Cambridge University Press.

Note

1. An expanded version of this example can be found in Zeller (2000).

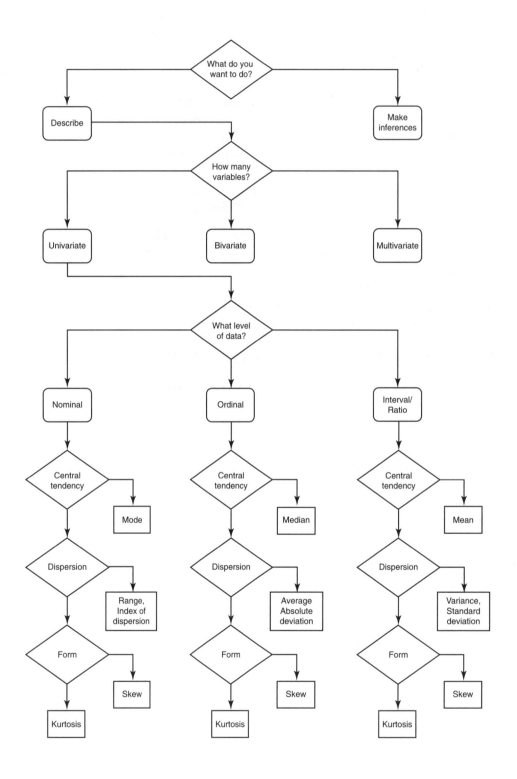

The Form of a Distribution

The final univariate descriptive statistic, the **form** of the distribution, ties together the central tendency and dispersion of the data. Three characteristics make up the form of the distribution: the number of modes, the symmetry, and the kurtosis. In addressing the form of a distribution, a polygon can generally be used to represent these characteristics visually.

■ Moments of a Distribution

In some statistics books and other places, distributions and the form of distributions are referred to in terms of the **moments of the distribution**. There are four moments that are considered important to a distribution. Moments are calculated as follows:

$$\frac{\Sigma(X - \overline{X})^i}{N}$$

where $(X - \overline{X})$ represents the deviations from the mean (as has been the case in Chapters 4 and 5), N is the total number of cases in the distribution, and i is the moment being calculated.

Since the sum of the deviations around the mean is always zero, the first moment is always zero. If X is taken to the second power in the formula above, you can see that this is the formula for the variance—thus, the second moment is the variance. The third moment is usually associated with the skew of the distribution, although the exact formula is to divide the formula for the third moment by the variance to the power of 1.5. Similarly, the kurtosis of a distribution is associated with the fourth moment, although the exact formula is to divide the formula for the fourth moment by the variance squared. The mean and variance were discussed in Chapters 4 and 5. The skew and kurtosis of a distribution are discussed in this chapter, together with the third measure of the form of a distribution: the number of modes.

■ Number of Modes

The first measure of the form of a distribution is the **number of modes**. The number of modes is important to higher-order analyses because it is indicative of the normality of the distribution. To use many bivariate and multivariate statistical procedures, a unimodal distribution is preferred.

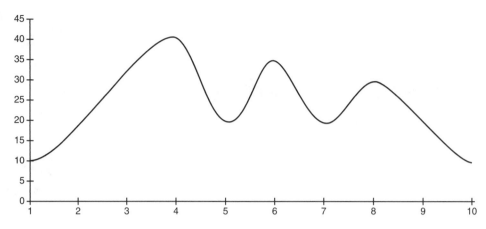

Figure 6-1 Polymodal Distribution

In determining the number of modes, a slight deviation from determining *the* mode may be necessary. Recall from the discussion of central tendency that it is common to count only the highest frequency in a distribution as the mode, even though some people argue that all peaks of a distribution should be considered. For determining the number of modes in an analysis of form, it may be more beneficial to look at peaks rather than to find the one, highest value. Consider, for example, the distribution in **Figure 6-1**. Even though there is only one highest value, there are three peaks in the distribution. These peaks may make the data unsuitable for certain statistical procedures unless transformations are made. In this distribution, all three modes should probably be counted in evaluating the form of the distribution even though the mode is actually only 4.

■ Skewness

The next characteristic of the form of the distribution is the degree of **symmetry** (skewness) of the distribution. This measure of the form of a distribution has three categories: symmetrical, positively skewed, and negatively skewed. A fully symmetrical distribution has mirror-image sides such that the distribution could be split at the mean and the sides folded over each other for a perfect match. In **Figure 6-2**, it is easy

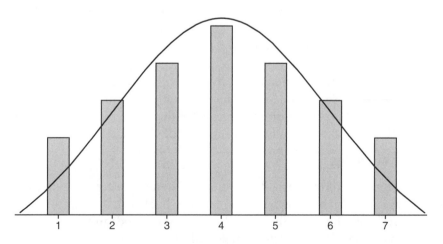

Figure 6-2 Histogram and Normal Curve for a Symmetrical Distribution

to see the symmetry in the distribution. This is the histogram from Figure 4-7. The frequencies displayed in this distribution are very balanced: categories 1 and 7 have the same frequency, as do 2 and 6, and 3 and 5. Category 4 has the highest frequency level. It is easy to see that this distribution could be folded in half and the two sides would match perfectly. This distribution is therefore, a perfectly symmetrical distribution. In actual research, however, it is not common to see a perfectly symmetrical distribution. More typically, the distribution will either be only close to symmetrical or not at all symmetrical.

It should be noted here that the number of modes does not necessarily affect the skew of the distribution. A distribution that is bimodal could still be cut in half where the distribution mirrors itself. The only difference in this case is that the mode and other measures of central tendency would not be the same.

Analysis of Skew

If a distribution is such that one side is different from the other, it is said to be **skewed**. In skewed distributions, there is no point that can be drawn in the polygon where it could be divided into two similar parts. If the point of the curve is to the left of the graph, it is said to be **negatively skewed** (the tail of the graph points to the negative end of the scale, smaller positive numbers). In **Figure 6-3**, "children" is an example of a negatively skewed distribution. Here, the point of the curve is toward category 1 or the left of the graph. If the point of the curve is to the right of the graph, it is said to be **positively skewed** (the tail of the graph points toward the positive end of the scale, larger positive numbers). In Figure 6-3, "gun-wher" is an example of a positively skewed distribution. Here, the point of the curve points toward category 12.5 or the right of the graph.

SPSS provides measures of skew in frequency output. A value of 0 means there is no skew to the data. Skew values of zero are almost never obtained, however, and a distribution is considered symmetrical if the skew value in SPSS is between −1 and 1.[1] A distribution is generally considered skewed if it has a skew greater than +1.00 or less than (a greater negative number) −1.00. The magnitude of the number will represent the degree of skew. When conducting research, it is desirable to obtain a distribution that has a skew as close to zero as possible. If the skew is outside +1 to −1, the distribution may be too skewed to work with, and efforts should be made to get the distribution closer to normal. This is done through transformations, which is addressed in the chapter on regression.

The frequency distribution that has been used with the other univariate measures is shown in **Table 6-1**. Here the value of skew is −0.477, which means that the distribution is not perfectly symmetrical but that it exhibits an acceptable level of skew (it is

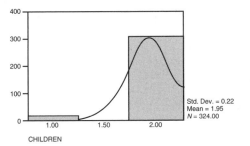
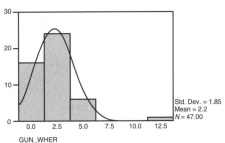

Figure 6-3 Negatively and Positively Skewed Distributions

TABLE 6-1   SPSS Output of Measures of Form

What is your highest level of education?

Value Label	Value	Frequency	Percent	Valid Percent	Cumulative Percent
Less than High School	1	16	4.6	4.8	4.8
GED	2	59	17.0	17.6	22.3
High School Graduate	3	8	2.3	2.4	24.7
Some College	4	117	33.7	34.8	59.5
College Graduate	5	72	20.7	21.4	81.0
Post Graduate	6	64	18.4	19.0	100.0
Missing		11	3.2	Missing	
Total		347	100.0	100.0	

N	Valid	336
	Missing	11
Mean		4.08
Median		4.00
Mode		4
Std. Deviation		1.460
Variance		2.131
Skewness		−.477
Std. Error of Skewness		.133
Kurtosis		−.705
Std. Error of Kurtosis		.265

within the acceptable range of 0 to −1.00). There is some negative skew to this distribution, as exhibited by the negative value, but it is not enough to warrant additional analyses or give cause for concern. If this value had been less than −1.00 (e.g., −2.77), it might have been necessary to transform the distribution.

Although quantitative measures of skewness and kurtosis will almost always be available when conducting actual research, an estimate of the skew of a distribution can be made even without a skew calculation. If the mean and the median are different, the distribution is at least somewhat skewed, although it is not possible to tell if it is beyond +1 or −1. Additionally, the skew is in the direction of the mean. For example, if the distribution is positively skewed, the mean will be larger than the median, but if the skew is negative, the mean should be smaller than the median. It should also be noted that the mode is generally on the opposite side of the median from the mean in skewed distributions. This is not always the case, however, and should not be treated as a rule. MacGillivray (1981) discusses the conditions under which each of these examples would fall.

Skew in direction of mean

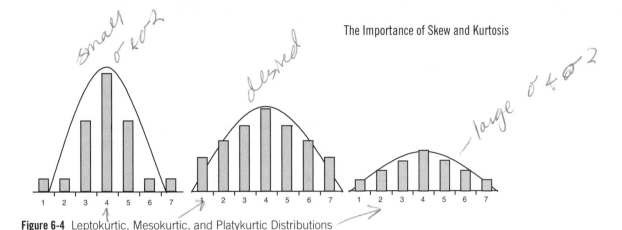

handwritten annotations: small σ of 1 or 2; desired; large σ of 4 or 2

Figure 6-4 Leptokurtic, Mesokurtic, and Platykurtic Distributions

■ Kurtosis

The last characteristic of the form of a distribution is the **kurtosis**. For kurtosis, think again of stacking blocks (or beer cans) on top of each other to represent the frequency of the categories in a histogram. The kurtosis is the extent to which cases are piled up around the measure of central tendency or in the tails of the distribution. If most of the values in the distribution are very close to the measure of central tendency, the distribution is said to be **leptokurtic** (as shown on the left in **Figure 6-4**). If most of the values in the distribution are out in the tails, the distribution is said to be **platykurtic**, as shown on the right in Figure 6-4. If the values in the distribution are such that they represent a distribution such as that shown in Figure 6-2, the distribution is said to be **mesokurtic**, as shown in the center of Figure 6-4. It is desirable to have a mesokurtic distribution in research; otherwise, the data may have to be transformed.

The shape of these curves also offers an opportunity to talk about variance and standard deviation. As discussed in Chapter 5, the variance and standard deviation dictate the shape of the distribution. In a leptokurtic distribution, the variance and standard deviation would be smaller than in a mesokurtic distribution. The variance and standard deviation of a platykurtic distribution would be larger than either a mesokurtic or leptokurtic distribution. This is one application of the variance and standard deviation. A more expanded discussion of this application is presented later in the section on the normal curve.

Analysis of Kurtosis

In SPSS, kurtosis is measured in the same way as skew. A value between +1 and −1 represents a mesokurtic distribution. Positive numbers greater than 1 represent leptokurtic curves. Negative numbers less than −1 (a greater negative number) represent platykurtic curves. As with skew, it is desirable to get the kurtosis as close as possible to zero, using transformations if necessary. Examining the kurtosis value in Table 6-1 shows that the distribution is mesokurtic because the value (-0.705) is between -1.00 and 0. If this value had been -1.705, the distribution would have been platykurtic.

■ The Importance of Skew and Kurtosis

It is important to know the skew and kurtosis because some statistical procedures do not work well with skewed data or data that is not mesokurtic. If data in a research project is found to be skewed or kurtose, it may be necessary to transform the data.

handwritten margin notes: 1 to −1 is mesokurtic; # > +1 = leptokurtic; # < −1 = platykurtic

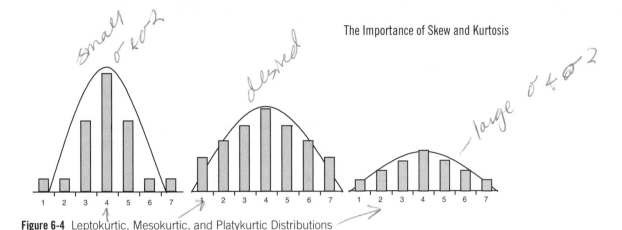

Skewed or kurtose = Non-Normal Curves.

This is covered in greater detail later in the book, but for now, remember two things about transformation. First, if the data is not within acceptable tolerances for skew and kurtosis, the data will need to be transformed prior to using some statistical procedures. Second, after making transformations, recheck both the skew and kurtosis. Transforming the data may bring one of these within acceptable tolerances but may make the other unacceptable. If that happens, you will need to choose another transformation. You should then recheck the skew and kurtosis again. This process should continue until you reach a point where both the skew and kurtosis are acceptable. If it is not possible to get both the skew and kurtosis in an acceptable range, you may need to consider using a different analysis procedure that is not susceptible to nonnormal curves.

■ Design of the Normal Curve

Extending the concepts of the frequency distribution, graphical representation of data, and measures of central tendency, dispersion, and form brings us to the point of discussing a key concept in statistical analysis, the **normal curve**. At this point, you should not be concerned with applications of the normal curve; that is covered in more detail in the chapters of the book on inferential analysis. The purpose of the present discussion is to introduce the properties of the normal curve.

An introduction to the normal curve is included in descriptive analyses rather than inferential analyses for two reasons. First, the normal curve can be used to provide an interpretation of the variance and standard deviation. Second, the normal curve is important to a number of statistical procedures that will be discussed before reaching information on inferential statistical procedures.

An example of relatively normally distributed data can be shown in grades in a course (see **Figure 6-5**). Say that most people taking the course score a C on the first test. This would be the modal grade (the top part of the curve). There are those who receive high A's, but there would be only a few of these; they would be at the positive end of the curve. There are also those who receive very low F's, but these are also few; they would be at the negative end of the curve. Most people would be in between these

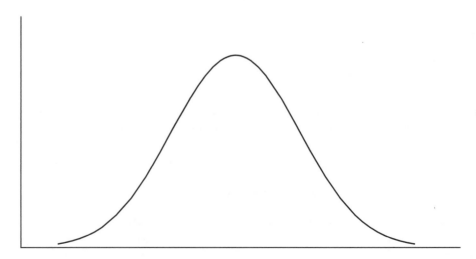

Figure 6-5 Normal Curve

two extremes, with more people making scores around C's than other grades, and more people making B's and D's than A's and F's. For the sake of argument, though, say that no one received a 100 and there were a few people who did not take the test. Therefore, the ends of the tails will never completely touch the baseline.

This type of data represents a special form of distribution called the normal curve. This type of curve or distribution is very much like those that have been used in this chapter and at the end of Chapter 4. A normal curve is special because it has certain characteristics. First, a normal curve is symmetrical in that it can be folded in half, and both sides would be exactly the same (as in Figure 6-2, where the frequencies of category 1 and 3 are the same as those of category 5 and 7, respectively). Note, though, that this does not mean this curve cannot be kurtose. Some symmetrical distributions are leptokurtic or platykurtic. This is shown in Figure 6-4, where each of those distributions was symmetrical even if it was kurtose. Also, a normal curve is unimodal; there is one, and only one, peak. This peak is at the maximum frequency of the data distribution, so that the mean, median, and mode all have the same value. The normal curve shown in Figure 6-5 has only one mode. From the peak, the tails of a normal curve fall off on both ends and extend to infinity, always getting closer to the baseline but never touching it. This is shown in Figure 6-5, where the bottom part of the curve straightens out and runs relatively parallel to the X axis. You may say this makes no sense; all distributions have an end, so why would the normal curve not have an end? The answer lies in the scientific process. Take the example of computers. Less than 20 years ago, scientists and engineers thought they had achieved the ultimate when they were able to reach 640K of random access memory (RAM) in a computer. They felt that this was the maximum that could be achieved and all that anyone would ever need. To them, the distribution limits were set. We now know, of course, that 640K was only the beginning and that computers are far beyond that now. It would have been foolish, then, to have the curve touch the line at 640K; it should not touch the line because we do not know what will come in the future. A final characteristic of the normal curve that merits discussion is that the area under a normal curve is always the same, regardless of the data set. The area under the normal curve is 1.00, or 100% of all values in the distribution. This is extremely important in the section of this book concerning inferential analyses because of its importance in estimating the placement of a sample distribution within a population or another sample.

The area under the normal curve also offers the opportunity to put the variance and standard deviation into practice. Say, for example, that a researcher was examining the time prisoners were out on parole before they committed another crime or returned to prison on a technical violation. If the time each parolee took before being reincarcerated was plotted, it might look as in **Figure 6-6**. There were a few people who returned to prison right away, most of the parolees who returned to prison did it within two to four years, and some took longer. Some had not recidivated at the time of the research, so the end of the distribution is open.

An analysis of the central tendency would put the mean of this distribution at 36 months, which is represented by the vertical line. This is good information: The average length of time for parolees to be reincarcerated is three years. It is obvious from this distribution, however, that not all of the parolees were reincarcerated at the same time. The span of time runs from a couple of months to more than five years. To get a more accurate picture of the distribution of parolees, we might want to know, on average, how far each of them is from the mean.

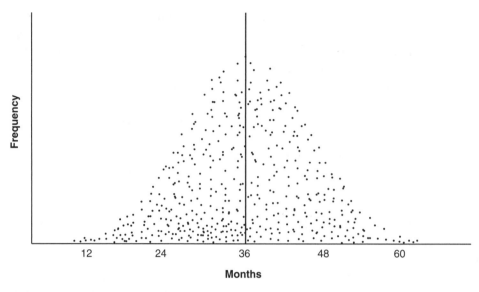

Figure 6-6 Distribution of Time to Reincarceration for Parolees

To calculate this, the procedure would be to take each person (each dot in Figure 6-6) and determine how far from the mean it is. This could be completed by measuring the distance with a ruler, but since this is a numeric scale, it could also be completed by subtracting each value from the mean. This operation would be noted as $X - \overline{X}$. Summing all of these values, represented by $\Sigma(X - \overline{X})$, would provide us with the total of the distance from each value to the mean. But remember that the sum of each value subtracted from the mean is zero, so that does not help us. A solution to this is to square each value before summing it. This calculation, $\Sigma(X - \overline{X})^2$, will give a positive value. Knowing the total distance between the mean and each value is good, but a simpler value would be the average of the distance from each value to the mean. The calculation for this procedure is:

$$\frac{\Sigma(X - \overline{X})^2}{N}$$

The only problem with this calculation is that the value that would be obtained is not on the same scale as the original data. To return this value to the original scale, take the square root of the value. Of course, you recognize that this is the procedure for calculating the variance and, ultimately, the standard deviation for the distribution. Although this procedure can be completed for any distribution, it is particularly important for a normal curve because of what the standard deviation represents.

Because the area under the normal curve is always 100% and since standard deviations represent standard distances in relation to the mean of the distribution, standard deviations can be used to calculate the area under the normal curve for particular values. For example, between the mean and 1 standard deviation under a normal curve lies 34.13% of all the data. Between the mean and 2 standard deviations is 47.72% of all the data. Between the mean and 3 standard deviations is 49.87% of all the data. That covers 49.87% of the possible 50% of one half of the curve. Since the normal curve is

68.26 1SD
95.44 2SD
99.74 3SD

symmetrical, the same values can be obtained whether counting from the left or right of the mean. This also means that the values could be doubled from one side and the area under the whole normal curve could be determined for values plus or minus a certain number of standard deviations from the mean. If the figures above are doubled, the result would be 68.26% of the data between -1 and 1 standard deviation, 95.44% of the data between -2 and 2 standard deviations, and 99.74% of the data between -3 and 3 standard deviations.

This is very useful to know because it is now possible to determine what percentage of the scores lie between the mean and any standard deviation within the distribution. The problem is that researchers are not always interested in determining values (or determining the area under a normal curve) only for values directly on a particular standard deviation. Often, calculations need to be made that allow *any* value in the distribution to be converted to standard deviation units such that the area under the normal curve can be calculated for that value. The procedure for doing this is to calculate a **Z score** for that value. Z scores convert any value in the distribution to standard deviation units. The formula for calculating a Z score is

$$Z = \frac{X - \overline{X}}{s}$$

standard deviation of the distribution.

where X is the number to be converted to a Z score, $\overline{X}$ the mean of the distribution, and s the standard deviation of the distribution.

The process of converting values to Z scores is simply a method of deriving standard scores. To be able to make comparisons between things, it is necessary to set a standard on which both items may be measured. If measuring distance, we could use feet or meters; weight could be in pounds or kilograms. But what about social data such as crime? Crime is generally discussed in terms of crime rates. Calculating crime rates is simply a process of standardizing crime in terms of the population of a city— taking numbers that may be difficult to compare between different things and making them comparable. The same can be done with the mean, standard deviation, and normal curve through Z scores. This standardizes scores based on the normal curve.

What this means is that if a researcher knows the mean and standard deviation of a population, Z scores can be used to calculate the distance between any value and the mean. Using the area under the normal curve allows inferential analyses (see the chapters in this book on inferential analyses) to be used to determine the chances that any number in a distribution will be in a sample drawn from that population.

The process of calculating a Z score, determining the area under the normal curve between that value and the mean, and examining how that score relates to the mean, is rather simple. First, subtract the raw score from the mean. This determines whether the number is above or below the mean: Negative numbers are below the mean, positive numbers above the mean. Then divide the number by the standard deviation to determine how many standard deviation units the number is above or below the mean. The answer obtained from this calculation is the score's deviation from the mean in standard units (the Z score in standard deviation units). Note that you may get a negative number; all this means is that the Z score is to the left of the mean. It does not affect the calculations at all and can be dropped in the remainder of the procedure.

The next step is to take this number and turn to Table B-1 in the appendix. The row to use corresponds to this number in column a of the table. The columns to use within this row depend on what area under the normal curve is being examined. If you are looking for the area between the score and the mean, look to column b in the table; if examining the area beyond that score, which is farther into the tail of the distribution, use column c.

Suppose that the number of complaints against a police department averaged 12 complaints a month with a standard deviation of 1.5. Then suppose that the chief wants you to determine what percentage of the scores fell between the mean and the current month's score of 14. This would require determining the area between the mean and 14. To do this, first calculate a Z score, as shown below:

$$Z = \frac{X - \overline{X}}{s}$$

$$= \frac{14 - 12}{1.5}$$

$$= \frac{2}{1.5}$$

$$= 1.33$$

Then look at column a in Table B-1 and find 1.33. You are looking for the area between the score and the mean, so you would use column b to get the percentage of the area under the normal curve that falls between the mean and 14. This value is 0.4082, so the area in this distribution between 12 and 14 is 40.82%. What does this mean? It means that almost 41% of the months in the current year had fewer complaints recorded than those in the current month.

But what if the chief wanted to know how many months had more complaints recorded than the current month? To determine the percentage of scores greater than 14, you would calculate the Z score exactly as before. That number, 1.33, is then found in column a of Table B-1. Since you are trying to determine the number of scores greater than 14, column c in the table would be used. The value in column c that corresponds to a Z score of 1.33 is 0.0918, so 9.18% of the scores in the distribution are greater than 14, which means that almost 10% of the months had more complaints recorded than the current month. Actually, because the area between the mean and 14 was known, it was not necessary to complete the second procedure. Because it was already determined that the percentage of scores between the mean and 14 was 40.82%, and because the area under the normal curve is always 100% and the area under half of the normal curve is 50%, the 40.08 could have been subtracted from 50 to arrive at the score of 9.18%—the area beyond that value.

These procedures work whether using scores above the mean or scores below the mean. As stated previously, -1 standard deviation is the same as 1 standard deviation; thus, it would be the same percentage if we were looking for the area under the normal curve representing less than 10. But what if you wanted to know how many scores were higher than one score and lower than another: For example, how many people scored higher than 90 and lower than 62 on a test. This case would require determining the percentage greater than the score of 90 and the percentage less than the score of 62. The final pieces of information needed here are that the mean of the distribution is 76 and the standard deviation is 10.5. The areas beyond these two scores can be determined simply by adding together the percentages in Table B-1. The Z score calculations for these two values would be as follows:

$$Z = \frac{X - \overline{X}}{s} \qquad Z = \frac{X - \overline{X}}{s}$$

$$= \frac{90 - 76}{10.5} \qquad = \frac{62 - 76}{10.5}$$

$$= \frac{2}{1.5} \qquad = \frac{-2}{1.5}$$

$$= 1.33 \qquad = -1.33$$

The results of these calculations would be taken to Table B-1. Finding 1.33 in column a and wanting to determine the area under the normal curve beyond this value, we would look in column c. The value found in column c is 0.0918. This means that 9.18% of the scores are higher than 90. Since 62 has the same value, but negative, it also has a value in column c of 0.0918, so 9.18% of the scores are lower than 62. Adding these two values together would establish what percentage of people scored greater than 90 and less than 62:

$$0.0918 + 0.0918 = 0.1836$$

So 18.36% of the people in the class scored higher than 90 (making an A) or lower than 62 (making an F). It is also possible to determine the percentage of scores that fall between two scores. In the example we have been using, the percentage of scores that fall between 90 and 62 can be determined by adding together the areas between them and the mean. Looking at column b in Table B-1, the percentage of scores between the mean of 76 and a score of 90 (the Z score of 1.33) is 40.82%. Since the score of 62 has the same Z score value, the percentage of scores between 62 and 76 is also 40.82%. Adding those percentages together produces a percentage of 81.64% as shown below:

$$0.4082 + 0.4082 = 0.8164$$

So 81.64% of the class scored a B, C, or D.

Points to Remember About the Normal Curve

Two Z scores that will become important later are 1.96 and 2.58. These correspond to 95% and 99% of the area under the normal curve, respectively. These are important because researchers often want to know where 95% or 99% of the values in the distribution fall, or they may want to compare two values and see if they fall within 95% or 99% of the values in the distribution.

Another point you should remember is that the normal curve is a theoretical ideal. Normal distributions do not really exist, and the best we can do is to gather data that is close to a normal curve but not exact. The conclusions drawn when using the theory of the normal curve, then, will only be estimates, not hard facts.

Finally, you should realize that not all data sets even come close to a normal curve. As discussed in the chapter on measures of central tendency, some distributions are multimodal and almost jagged looking. There are many distributions that are so skewed they are J shaped and do not conform at all to a normal distribution. There are even a few distributions that are absolutely flat. Each of these distributions requires special treatment that is discussed in the chapters on bivariate and multivariate analysis.

■ Conclusion

This chapter completes the description of one variable at a time from a distribution—univariate descriptive analysis. You have learned how to describe a variable such that it can be relayed to another person with some accuracy and in a form that allows the person to get a mental image of the distribution. Describing single variables can take many forms: frequency distributions and graphs, measures of central tendency, measures of dispersion, and the form of the distribution. These all have the goal of describing the attributes of the distribution and determining if a variable is suitable for further analyses.

In the next chapters, we put together two variables—bivariate descriptive analyses—and describe what relationship might exist between them. The success of those analyses depends on the successful univariate description of data.

■ Key Terms

form

kurtosis

leptokurtic

mesokurtic

moments

negatively skewed

normal curve

number of modes

platykurtic

positively skewed

skew

symmetry

Z score

■ Summary of Equations

Moments of a Distribution

$$\frac{\Sigma(X - \overline{X})^i}{N}$$

Z Score

$$Z = \frac{X - \overline{X}}{s}$$

■ Exercises

1. Use the following frequency tables (from the gang database):
   a. To discuss the number of modes.
   b. To determine the skew and kurtosis and discuss whether the distribution is positively skewed or negatively skewed and whether it is leptokurtic, mesokurtic, or platykurtic.

HOME: What type of house do you live in?

Value Label	Value	Frequency	Percent	Valid Percent	Cumulative Percent
House	1	280	81.6	82.4	82.4
Duplex	2	3	.9	.9	83.2
Trailer	3	34	9.9	10.0	93.2
Apartment	4	21	6.1	6.2	99.4
Other	5	2	.6	.6	100.0
Missing		3	.9	Missing	
Total		343	100.0	100.0	

N	Valid	340
	Missing	3
Mean		1.41
Std. Error of Mean		.051
Median		1.00
Mode		1
Std. Deviation		.945
Variance		.892
Skewness		2.001
Std. Error of Skewness		.132
Kurtosis		2.613
Std. Error of Kurtosis		.264
Range		5

ARREST: How many times have you been arrested?

Value	Frequency	Percent	Valid Percent	Cumulative Percent
0	243	70.8	86.2	86.2
1	23	6.7	8.2	94.3
2	10	2.9	3.5	97.9
3	3	.9	1.1	98.9
5	2	.6	.7	99.6
24	1	.3	.4	100.0
Missing	61	17.8	Missing	
Total	343	100.0	100.0	

N	Valid	282
	Missing	61
Mean		.30
Std. Error of Mean		.093
Median		.00
Mode		0
Std. Deviation		1.567
Variance		2.455
Skewness		12.692
Std. Error of Skewness		.145
Kurtosis		187.898
Std. Error of Kurtosis		.289
Range		24

TENURE: How long have you lived at your current address (months)?

Value	Frequency	Percent	Valid Percent	Cumulative Percent
1	14	4.1	4.3	4.3
2	6	1.7	1.8	6.1
3	4	1.2	1.2	7.3
4	4	1.2	1.2	8.6
5	6	1.7	1.8	10.4
6	6	1.7	1.8	12.2
7	1	.3	.3	12.5
8	3	.9	.9	13.5
9	2	.6	.6	14.1
10	1	.3	.3	14.4
11	1	.3	.3	14.7
12	11	3.2	3.4	18.0
14	1	.3	.3	18.3
18	5	1.5	1.5	19.9
21	1	.3	.3	20.2
24	30	8.7	9.2	29.4
30	1	.3	.3	29.7
31	1	.3	.3	30.0
32	1	.3	.3	30.3
36	22	6.4	6.7	37.0
42	1	.3	.3	37.3
48	12	3.5	3.7	41.0

Value	Frequency	Percent	Valid Percent	Cumulative Percent
60	24	7.0	7.3	48.3
72	14	4.1	4.3	52.6
76	1	.3	.3	52.9
84	8	2.3	2.4	55.4
96	18	5.2	5.5	60.9
108	4	1.2	1.2	62.1
120	9	2.6	2.8	64.8
132	11	3.2	3.4	68.2
144	21	6.1	6.4	74.6
156	13	3.8	4.0	78.6
168	11	3.2	3.4	82.0
170	5	1.5	1.5	83.5
180	7	2.0	2.1	85.6
182	2	.6	.6	86.2
186	1	.3	.3	86.5
192	14	4.1	4.3	90.8
198	1	.3	.3	91.1
204	24	7.0	7.3	98.5
216	3	.9	.9	99.4
240	2	.6	.6	100.0
Missing	16	4.7	Missing	
Total	343	100.0	100.0	

N Valid	327	
Missing	16	
Mean	88.77	
Std. Error of Mean	3.880	
Median	72.00	
Mode	24	
Std. Deviation	70.164	
Variance	4923.055	
Skewness	.365	
Std. Error of Skewness	.135	
Kurtosis	−1.284	
Std. Error of Kurtosis	.269	
Range	239	

SIBS: How many brothers and sisters do you have?

Value	Frequency	Percent	Valid Percent	Cumulative Percent
0	39	11.4	11.5	11.5
1	137	39.9	40.5	52.1
2	79	23.0	23.4	75.4
3	39	11.4	11.5	87.0
4	17	5.0	5.0	92.0
5	13	3.8	3.8	95.9
6	6	1.7	1.8	97.6
7	4	1.2	1.2	98.8
9	1	.3	.3	99.1
10	1	.3	.3	99.4
12	1	.3	.3	99.7
15	1	.3	.3	100.0
Missing	5	1.5	Missing	
Total	343	100.0	100.0	

N	Valid	338
	Missing	5
Mean		1.94
Std. Error of Mean		.098
Median		1.00
Mode		1
Std. Deviation		1.801
Variance		3.245
Skewness		2.664
Std. Error of Skewness		.133
Kurtosis		12.027
Std. Error of Kurtosis		.265
Range		15

2. For a distribution with a mean of 50 and a standard deviation of 10:
   a. Calculate the Z score for a score of 40.
   b. Determine the area under the normal curve between the mean and this value.
   c. Determine the area under the normal curve beyond this value.
   d. Calculate a Z score for a score of 65.
   e. Determine the area under the normal curve between the mean and this value.

    f. Determine the area under the normal curve beyond this value.

    g. Determine the area under the normal curve between 40 and 60.

    h. Determine the area under the normal curve outside 40 and 60.

3. Say a parole board has a policy that it will only release prisoners who meet a minimum amount of time served, have a minimum number of good-time points, and have made an acceptable score on tests in their drug awareness class. If the mean of this distribution is 90, the minimum acceptable score on these criteria is 70, and the standard deviation of scores is 15:

    a. Calculate the Z score for a score of 90.

    b. Calculate the Z score for a score of 50.

    c. Determine the area under the normal curve between the mean and the values in parts a and b.

    d. Determine the area under the normal curve beyond these values.

    e. Calculate a Z score for a score of 70.

    f. Determine the area under the normal curve between the mean and this value.

    g. Determine the area under the normal curve beyond this value.

References

MacGillivray, H. L. (1981). The mean, median, mode inequality and skew for a class of densities. *Australian Journal of Statistics*, 23:247.

For Further Reading

Pearson, K. (1894). On the dissection of asymmetrical frequency-curves: General theory. *Philosophical Transactions of the Royal Society of London*, Series A, Vol. 185. London: Cambridge University Press.

Pearson, K. (1895). Classification of asymmetrical frequency curves in general: Types actually occurring. *Philosophical Transactions of the Royal Society of London*, Series A, Vol. 186. London: Cambridge University Press.

Note

1. SPSS uses the convention of +1 to −1 as the measure of acceptable skew. This is based on a particular formula that "standardizes" skew and kurtosis scores. There is a similar (and popular) formula that calculates skew and kurtosis where the acceptable range is +3 to −3. Programs such as SAS use this formula. It is important, then, that you understand which formula is being used (what the acceptable range is) for these values before making a judgment about them.

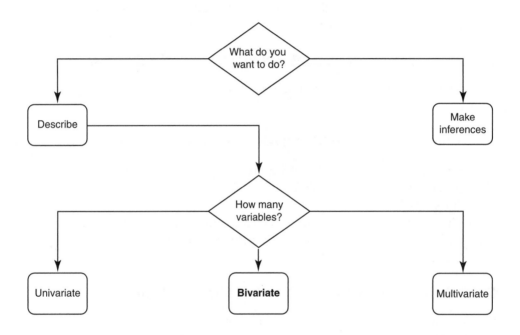

Introduction to Bivariate Descriptive Statistics

In Chapter 3, we introduced frequency tables as a method of examining the characteristics of a particular variable. It is possible to use those frequency tables to examine the **relationship** between two variables. As an example, take the two variables *sex* and *type of victimization*. The combination tables from SPSS for each variable are shown in **Table 7-1**. The univariate statistics of each of these variables can be determined from these frequency tables. These show that there are more than double the number of female respondents than males and that the most frequent victimization was for burglary, closely followed by vandalism and rape. The table for victimization also contains 229 people who indicated on a previous question that they had been victimized but who did not respond as to the actual crime in this question.

It is difficult, however, to make any determinations about the possible sex of victims from these tables. For instance, it is impossible to determine from *these* tables how many males and how many females were victims of burglary. However, with some additional work and analyses, the tables can be arranged by the sex of the respondent. This allows statements to be made about male and female victims. The data is shown in **Table 7-2**. Here, it can be shown that females had a higher number and percentage of vandalism, car theft, assault, and rape, and that males had a higher percentage of burglaries, robberies, and other victimizations (although there were still more females in each of these categories except burglary—which had the same number of males and females—and no response). This is a somewhat complicated process, however, and requires looking back and forth between tables. It would be much easier if we could make more direct, or side-by-side, comparisons of the variables. A bivariate table does just that.

■ Bivariate Tables and Analysis

A **bivariate table** displays the joint frequencies of two variables and attempts to show an association between the variables by determining how many times they have paired occurrences in their categories, such as females who have been victims of burglary, versus the differences in those categories. This table is commonly referred to as a *cross-tabulation table*, or **crosstab**. A crosstab for the victimization and sex tables is shown in **Table 7-3**. This table presents the categories—frequencies and/or percentages—of one variable as they are compared to the categories of another variable. This is usually presented in a format such as females who have been victims of a crime versus males who have been victims.

TABLE 7-1   Combination Tables for Sex and Victimization

Sex of respondent

Value Label	Value	Frequency	Percent	Valid Percent	Cumulative Percent
Male	1	99	28.5	29.1	29.1
Female	2	241	69.5	70.9	100.0
	Missing	7	2.0	Missing	
	Total	347	100.0	100.0	

Valid cases, 340; missing cases, 7

Have you been a victim of any of these crimes in the last six months?

Value Label	Value	Frequency	Percent	Valid Percent	Cumulative Percent
No response	0	7	2.0	5.9	5.9
Burglary	1	30	8.6	25.4	31.4
Vandalism	2	28	8.1	23.7	55.1
Car theft	3	14	4.0	11.9	66.9
Robbery	4	9	2.6	7.6	74.6
Assault	5	3	0.9	2.5	77.1
Rape	6	22	6.3	18.6	95.6
Other	7	5	1.4	4.2	100.0
Missing		229	66.0	Missing	
Total		347	100.0	100.0	

Valid cases, 118; missing cases, 229

TABLE 7-2   Data for Male and Female Victims of Crime

Type of Crime	Male Victim		Female Victim	
	Frequency	Percent	Frequency	Percent
No response	1	2.9	4	5.1
Burglary	15	42.9	15	19.0
Vandalism	5	14.3	21	26.6
Car theft	3	8.6	11	13.9
Robbery	3	8.6	6	7.6
Assault	0	0	3	3.8
Rape	5	14.3	17	21.5
Other	3	8.6	2	2.5

Statistical Tables versus Presentation Tables

Tables may be broken into two general types: statistical tables and presentation tables. *Statistical tables* are those represented by the output from statistical programs such as SPSS, and are used in this book. *Presentation tables* are those found in journals and books. These tables are generally made by taking the statistical tables and formatting them to fit (1) a particular editorial style, such as Turabian or APA; (2) the format

TABLE 7-3   Crosstab of Victim by Sex

			Sex of respondent		
			Male	**Female**	**Total**
Have you been a victim of any of these crimes in the last 6 months?	Other	Count	3	2	5
		Expected Count	1.5	3.5	5.0
		% within Have you been a victim of any of these crimes in the last 6 months?	60.0%	40.0%	100.0%
		% within Sex of respondent	8.6%	2.5%	4.4%
		% of Total	2.6%	1.8%	4.4%
	Rape	Count	5	17	22
		Expected Count	6.8	15.2	22.0
		% within Have you been a victim of any of these crimes in the last 6 months?	22.7%	77.3%	100.0%
		% within Sex of respondent	14.3%	21.5%	19.3%
		% of Total	4.4%	14.9%	19.3%
	Assault	Count	0	3	3
		Expected Count	.9	2.1	3.0
		% within Have you been a victim of any of these crimes in the last 6 months?	.0%	100.0%	100.0%
		% within Sex of respondent	.0%	3.8%	2.6%
		% of Total	.0%	2.6%	2.6%
	Robbery	Count	3	6	9
		Expected Count	2.8	6.2	9.0
		% within Have you been a victim of any of these crimes in the last 6 months?	33.3%	66.7%	100.0%
		% within Sex of respondent	8.6%	7.6%	7.9%
		% of Total	2.6%	5.3%	7.9%

(continues)

TABLE 7-3   Crosstab of Victim by Sex (*continued*)

| | | | Sex of respondent | | |
			Male	Female	Total
Car theft		Count	3	11	14
		Expected Count	4.3	9.7	14.0
		% within Have you been a victim of any of these crimes in the last 6 months?	21.4%	78.6%	100.0%
		% within Sex of respondent	8.6%	13.9%	12.3%
		% of Total	2.6%	9.6%	12.3%
Vandalism		Count	5	21	26
		Expected Count	8.0	18.0	26.0
		% within Have you been a victim of any of these crimes in the last 6 months?	19.2%	80.8%	100.0%
		% within Sex of respondent	14.3%	26.6%	22.8%
		% of Total	4.4%	18.4%	22.8%
Burglary		Count	15	15	30
		Expected Count	9.2	20.8	30.0
		% within Have you been a victim of any of these crimes in the last 6 months?	50.0%	50.0%	100.0%
		% within Sex of respondent	42.9%	19.0%	26.3%
		% of Total	13.2%	13.2%	26.3%
No Response		Count	1	4	5
		Expected Count	1.5	3.5	5.0
		% within Have you been a victim of any of these crimes in the last 6 months?	20.0%	80.0%	100.0%
		% within Sex of respondent	2.9%	5.1%	4.4%
		% of Total	.9%	3.5%	4.4%
Total		Count	35	79	114
		Expected Count	35.0	79.0	114.0
		% within Have you been a victim of any of these crimes in the last 6 months?	30.7%	69.3%	100.0%
		% within Sex of respondent	100.0%	100.0%	100.0%
		% of Total	30.7%	69.3%	100.0%

TABLE 7-4   Presentation Table of Table 7-3: Crime Type Victimization by the Sex of the Offender[a]

Type of Crime	Male		Female	
	Frequency	Percent	Frequency	Percent
Other	3	2.6	2	1.8
Rape	5	4.4	17	14.9
Assault	0	0	3	2.6
Robbery	3	2.6	6	5.3
Car theft	3	2.6	11	9.6
Vandalism	5	4.4	21	18.4
Burglary	15	13.2	15	13.2
No response	1	0.9	4	3.5
Total	35	30.7	79	69.3

[a]Format of table taken from *Criminology*. Percents are percentages of the total ($N = 144$).

required by a particular publication; or (3) the aesthetic notions of the author, that is, what he or she thinks looks best.

It is important to be able to understand and interpret both types of tables. Much of what you will see as students will be presentation tables in journal articles. Most such tables are fairly easy to understand because the reason they are created the way they are is to make them easy to read. What is probably more important, however, is the ability to read statistical tables. If you cannot read statistical tables, it is not possible to analyze the data or to convert the statistical tables to presentation tables. For this reason, in this book we focus on statistical tables. The tables contained here are direct output from SPSS. This will allow you to begin to understand statistical output, to be able to analyze that output, and to be able to put that analysis in writing. To show the difference between statistical tables and presentation tables, the information in Table 7-3 is shown as a presentation table in **Table 7-4**.

Some tables in this book will be shown as both statistical tables and presentation tables to show the differences and similarities. The focus of the tables, however, is on statistical tables and analysis.

■ Constructing Bivariate Tables

There are many different ways to construct a bivariate table. Some conventions have been developed, however, that make the analysis of bivariate tables somewhat easier and more uniform. The basic format of a bivariate table is to put the variables in rows and columns with the dependent variables in the rows and the independent variables in the columns. Although this ordering can be reversed, some bivariate procedures assume this ordering, and consistency does make analysis easier. In Table 7-3, *victimization* is the dependent variable and *sex* is the independent variable; the probability of persons becoming victims may be influenced by whether they are male or female.

A bivariate table is classified by the number of rows and columns, using $Y \times X$ as the model. For example, a 4×3 table is a bivariate table with the dependent variable, Y, divided into four categories, and the independent variable, X, divided into three categories.[1] Table 7-3 is a 7×2 table: that is, seven categories of the dependent variable (Y) and two categories of the independent variable (X).

Each of the boxes of a table is called a **cell**. The frequency of a cell is determined by calculating the number of occurrences of a pair of values from the independent and dependent variables. This number is called the **cell frequency** and is usually denoted by the symbol n_{ij} or n_{rc}, where ij and rc stand for the row and column in which the cell is located, respectively. For example, in Table 7-3 the cell that contains the joint frequencies for males who had *no response* is cell 1–1 (row 1, column 1). The frequency for this cell is 3, which means that there were three males who had no response. In written form, this frequency might be represented in the form $n_{11} = 3$.

The type of bivariate table is taken from the type of data it contains, the same as with identifying the type of frequency table. If the values in the cells are data values, the table is called a *frequency table;* if the values are percentages, the table is called a *percentage table;* if there are both, it is called a *combination table.*

The level of the bivariate table is taken from the level of data it contains; that is, ordinal level tables contain ordinal level variables, nominal level tables contain nominal level variables. The level of data in the table, and thus the table itself, is an important consideration in analysis because the level of data will determine the type of statistical procedures that can be used to analyze the data. Because the development of an ordinal level table is somewhat more structured, it is discussed in detail and discussed first. Then, construction of a nominal level table is covered, based on the understanding of ordinal level table design.

Ordinal Level Table Construction

For use of ordinal measures of association, both variables must be ordinal level, dichotomized nominal level (such as male/female),[2] or categorized interval level variables (which is interval level data such as income that has been placed into categories after data collection to ease analysis or presentation). Ordinal bivariate tables can set the foundation for analyses identifying the existence of an association, as well as strength, direction, and nature (see the chapters that follow).

How do you do that? Constructing a Bivariate Table in SPSS

1. Open a data set.
   a. Start SPSS.
   b. Select *File*, then *Open*, then *Data*.
   c. Select the file you want to open, then select *Open*.
2. Once the data is visible, select *Analyze*, then *Descriptive Statistics*, then *Crosstabs*.

3. Select the variable you wish to include as your dependent variable and press the ▶ next to the *Row(s)* window.

4. Select the variable you wish to include as your independent variable and press the ▶ next to the *Column(s)* window.

5. For now, do not worry about the *Exact* and *Statistics* boxes at the bottom of the window.

6. Select the *Cells* button and check any of the boxes for information you may want in your crosstab. For a crosstab with full information, select *observed*, *expected*, *row*, *column*, and *total*; then press *Continue*.

7. IMPORTANT. Select the *Format* button, and check the box marked *Descending*. If you do not do this, your table will not be formatted as discussed in this chapter.

An output window should appear containing a distribution similar in format to Table 7-3.

The first step in constructing an ordinal level table is to set up the title. The title is created in the form *dependent variable (Y) by independent variable (X)*. As shown in **Table 7-5**, the title is Walking at Night (WALK_NIT) by Perceptions of Crime Problem (CRIM_PRB), where *Walking at Night* is the dependent variable and *Perceptions of Crime Problem* is the independent variable. Columns contain independent variables, so the independent variable title is across the top of the table. Column headings should be arranged in ascending order (from low to high) as you move from left to right across the table. In the example shown in Table 7-5, the categories of *CRIM_PRB* move from No Problem (category 1) to Big Problem (category 3) as you move across the top of the table. Rows contain the dependent variables, so the dependent variable title is to the left of the table. Row headings should be arranged in descending order (from high to low) as you move from the top to the bottom of the table. In the example shown in Table 7-5, the rows for *WALK_NIT* move from Often (category 3) to Never (category 1) as you move from top to bottom. Arranging the dependent variable so it moves from high values at the top of the table to low values at the bottom of the table will ensure the signs of analyses will be correct and the direction is visually attainable by how the values are arranged.

Ordering the variables and categories as described above is a convention, but one that is not always followed. Some statistical programs, including SPSS, do not automatically set the variables in the position described here, nor do they automatically order the categories as outlined in this chapter. Because of the need for flexibility to work with data, most programs allow a researcher to specify variable arrangement and category order. When conducting research, then, it is important to ensure that a table be set up in the prescribed manner before additional analyses are conducted; otherwise, the analyses described in this book will not work. It is not that you could not analyze the data, it is just that you would have to use different methods than those described here because the table would be backward to how the analysis is addressed in this book.

To fill out the table, place the proper numbers in each cell by counting how many times each pair occurs. The data that would be contained in the top row of Table 7-5 is shown below. For each of the two variables, the cell frequencies would be determined by taking each pair of scores and hash-marking them in the appropriate cell. Once all

TABLE 7-5   Crosstab of Walking at Night by Perceptions of Crime Problem

How often do you walk in your neighborhood at night? * How big a problem is crime? Crosstabulation

			How big a problem is crime?			
			No Problem	Small Problem	Big Problem	Total
How often do you walk in your neighborhood at night?	Often	Count	2	2	0	4
		Expected Count	2.7	1.2	0.2	4.0
		% within How often do you walk in your neighborhood at night?	50.0%	50.0%	0.0%	100.0%
		% within How big a problem is crime?	1.2%	2.8%	0.0%	1.6%
		% of Total	0.8%	0.8%	0.0%	1.6%
	Occasionally	Count	20	13	3	36
		Expected Count	23.9	10.4	1.7	36.0
		% within How often do you walk in your neighborhood at night?	55.6%	36.1%	8.3%	100.0%
		% within How big a problem is crime?	12.1%	18.1%	25.0%	14.5%
		% of Total	8.0%	5.2%	1.2%	14.5%
	Never	Count	143	57	9	209
		Expected Count	138.5	60.4	10.1	209.0
		% within How often do you walk in your neighborhood at night?	68.4%	27.3%	4.3%	100.0%
		% within How big a problem is crime?	86.7%	79.2%	75.0%	83.9%
		% of Total	57.4%	22.9%	3.6%	83.9%
Total		Count	165	72	12	249
		Expected Count	165.0	72.0	12.0	249.0
		% within How often do you walk in your neighborhood at night?	66.3%	28.9%	4.8%	100.0%
		% within How big a problem is crime?	100.0%	100.0%	100.0%	100.0%
		% of Total	66.3%	28.9%	4.8%	100.0%

of the pairs of values have been determined, these hash marks could be converted to frequencies. Here, there were two respondents who stated that they often walked at night and that crime was a small problem in their neighborhood, and there were two people who said that they often walked at night and that crime was a big problem in their neighborhood.

Respondent	WALK_NIT	CRIM_PRB
A	Often (3)	Small problem (2)
B	Often(3)	Big problem (3)
C	Often (3)	Big problem (3)
D	Often (3)	Small problem (2)

In this case, there were no *No Problem* responses, so there were no category 1 responses to hash mark. When these results are put in terms of a frequency, they would look like the top row in Table 7-5.

As discussed previously, it is often difficult to compare two or more variables using their scores because score values are difficult to interpret. It is a good practice, therefore, to convert frequencies to percentages so the variables can be compared more easily. This is accomplished by dividing each cell frequency by its column total, row total, or *N*, and then multiplying by 100. SPSS does this for you. When you check the box indicating *Row*, *Column*, and/or *Total*, as described above, SPSS will calculate the **row percentages**, **column percentages**, and/or **total percentages** of the data. How these are listed in the printouts is "% within" the variable in the row or column or as "% of total." For example, in row 2, cell 2–3 has a *row percentage* of 8.3% (3/36), cell 2–2 has a row percentage of 36.1 % (13/36), and cell 2–1 has a row percentage of 55.6% (20/36). You know these are row percentages because they are labeled "% within WALK_NIT. How often do you walk in your neighborhood at night," which is the variable in the rows. The same can be completed for *column percentages*. In column 1 of Table 7-5, cell 1–1 would have a column percentage of 1.2% (2/165), the column percentage of cell 2–1 would be 12.1% (20/165), and the column percentage of cell 3–1 would be 86.7% (143/165). You know these are column percentages because they are labeled "% within CRIM_PRB. How big a problem is crime?", which is the variable in the columns. Finally, the *total percentages* can be calculated by taking each cell and dividing by *N*. For example, in column 2, the total percentage for cell 1–2 is 0.8% (2/249), the total percentage for cell 2–2 is 5.2% (13/249), and the total percentage for cell 3–2 is 22.9% (57/249).

The final piece of information available from SPSS is the **Expected Count**. This is the frequency that would be expected if the two variables were not related at all. This is an important part of bivariate analysis, and will be addressed in Chapter 8.

The numbers on the outside of the table are called the **marginals**. These are simply the total frequency counts or percentages of the data in the rows or columns. For example, *Row Total* contains the total cell frequency for all scores in that row, and the same is true for *Column Total*. In Table 7-5, the row total for row 1 is 4 because there were two cells with frequencies of 2 and one cell with a frequency of 0. Similarly, the column total for column 3 is 12 because there were three cells with frequencies of 0, 3, and 9, respectively. The percentage values in the marginals are the contribution of that row or column to the total. For example, in Table 7-5 the value of 1.6% in row 1 means that the score *Often* contributes 1.6% of the total scores for all values of crime problem (with *Occasionally* and *Never* contributing the other 98.4%). If these marginals were compared to the original frequency or percentage tables, the numbers would match for each category.

The values in the bottom right-hand corner of the table represent the totals for the table. In Table 7-5 the total frequency (N) is 249, and the 100% represents 100% of the row totals and 100% of the column totals.

Nominal Level Table Construction

If one of the variables to be studied is nominal level (excluding dichotomized nominal), a nominal table and nominal statistics should be used to measure the association. Nominal tables provide less information than do ordinal level tables because of the type of data. Nominal level tables can provide information on the existence and strength of the relationship. If there are enough categories in each variable, nominal level tables can provide some information about the nature of the relationship, but no determination can be made concerning the direction of the data because nominal level data cannot be ordered.

A nominal table is created the same way as an ordinal table. The only difference is that the ordering of categories is not important. For these tables, the variable categories should be arranged in a way that makes the most logical sense. Even though no direction can be determined from a nominal table, any ordinal level variables or any categorized interval level variables used should be put in order according to the instructions for an ordinal level table. This assists in examining the data and standardizes the way the tables are constructed.

An example of a nominal table is shown in Table 7-3. Here both sex and type of victimization are nominal level. Note that it is probably possible to make type of victimization ordinal level if the degree of seriousness is considered. In this case, however, no such distinctions were made, so the variable is considered nominal.

■ Analysis of Bivariate Tables

Constructing bivariate tables is often the first step in bivariate analyses. Bivariate analyses are covered in the next three chapters, but a short introduction is included here to show the link with table construction. Bivariate tables and the statistics associated with them are one of the most common forms of analysis for nominal and ordinal level data. These tables are seldom used for interval and ratio level data, however, because there are often so many categories with this level of data that the table would be too large and complex to be useful. Furthermore, the statistics for interval and ratio level data are much stronger if the data is not categorized, so the data is usually left ungrouped rather than putting it in tables.

Bivariate analyses examine the relationship between two variables, or the differences between categories of one variable as they relate to categories of a second variable. There are four steps involved in bivariate analysis. These steps examine or determine the characteristics of an association (see Chapters 8, 9, and 10). For nominal and ordinal level data, these analyses are conducted using bivariate tables and the statistical procedures that go with them. For interval and ratio level data, the analyses are generally conducted without the use of bivariate tables.

The first step in a bivariate analysis is testing for the **existence** of an association (see Chapter 8). An association is said to exist between two variables if the distribution of one variable differs in some respect between categories of the variable. The next step is measuring the **strength** of an association (see Chapter 9). If an association exists, this determines how closely the two variables are associated. The third step is determining the **direction** of an association (see Chapter 10). If the higher values of one variable are associated with higher values of the other, the association is said to be *positive*. If the higher values of one variable are associated with lower values of the other, the association is said to be *negative*. The final step is determining the **nature** of an association (see Chapter 10). Patterns of an association may be irregular. If an increase in one increment in one variable is always related to a constant increase or decrease of a certain number of increments in the other variable, the nature of the association is said to be *linear*. Some patterns are curvilinear or nonlinear, however, and may influence other analyses of the variables.

Not all of these determinations can be made with the statistics available to all levels of data. For example, since nominal level data has no ordering, there is no way to determine direction. Proper bivariate analysis strives, however, to provide indicators of each of these steps so a complete summarization of the data and relationship can be made. This process is discussed more fully in the next three chapters.

■ Conclusion

In this short chapter, we introduced the construction and contents of bivariate tables. We also provided a brief introduction to the analysis of two variables. After reading this chapter, you should know how to properly construct a nominal and ordinal bivariate table and the four analyses that make up bivariate analysis. The techniques for bivariate analysis—existence, strength, direction, and nature—are discussed in the following three chapters.

■ Key Terms

bivariate table	marginals
cell	nature
cell frequency	relationship
column percentage	row percentage
crosstab	strength
direction	total percentage
existence	
expected count	

■ Exercises

1. For the variables *sex* (male, female) and *victim* (yes, no):
   a. Fill in the information in the blanks to properly format the table.

   _____ by _____

	_____ 1	_____ 2	
_____ 2	20	37	57
_____ 1	27	48	75
	47	85	132

   b. Is this a nominal or an ordinal table? Why?
2. a. From the following information, construct the appropriate table for use in a bivariate analysis.

Fear of Walking Alone During the Day	How Well Police Perform Their Duties
Great fear	Very well
Medium fear	Average
Small fear	Below average
No fear	Not at all

b. Is this a nominal or an ordinal level table? Why?

3. A survey of prison inmates was conducted in which inmates were asked about their criminal careers and their income at the time of their arrest. The results of the survey follows.

Crimes per Month	Income	Crimes per Month	Income
15	$18,000	15	$24,000
5	21,000	15	18,000
15	23,000	15	21,000
10	22,000	15	31,000
10	25,000	20	52,000
5	14,000	7	21,000
20	23,000	7	17,000
20	46,000	20	48,000
5	15,000	20	45,000
5	17,000	10	19,000

a. In the spaces below, construct an ordinal bivariate table for *crimes per month* (dependent variable) and income (independent variable). For *crimes per month*, use the categories "10 and less" and "greater than 10." For income, use the categories "$21,000 and more" and "less than $21,000."

_____ by _____

_____

_____ _____   _____

_____ _____   _____ _____

_____ _____   _____ _____

_____ _____   _____ _____

b. Is this a nominal or an ordinal table? Why?

4. For the table below, do the following exercises:
   a. State whether it is nominal or ordinal level, and why.
   b. Identify the row percentage of cell 2–3.
   c. Identify the column percentage of cell 1–2.
   d. Identify the total percentage of cell 2–2.
   e. Identify the row percentage for row 1.
   f. Identify the column percentage for column 2.

Crosstab of Whether a Person Considered Moving in the Preceding Six Months by the Perception of Fear of Crime During the Day

Is crime a serious enough problem that you have considered moving in the past 12 months? * How much has fear of crime affected your decision to walk during the daytime? Crosstabulation

Is crime a serious enough problem that you have considered moving in the past 12 months?			How much has fear of crime affected your decision to walk during the daytime?			Total
			1 No Effect	2 Small Effect	3 Great Effect	
2 Yes		Count	83	48	22	153
		% within Is crime a serious enough problem that you have considered moving in the past 12 months?	54.2%	31.4%	14.4%	100.0%
		% within How much has fear of crime affected your decision to walk during the daytime?	69.2%	44.0%	24.7%	48.1%
		% of Total	26.1%	15.1%	6.9%	48.1%
1 No		Count	37	61	67	165
		% within Is crime a serious enough problem that you have considered moving in the past 12 months?	22.4%	37.0%	40.6%	100.0%
		% within How much has fear of crime affected your decision to walk during the daytime?	30.8%	56.0%	75.3%	51.9%
		% of Total	11.6%	19.2%	21.1%	51.9%
Total		Count	120	109	89	318
		% within Is crime a serious enough problem that you have considered moving in the past 12 months?	37.7%	34.3%	28.0%	100.0%
		% within How much has fear of crime affected your decision to walk during the daytime?	100.0%	100.0%	100.0%	100.0%
		% of Total	37.7%	34.3%	28.0%	100.0%

Notes

1. Note that tables can be multivariate in the form $Y \times X \times Z$, where Z is a control variable. The procedure is to hold the first cell constant and compare X and Y, then repeat the procedure for each cell. This produces a number of bivariate, "partial" tables, which are then compared.
2. Note that it is not particularly proper to consider a table ordinal if both variables are dichotomized nominal level. It is appropriate to use dichotomized nominal level data in combination with a true ordinal or categorized interval level variable only to consider the table ordinal level.

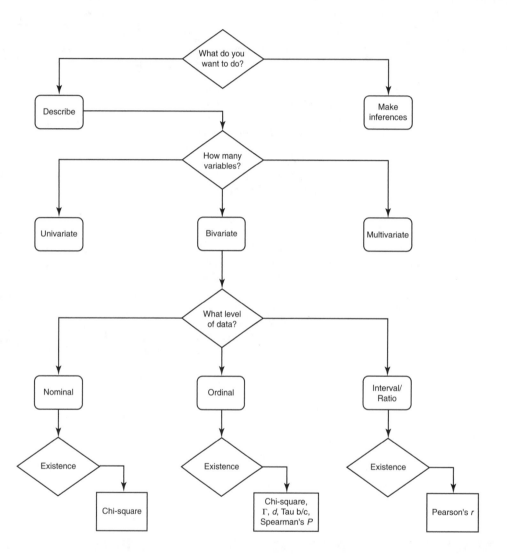

Measures of Existence and Statistical Significance

In Chapter 7 we described how to construct a bivariate table, often the first step in conducting a bivariate analysis. Once the data has been put into a bivariate table, typically the next step is to determine if there is any real relationship between the variables or if the relationship could have occurred by chance. This is a process of determining the **existence** of a relationship—an important step because it may prevent needless effort expended on other bivariate analyses. If there is no relationship between the variables, or if that relationship is probably occurring by chance, there may be no need to conduct other analyses. In this chapter we discuss how to determine the existence of a relationship and whether or not such an existence might be due to chance. The key to establishing existence is the statistical significance of the relationship.

■ Nominal Level Measures of Existence

At the most basic level, an **association** is said to exist between two variables when the distribution of the categories of one variable differs from some of the categories of the other variable. The greater the difference, the greater the association. This may sound counterintuitive; it may seem to make more sense that if two variables are associated, they should have the same rather than different values. What you must keep in mind is that bivariate analyses look for *change* in one variable that may be associated with *change* in the other variable. In a perfect world, an increase of one unit in an independent variable would be associated with a one-unit increase (or decrease) in the dependent variable. For example, if we were studying drug use and crime, ideally we would like to see that for every instance of drug use, the probability of committing a crime rose by 1%. If changes in the independent variable reflect no changes in the dependent variable (i.e., increases in categories of one variable show no change in the categories of the other variable), nothing has been gained by including the independent variable. In drug use and crime, for example, if criminal behavior did not increase no matter how much a person's drug use increased (or if the changes were so random that you could not tell if an increase in drug use produced an increase or decrease in crime), there would be no value in collecting data on drug use to explain criminal behavior. The way this change is examined is through bivariate tables, bivariate analyses, and establishing the statistical significance of the relationship.

■ Tables, Percentages, and Differences

As discussed in Chapter 1, the best way to make preliminary comparisons between variables, especially if they have different *N*s, is to convert the frequencies to percentages. This translates to bivariate analysis by constructing a *percentaged table*, which allows direct comparison between categories of the variables.

There are three types of percentaged tables (corresponding to the percentages contained in the crosstabs discussed in Chapter 7). A **row percentaged table** is based on the rows. This means that the rows will equal 100% but the columns may not. A **column percentaged table** is based on the columns, so the columns will total 100% but the rows may not. A **total percentaged table** is constructed by dividing each cell frequency by *N*. Here, the total percentages of rows and columns will total 100%, but probably none of the individual rows or columns will.

The type of table that is constructed is important in determining the existence of a relationship between two variables because *comparisons are made in the opposite direction from the way in which percentages are calculated.* If rows are used as the base of percentages (a row percentaged table), comparisons will be made on the dependent variable. If columns are used as the base of percentages (a column percentaged table), comparisons will be made on the independent variable. A column percentaged table is the desired way to make comparisons because the goal is to determine what influence the independent variable is having on the dependent variable. To accomplish this requires examining the influence of the independent variable on the dependent variable. This requires removing the effect of the frequencies in the categories of the independent variable, which is completed by totaling those columns to 100%. In **Table 8-1**, the column percentages are highlighted. These are the percentages that would be used in this analysis.

The example in Table 8-1 shows why column percentages should be used in these analyses. As can be seen in this crosstab, the number of females in the distribution far exceeds the number of males (241 females to 99 males). As a result, the frequency of female victimizations exceeds that of males for all categories. This misrepresents the relationship between the variables, however. As shown in **Table 8-2**, if column percentages are used for the comparison, thus controlling for the larger number of females, the cells are much closer. The percentage of persons not victimized is within one percentage point between male and female. Additionally, the relationship of males to females victimized is actually reversed when their proportion of the respondents is taken into account (31.3% to 30.3%). A properly percentaged table, then, using the categories of the independent variable, is important for proper comparison of the variables.

Once a table is set up properly, a difference between the variables (denoting the existence of a relationship) can be determined by calculating an **Epsilon**, symbolized by the Greek letter ε. Epsilon is easily calculated by determining the differences in the column percentages across the rows.

Table 8-2 shows an example of a table set up to examine Epsilon for the existence of the relationship between victimization and sex. Here the comparison is by rows, so the table is totaled by columns, as evidenced by both columns totaling to 100%, with the rows totaling to 69.4% and 30.6%, respectively. Comparisons are then made based on the differences in column percentages by row. In Table 8-2, we want to know how a person's sex may have influenced whether the person was victimized, so we will look in

TABLE 8-1 Crosstab of Victim by Sex of Respondent

Recoded VICTIM to yes and no * Sex of respondent Crosstabulation

			Sex of respondent		
			Male	Female	Total
Recoded VICTIM to yes and no	No	Count	68	168	236
		Expected Count	68.7	167.3	236.0
		% within Recoded VICTIM to yes and no	28.8%	71.2%	100.0%
		% within Sex of respondent	68.7%	69.7%	69.4%
		% of Total	20.0%	49.4%	69.4%
	Yes	Count	31	73	104
		Expected Count	30.3	73.7	104.0
		% within Recoded VICTIM to yes and no	29.8%	70.2%	100.0%
		% within Sex of respondent	31.3%	30.3%	30.6%
		% of Total	9.1%	21.5%	30.6%
Total		Count	99	241	340
		Expected Count	99.0	241.0	340.0
		% within Recoded VICTIM to yes and no	29.1%	70.9%	100.0%
		% within Sex of respondent	100.0%	100.0%	100.0%
		% of Total	29.1%	70.9%	100.0%

TABLE 8-2 Column Percentaged Table: Victim by Sex

	Sex		
Victim	Male	Female	Total
No	68.7	69.7	69.4
Yes	31.3	30.3	30.6
Total	100.0	100.0	100.0

the "Yes" row of the table. The difference between male and female victimizations, the ε, is the difference of the first row (31.3% to 30.3% = 1%). Actually, the existence of an Epsilon can be found if there is a difference in *any* row of the table; that is, if there is a nonzero ε, there is a relationship. An Epsilon of 1% means that there is *some* difference between the variables and that a relationship does exist.

The problems with using ε as a measure of the existence of a relationship between two variables is that Epsilon cannot determine if the difference is by chance or if there is a real difference (is a 1 % difference between the variables meaningful?). In social science research, there will always be some difference, but this difference may be in the way the data is arranged, bad luck, or some other factor. What is needed is to be able to tell if those differences are meaningful or if the way the data is arranged is because of chance.

A more precise method of examining the relationship is determining if it is **statistically significant**. This is important because if the numbers' arrangement is by chance (i.e., if they are not statistically different enough to be considered a true difference), they may not be of any importance. It is generally accepted that it is worth the effort to conduct other bivariate analyses only if a relationship is significant. Some will argue that the discussion here is more about inferential analyses (see the section in this book on inferential analyses) than descriptive analyses. It is certainly true that examining the existence of a difference between two variables is an issue of inferential analysis. It is, however, also an issue of descriptive analyses. Even if you are examining two variables that are assumed to represent the population of data, which descriptive analyses do, you still want to know if they are related statistically. If there is no true relationship—if the data is arranged this way only by chance or because of the way the data was gathered—it may not be worthwhile to spend a lot of time describing the variables further. If they are related by chance, collecting additional data will probably change the way the relationship is described, and other researchers will probably not find the same results. It is important, then, even in descriptive analyses, to determine the existence of a relationship before proceeding to other analyses. Also, even though descriptive analyses are meant to describe a population, the population being described is often quite small: for example, criminals in Kansas City. Even though the findings of this research should be applied only to this population, researchers often want to make the supposition that the findings *could* be applied to a much larger group, such as all criminals in the United States. Using the tests of existence discussed in this chapter (particularly Chi-square) facilitates such a supposition because some conclusions (or inferences) can be drawn about how this data might actually be generalized to a larger group (population).

A bivariate table can be tested to see if the relationship is significant by computing a Chi-square Test of Independence. This is a test based on the Delta statistic, which tests the hypothesis that there is *no* relationship between the two variables (see the description of the null hypothesis in Chapter 16). **Delta (Δ)** is calculated by determining what the values in the cells would be if there was no association between the variables ($\varepsilon = 0$ for all categories) and then comparing those values with the frequencies in the table. The first step in this process is determining the **expected value** of each cell (called the expected count in some versions of SPSS), which is symbolized by f_e, the expected frequency of values. The expected value is the value we would anticipate if there was absolutely no association between the variables. This occurs, and the variables are said to be *independent* or have no association, if the probability that a case falls into a given cell is simply the product of the marginal values (row total and column total) of the two categories that make up the cell. If the expected value (expected count) is not given in a crosstab printout, it can be calculated by multiplying the row total for that row by the column total for that column and dividing by the N of the distribution, as shown below:

$$f_e = \frac{\text{row total} \ \times \ \text{column total}}{N}$$

TABLE 8-3   Column Percentaged Table: Victim by Sex

Victim	Sex		Row Total
	Male	**Female**	
No	$f_e = \dfrac{(236)(99)}{340}$ $= 68.71$	$f_e = \dfrac{(236)(241)}{340}$ $= 167.29$	236
Yes	$f_e = \dfrac{(104)(99)}{340}$ $= 30.28$	$f_e = \dfrac{(104)(241)}{340}$ $= 70.72$	104
Column total	99	241	340

This process is completed for each cell in the crosstab.[1] This would be accomplished for the data in Table 8-1 with the calculations shown in **Table 8-3**. The result of these calculations produces a value that can be compared to the frequency for that category. If the frequency is something other than the expected value, the variables are said to be *associated*, at least in some way. This distribution of expected values is called **model of no association**. These values (68.71, 167.29, 30.28, and 70.72) represent what would be expected for **observed values** (frequency) if there was no association between the two variables. Deviation from this model of no association represents some degree of relationship between the two variables.

The next step in examining existence is to determine the **observed value** of each cell in the distribution, which is symbolized by f_o, the frequency of observed value. These are simply the cell frequencies as calculated in Chapter 7. The cell frequencies for Table 8-1 are 68, 168, 31, and 73. Notice here the switch from percentages to frequencies. This is an important change. For Epsilon or other crude comparisons, percentages are valuable because they standardize the data. More powerful statistical procedures, however, benefit from the more precise measure of frequency.

Delta is calculated by taking the difference between the observed (f_o) and the expected (f_e) values for the cells. This formula will determine if there is any association between the variables. Delta is calculated as follows:

$$\Delta = (f_{o11} - f_{e11}), (f_{o12} - f_{e12}), (f_{o21} - f_{e21}), (f_{o22} - f_{e22}), \ldots$$

If there is a nonzero value of Δ in one of the categories, there is some association between the variables. In the crosstab in Table 8-1, Delta would be calculated as follows:

$$\Delta = (68 - 68.7)\ (168 - 167.3)\ (31 - 30.3)\ (73 - 73.7)$$
$$= \quad (-0.7) \qquad (0.7) \qquad (-0.7) \qquad (0.7)$$

This shows that there is an association between the variables because each calculation had a nonzero value for Delta. Although there were differences in each of the Deltas calculated here, it is not mandatory that all Deltas show a difference; if one has a nonzero value, the Delta is said to show that a relationship exists.

The finding that there is some difference between the observed and expected values here shows that there is some association between the variables. There are two problems with Delta, however. First, Delta can determine only if there is some difference between one category in the distribution at a time. A better procedure would be to check all of the differences between the expected and observed values and provide one summary of the relationship between the two variables in one measure. Delta alone cannot do this. Suppose, however, that the Deltas for the various categories of the two variables above could be added together into one value. This could be done using the formula

$$\Delta = \Sigma(f_o - f_e)$$

Extending this formula would produce the following derivatives:

$$\Delta = \Sigma(f_o - f_e)$$
$$= \Sigma f_o - f_e$$
$$= N - N$$
$$= 0$$

This is the result that would occur if the Deltas in the example above were added together $(-0.7 + 0.7 + -0.7 + 0.7 = 0)$. Since Delta is based on the model of no association, summing Deltas always produces a zero value. This does not aid in determining the total relationship of the two variables. This problem can be overcome, however, by squaring the values of the sum of $(f_o - f_e)$, the same procedure used when making the variance nonzero. The following formula, then, results in a zero value only if there is a true fit between the two distributions and no difference exists between the observed and expected values:

$$\Delta = \Sigma(f_o - f_e)^2$$

The other problem with Delta is that it does not account for differences in the magnitude of frequencies in the cells. For example, in simply measuring the difference between the expected and observed values, Delta does not recognize that there are twice as many females as males in the distribution. Overcoming this problem simply requires standardizing the formula above by dividing it by the expected values. The formula that would result is

$$\Sigma\frac{(f_o - f_e)^2}{f_e}$$

This is the formula for the statistical procedure Chi-square.

■ Chi-Square

The **Chi-square test**, symbolized by χ^2, is the primary method of establishing the statistically significant existence of a relationship between two nominal or, sometimes, higher-level variables. Chi-square allows researchers to move from simply checking the existence of a relationship to testing the statistical significance of the relationship through the null hypothesis (see the chapters in this book on inferential analyses). The null hypothesis being tested here is that there is no statistically significant relationship between the variables being studied: specifically, that there is no difference between the values in the rows and the values in the columns of the table being examined.[2] If that null hypothesis can be rejected, it can be said with a particular degree of certainty that there is a relationship between the two variables.

Chi-square is generally traced to Karl Pearson (1900). There are a number of statistical procedures involving Chi-square that deal with the relationship between distributions. **Chi-square** is also used for a family of sampling distributions. Two of the more popular statistical procedures are the **Chi-square Test of Goodness of Fit**, which compares a distribution to a theoretical distribution to determine the degree of fit between the two, and the **Chi-square Test of Independence**, which compares two distributions to determine if there are significant differences between them. These use the same formula but have different applications. The procedure used here will be the Pearson Chi-square Test of Independence, which we use to examine whether or not the observed frequencies resemble those that would be found if the two variables were actually independent of each other.

In testing for the existence of a relationship, Chi-square examines the difference between the observed and expected frequencies. Differences between these two imply one of two conclusions: Either the difference is truly due to chance and the null hypothesis is correct, or the values are so different that it cannot be said that the difference is due to chance and the null hypothesis must be rejected. Although Chi-square actually tests whether the two variables of a sample are independent in the *population,* it is often extended as an approximate measure of whether the variables are independent in the sample (or population if the analysis is purely descriptive).

Chi-square sums the values of the Delta statistic squared, divided by the expected value for each cell. The formula for chi-square is

$$\chi^2 = \sum \left[\frac{f_o - f_e)^2}{f_e} \right]$$

Essentially, this calculation measures the extent to which the observed frequencies in the table differ from those that would be expected if the hypothesis of no association were true. The result of the formula is the absolute value of the divergence between the observed and expected values. When interpreting the results of this calculation, the larger the differences between the observed and expected frequencies, the larger the Chi-square value. Chi-square values can run from zero into the hundreds.[3] Chi-square will be zero if all observed and expected values are equal. In the example from Table 8-1, Chi-square would be calculated as follows:

$$\chi^2 = \sum \frac{(f_o - f_e)^2}{f_e}$$

$$\chi^2 = \sum \frac{(68 - 68.7)^2}{68.7} + \frac{(168 - 167.3)^2}{167.3} + \frac{(31 - 30.3)^2}{30.3} + \frac{(73 - 73.7)^2}{73.7}$$

$$\chi^2 = \sum \frac{(-0.7)^2}{68.7} + \frac{(0.7)^2}{167.3} + \frac{(0.7)^2}{30.3} + \frac{(-0.7)^2}{73.7}$$

$$\chi^2 = \sum \frac{0.49}{68.7} + \frac{0.49}{167.3} + \frac{0.49}{30.3} + \frac{0.49}{73.7}$$

$$\chi^2 = 0.0071 + 0.0029 + 0.0162 + 0.0066$$

$$\chi^2 = 0.0328$$

Here, all we have done is to calculate Delta $(f_o - f_e)$ for each cell in the table, squared that value, divided by f_e for that cell, and then summed all of those scores.

Computing Chi-square in this manner provides a value for the likelihood of the data arrangement occurring by *chance*. But some method of interpreting this value is needed. This is done by determining the degrees of freedom for the variables and making comparisons with a table of Chi-square values.

The next step, then, is to determine the degrees of freedom for the two variables. Calculating the **degrees of freedom** is a statistical method of compensating for error that arises when samples are used rather than populations. In descriptive analyses, it is also a compensation for small data sets, where some of the data may have occurred by chance even if all of the data did not. Since researchers can rarely be certain of the accuracy of all of the values, they must compensate for differences between the data drawn and the true nature of the distribution, even assuming it is a population. When using a crosstab, degrees of freedom are essentially the number of values that would be required to be given to you before you could figure out the rest of the table by using the marginals. Degrees of freedom are calculated as follows:

$$df = (\text{row categories} - 1)(\text{column categories} - 1)$$

For a 2×2 table, the degrees of freedom would be $(2 - 1)(2 - 1)$, or 1. This means that we would only need to have one value given to us to be able to calculate the rest of the values for the other cells. For example, in the crosstab in **Table 8-4** (the same example as in Table 8-1), we need only one value to be able to determine the other three because we could subtract that value from the row or column totals to obtain the other values. Here, knowing that the value in cell 1–1 is 68 and knowing that the row total is 236, the value of cell 1–2 can easily be calculated as 168 ($236 - 68 = 168$). Similarly, knowing that cell 1–1 is 68 and that the column total is 99, we know that the value of cell 2–1 is 31 ($99 - 68 = 31$). Finally, knowing that the value of cell 2–1 is 31 and that the row total is 104, we know that the value of cell 2–2 is 73 ($104 - 31 = 73$).

The greater the number of degrees of freedom, the closer we are to the true population or to the true nature of the distribution. Smaller numbers mean that the data may not be an accurate measure of the true nature of the data distribution and for inferential analyses that the sample drawn is probably only a small portion of the population and may not be representative.

The last element of Chi-square to be determined before consulting the table is the level of confidence, or *statistical significance*, desired to make a conclusion about the data. In other words, how sure do we want to be that the relationship does not occur by

TABLE 8-4   **Explanation of Degrees of Freedom for Crosstabs**

Victim	Sex		Column Total
	Male	Female	
Yes	68	—	236
No	—	—	104
Row total	99	241	

chance? In statistical analyses, two levels of statistical significance are used most often: 95% and 99%. These mean that there is a 95% or 99% probability that the relationship between the variables did not occur by chance (and a corresponding 5% or 1% chance that it did). Other values that are often used are 99.9%, 90%, 80%, and 70%, as shown in Table B-2 in the appendix. The ultimate decision of what level of significance to choose is up to the researcher.

These three pieces of information are used to determine the *critical value* drawn from Table B-2. The *obtained* (*calculated*) *value* of Chi-square and the degrees of freedom are used to make comparisons to standard values of Chi-square in Table B-2 for the chosen level of significance. This will allow a determination of whether or not there is a statistically significant relationship between the two variables. To use these three pieces of information, turn to Table B-2 in the appendix. Select the row that has the degrees of freedom for the table being examined. Then select the column that contains the desired level of significance. Where the row and column intersect is the *critical value* of the χ^2. If the obtained value of χ^2 is *greater* than the critical value, the null hypothesis can be rejected and it can be concluded that there is a statistically significant relationship. If the obtained value is *less than* the critical value, we would fail to reject the null hypothesis, and it can be concluded that the relationship may have occurred by chance.

How do you do that? Obtaining Chi-square in SPSS

1. Open a data set.
   a. Start SPSS.
   b. Select *File*, then *Open*, then *Data*.
   c. Select the file you want to open, then select *Open*.
2. Once the data is visible, select *Analyze*, then *Descriptive Statistics*, then *Crosstabs*.
3. Select the variable you wish to include as your dependent variable and press the ▶ next to the *Row(s)* window.
4. Select the variable you wish to include as your independent variable and press the ▶ next to the *Column(s)* window.
5. Select the *Cells* button and check any of the boxes for information you may want in your crosstab. For a crosstab with full information, select *observed*, *expected*, *row*, *column*, and *total*; then press *Continue*.
6. Select the *Format* button, and check the box marked *Descending*.
7. Select the *Statistics* button at the bottom of the window.
8. Check the box next to "Chi-square."
9. Select *Continue*, then *ok*.
10. An output window should appear containing a distribution similar in format to Table 8-5.

Since the table in this case is a 2×2 table, there would be 1 degree of freedom, as calculated above. Next, say that we want to be 95% sure our relationship is not occurring by chance, so we would use this column in Table B-2. In Table B-2 under the 0.95 level of significance and at 1 degree of freedom, there is a value of 3.841. This means that for 1 degree of freedom, distributions that have a Chi-square value of 3.841 or more would, over time, occur by chance only about 5 times in 100. We could be confident, then, that a decision based on this value would be correct 95% of the time (and would be wrong 5 times in 100). The critical value in this case, 3.841, is definitely larger than the 0.0328 calculated using the Chi-square formula (the obtained value). We would, therefore, fail to reject the null hypothesis and conclude that the difference between the observed values and expected values occurs by chance.

To solidify this important statistical procedure, take one more example of a Chi-square calculation drawn from **Table 8-5** (see **Exhibit 8-1**). In this case, 12 sets of expected values would have to be calculated. These expected values would be calculated as

$$\chi^2 = \sum \frac{(f_o - f_e)^2}{f_e}$$

$$\chi^2 = \sum \begin{bmatrix} \dfrac{(13 - 5.8)^2}{5.8} + \dfrac{(5 - 8.0)^2}{8.0} + \dfrac{(1 - 5.1)^2}{5.1} + \dfrac{(26 - 20.2)^2}{20.2} \\[2mm] + \dfrac{(27 - 27.9)^2}{27.9} + \dfrac{(13 - 17.8)^2}{17.8} + \dfrac{(3 - 13.2)^2}{13.2} \\[2mm] + \dfrac{(24 - 18.2)^2}{18.2} + \dfrac{(16 - 11.6)^2}{11.6} + \dfrac{(0 - 2.8)^2}{2.8} \\[2mm] + \dfrac{(2 - 3.8)^2}{3.8} + \dfrac{(7 - 2.4)^2}{2.4} \end{bmatrix}$$

$$\chi^2 = \sum \frac{7.2^2}{5.8} + \frac{-3^2}{8.0} + \frac{-4.1^2}{5.1} + \frac{5.8^2}{20.2} + \frac{-0.9^2}{27.9} + \frac{-4.8^2}{17.8} + \frac{-10.2^2}{13.2}$$

$$+ \frac{5.8^2}{18.2} + \frac{4.4^2}{11.6} + \frac{-2.8^2}{2.8} + \frac{-1.8^2}{3.8} + \frac{4.6^2}{2.4}$$

$$\chi^2 = \sum \frac{51.84}{5.8} + \frac{9}{8.0} + \frac{16.81}{5.1} + \frac{33.64}{20.2} + \frac{0.81}{27.9} + \frac{23.04}{17.8} + \frac{104.4}{13.2}$$

$$+ \frac{33.64}{18.2} + \frac{19.36}{11.6} + \frac{7.84}{2.8} + \frac{3.24}{3.8} + \frac{21.16}{2.4}$$

$$\chi^2 = 40.22$$

Notice here that although these numbers are more complex than those shown earlier when calculating Delta, the numerators still sum to zero before squaring.

The next piece of information that is needed is the degrees of freedom. Since this is a 4×3 table, the degrees of freedom would be calculated as $(r - 1)(c - 1)$, or $(4 - 1)(3 - 1) = (3)(2) = 6$. So there would be 6 degrees of freedom for this table. If the desired significance (confidence) level is .99, Table B-2 can be consulted to determine if

TABLE 8-5 **Crosstab of COP-OK and TIME-OK**

In your neighborhood, how well do you think the police perform their duties? * How satisfied were you with the time it took the police to respond? Crosstabulation

In your neighborhood, how well do you think the police perform their duties?			How satisfied were you with the time it took the police to respond?			Total
			1 Not At All Satisfied	2 Somewhat Satisfied	3 Very Satisfied	
4 Very well	Count		13	5	1	19
	% within In your neighborhood, how well do you think the police perform their duties?		68.4%	26.3%	5.3%	100.0%
	% within How satisfied were you with the time it took the police to respond?		31.0%	8.6%	2.7%	13.9%
	% of Total		9.5%	3.6%	.7%	13.9%
3 Average	Count		26	27	13	66
	% within In your neighborhood, how well do you think the police perform their duties?		39.4%	40.9%	19.7%	100.0%
	% within How satisfied were you with the time it took the police to respond?		61.9%	46.6%	35.1%	48.2%
	% of Total		19.0%	19.7%	9.5%	48.2%
2 Below Average	Count		3	24	16	43
	% within In your neighborhood, how well do you think the police perform their duties?		7.0%	55.8%	37.2%	100.0%
	% within How satisfied were you with the time it took the police to respond?		7.1%	4.14%	43.2%	31.4%
	% of Total		2.2%	17.5%	11.7%	31.4%

(*continues*)

TABLE 8-5   Crosstab of COP-OK and TIME-OK (*Continued*)

In your neighborhood, how well do you think the police perform their duties?			How satisfied were you with the time it took the police to respond?			Total
			1 Not At All Satisfied	2 Somewhat Satisfied	3 Very Satisfied	
1 Not At All		Count		2	7	9
		% within In your neighborhood, how well do you think the police perform their duties?		22.2%	77.8%	100.0%
		% within How satisfied were you with the time it took the police to respond?		3.4%	18.9%	6.6%
		% of Total		1.5%	5.1%	6.6%
	Total	Count	42	58	37	137
		% within In your neighborhood, how well do you think the police perform their duties?	30.7%	42.3%	27.0%	100.0%
		% within How satisfied were you with the time it took the police to respond?	100.0%	100.0%	100.0%	100.0%
		% of Total	30.7%	42.3%	27.0%	100.0%

Chi-Square Tests

	Value	df	Asymp. Sig. (2-sided)
Pearson Chi-Square	39.874	6	.000
Likelihood Ratio	42.699	6	.000
Linear-by-Linear Association	33.303	1	.000
N of Valid Cases:	137		

Note: 3 cells (25.0%) have expected count less than 5. The minimum expected count is 2.43.

EXHIBIT 8-1   Sets of Expected Values for Table 8-5

$$F_{e1\text{-}1} = \frac{(19)(42)}{137} = 5.8 \qquad F_{e1\text{-}2} = \frac{(19)(58)}{137} = 8.0 \qquad F_{e1\text{-}3} = \frac{(19)(37)}{137} = 5.1$$

$$F_{e2\text{-}1} = \frac{(66)(42)}{137} = 20.2 \qquad F_{e2\text{-}2} = \frac{(66)(58)}{137} = 27.9 \qquad F_{e2\text{-}3} = \frac{(66)(37)}{137} = 17.8$$

$$F_{e3\text{-}1} = \frac{(43)(42)}{137} = 13.2 \qquad F_{e3\text{-}2} = \frac{(43)(58)}{137} = 18.2 \qquad F_{e3\text{-}3} = \frac{(43)(37)}{137} = 11.6$$

$$F_{e4\text{-}1} = \frac{(9)(42)}{137} = 2.8 \qquad F_{e4\text{-}2} = \frac{(9)(58)}{137} = 3.8 \qquad F_{e4\text{-}3} = \frac{(9)(37)}{137} = 2.4$$

the null hypothesis of no association between the two variables can be rejected. In Table B-2, for 6 degrees of freedom at .99, the critical value is 16.812. Since the obtained value in this case (40.22) is larger than the critical value, the null hypothesis can be rejected and an assumption can be made that these two variables are related; there is an association between a person's satisfaction with the time the police took to respond to a call and his or her perception of the quality of the job the police are doing.

Also notice in Table 8-5 that the value shown for the Pearson Chi-square is very close to the value obtained from the calculations above. Actually, these are the same values. The only difference is in the rounding and the fact that SPSS works with many more decimal places. The fact that the value obtained from the calculations above is the same as the value for Chi-square in SPSS is important. This means that the output of most statistical programs contains all the information that is needed to evaluate the significance of a relationship using Chi-square without resorting to the table. The speed and accuracy of the computer allows researchers to use an exact probability of the two distributions occurring by chance rather than having to rely on approximations. **Table 8-6** shows only the Chi-square portion of an SPSS output for the data in

TABLE 8-6   Chi-Square Output from SPSS

Chi-Square Tests

	Value	df	Asymp. Sig. (2-sided)	Exact Sig. (2-sided)	Exact Sig. (1-sided)
Pearson Chi-Square	.035	1	.853		
Continuity Correction	.003	1	.955		
Likelihood Ratio	.034	1	.853		
Fisher's Exact Test				.897	.475
Linear-by-Linear Association	.034	1	.853		
N of Valid Cases	340				

Notes: Computed only for a 2 × 2 table.

  0 cells (.0%) have expected count less than 5. The minimum expected count is 30.28.

Table 8-1. In this table, the Chi-square value is the same as that obtained in the calculation (.03). However, the obtained value of the Chi-square is not the important piece of information for evaluating the relationship. With this kind of output, the important piece of information is the **significance** for the *Pearson* (Chi-square). The goal is essentially the same as above, except in reverse. Whereas we would use 90%, 95%, 99%, and so on, in the table to establish significance, here we want to use .05, or .01, depending on how sure we want to be. If the number in the "Asymp. Sig (2-sided)" column is greater than the desired value, we would fail to reject the null hypothesis; if the output value is equal to or less than our desired cutoff value, we would reject the null hypothesis. In this case, the output value (.853) far exceeds any desired value (either .05 or .01) so we would fail to reject the null hypothesis.

Requirements for Using Chi-Square

Chi-square is a very powerful procedure for determining the statistical significance of a relationship. There are requirements, however, for its acceptable use. First, the N for the crosstab should be five times the number of cells. This ensures that there is ample data from which to draw conclusions concerning the relationship. It would not be desirable to have 30 cells in a table and an N of only 10—most of the cells would be blank, making it impossible to interpret the data. In the example in Table 8-1, the sample size should be at least 20 (4 cells $\times$ 5 = 20). Since this data set has an N of 340, this requirement is met. Similarly, in Table 8-5, the sample size should be at least 60 (12 cells $\times$ 5 = 60). Since this data set has an N of 137, this requirement is met.

If the N does not exceed five times the number of cells, it is possible to analyze the data by using Yates' Correction (1934) or Fisher's Exact Test (1922). These are two statistics that compensate for small Ns and essentially make it more difficult to reject the null hypothesis by lowering the obtained value of χ^2.[4]

Yates Correction adds a value of 0.5 to the numerator of the Chi-square formula and changes the difference between the observed and expected values to an absolute value, as shown below:

$$\chi^2_{\text{Yates}} = \sum \frac{(|f_o - f_e| - 0.5)^2}{f_e}$$

This has the effect of lowering the value in the numerator. Mathematically, this will lower the obtained value of Chi-square, making it more difficult to reject the null hypothesis. It is argued by many that a Yates Correction should always be used on 2 $\times$ 2 tables.

Fisher's Exact Test for small sample sizes uses a more precise measure of the data available and an alternative method of calculating the degrees of freedom to attempt to make a refined prediction of the relationship. The common conception that Fisher's Exact Test is an extension of Chi-square is actually not true. Fisher's Exact Test, in fact, uses a multinomial or hypergeometric calculation of the data to compute the exact probability of the data obtained representing the population. For example, suppose the population of probation officers can be divided into college graduates and nongraduates and can also be divided based on their correctional philosophy into those favoring rehabilitation and those opposed to it. This could be shown as in **Table 8-7**, with the distribution of the frequencies represented by the marginal totals $(a + c)$, $(b + d)$, $(a + b)$, and $(c + d)$. Fisher's Exact Test calculates the probability that in a sample of probation

TABLE 8-7   Population and Sample Characteristics for Fisher's Exact Test

	C_1	C_2	Total
R_1	a	b	$a + b$
R_2	c	d	$c + d$
Total	$a + c$	$b + d$	

officers chosen at random, exactly a probation officers fall into column C1 and b probation officers fall into column C_2.

Fisher's Exact Test in this case would use a hypergeometric to determine this probability, using the formula

$$\frac{(a + b)!\,(c + d)!\,(a + c)!\,(b + d)!}{N!\,a!\,b!\,c!\,d!}$$

In this equation, the letters correspond to Table 8-7, where N is the total number of cases and ! represents a factorial. This calculation requires that all possible combinations that would equal the marginal totals be tested. Obviously, this would require a great deal of work and is practical only for very small sample sizes or when a computer is used.

The result of this calculation is the probability that the values a, b, c, and d were distributed in the crosstab by chance. The value obtained from this calculation can be used as an approximate level of significance (i.e., values of less than .05 would be significant) after all calculations are made, or it can be compared to a Fisher's Exact Test table so that a smaller number of calculations can be used to approximate Fisher's Exact Test.

Fisher's Exact Test is automatically substituted for a Chi-square test in SPSS when the sample size requirement is violated or if one of the cells has a cell frequency of less than 5. This typically goes unnoticed in SPSS because it does not change the printout, but it is an important addition to any analysis because it helps to ensure that the assumptions of chi-square are met.

The second requirement of Chi-square is that 80% of the cells should have expected frequencies greater than 5.[5] Chi-square operates from the difference between observed and expected values. If there are a large number of expected values that are less than five, it makes calculations unstable. For example, in research including race, it is sometimes unwise to include some racial categories because there may or may not be any people of that race in the sample chosen. Because there can be such small observed or expected frequencies, the probability of a value occurring by **chance** is fairly large. If more than 20% of the cells of a table contain this kind of data, it can cause the significance testing to be very unstable. Whereas a sample size of at least five times the number of cells helps ensure there is an adequate supply of values from which to choose, the 80% rule helps ensure that the frequencies are sufficiently spread out in the table to make it useful for analysis. For example, even with a sample size of 300, it would be no help if all the values were in two cells of a 4×4 table. Like the requirement above, if this rule is violated in SPSS analyses, Fisher's Exact Test is used automatically.

It is common practice to collapse categories if this requirement is violated. This means that if there are five categories in a table and one of them has a very small expected or observed value, two of the categories can be combined to boost the observed and expected frequencies. Although this is common, its use should be approached with caution. With Chi-square, a random sample and categories defined in advance are assumed. Changing the categories after the data is drawn and the table constructed violates these assumptions, and any conclusions drawn may be erroneous. If there is some question about the veracity of combining categories, the procedure should be avoided because it may be better to violate an assumption of Chi-square than to risk not being able to make any interpretation at all.

The third requirement of Chi-square is that there be mutually exclusive and exhaustive categories. This is important because there is the assumption that the variables are independent. If the variables are measuring the same phenomena, the significance of the relationship is corrupted. Additionally, if there is not exhaustion of categories, the table will be incomplete or the variable will not be measured correctly, making calculations unreliable. These important properties can be shown using the bivariate table construction outlined in Chapter 7. There, the frequencies of two variables were placed in categories, with one variable represented in the rows of the table and another variable in the columns of the table. To fill in the cells properly, each joint frequency must fit into only one cell. If the attributes of one variable were such that they could fit into two of the three columns, it would be difficult or impossible to determine the proper column for the data. Furthermore, if the variable is not completely exhaustive, even after all the joint frequencies have been placed in the table, the variable will still not be measured accurately.

Limitations of Chi-Square

Although Chi-square is a valuable tool for determining the existence of a statistically significant relationship, it is not without problems. At least three limitations must be considered when using Chi-square as a measure of existence. First, Chi-square can determine if two variables are significantly related, but little else. It does not address whether the relationship is meaningful or what the nature of the relationship might be. As a result, it is generally considered more of a starting point than an end in itself.

This also shows the distinction between using Chi-square to analyze a relationship and using Analysis of Variance (ANOVA; see Chapter 18). In ANOVA, the numerical values themselves are used as the basis of comparison, whereas Chi-square simply uses the values to place them into categories. Analysis of variance bases the distinction between the variables on the means of the two groups, and conclusions can be drawn about the difference or distance between the two groups. This means that whereas Chi-square can only determine that a significant difference exists between the two variables, ANOVA can add how the variables differ and by how much they differ. This is why Analysis of Variance is preferred if the level of data (interval and ratio) and other criteria can be met.

Also, Chi-square is heavily influenced by sample size. This is a kind of a Catch-22. One of the requirements of Chi-square is that it have a fairly large N, otherwise, it is not reliable as a test of independence. If the N is much larger than that required for accurate results, however, it quickly reaches a point that almost any difference between the variables will show up as significant. This can be demonstrated with only a modicum of belief that almost everything on Earth is related in at least some manner, so enough data should be able to produce significant results in any variables compared.

This limitation can be also demonstrated by examining Table B-2 in the appendix. Notice that the table ends at 30 degrees of freedom. Although this deals with degrees of freedom rather than sample size, it is illustrative of the influence of a large amount of data on Chi-square. The table ends at 30 degrees of freedom because Chi-square reaches a point of diminishing returns.[6] Because Chi-square reaches this point at a relatively small value, large samples will be significant even if the variables are weakly related.

The final limitation of Chi-square is that the alternative hypothesis is not supported by rejection of the null hypothesis. Just because we can reject the null hypothesis that there is no relationship between the variables does not support the argument that a relationship definitely exists. Additional statistical analyses are required to further examine the relationship between the variables.

Final Note on Chi-Square

The procedure, analysis, and interpretation of Chi-square is very similar to that discussed in the chapters on inferential analysis. The difference is in what is being measured. Inferential analysis requires the variables to be normally distributed and interval level. This allows summary measures—most typically the mean or variance—to be compared between the two distributions. Ordinal and nominal level data cannot rise to that level of sophistication, however, and using the mode or range, for example, to make comparisons is mathematically useless. Chi-square uses the data available (the observed and expected values) to make comparisons not otherwise possible. This allows researchers to employ the theory and procedures of inferential analyses to lower-level data.

■ Tests of Existence for Ordinal and Interval Level Data

As stated above, Chi-square is a popular method of determining the existence of a relationship for nominal and ordinal level data. There are also derivatives of Chi-square specifically for use with ordinal level variables. Probably the most popular of these tests is the Kolmogorov–Smirnov Test, although these tests are typically eschewed in favor of either Chi-square or the analyses discussed in this section.

SPSS has also added (actually returned) a feature to the measures of association that aids in examining the statistical significance of those analyses. For Lambda and the ordinal measures of association used in SPSS, there is an **approximate significance** value included. This value, as its name implies, provides an approximation of the statistical significance for that procedure. Technically, however, these values are not measuring the same thing as Chi-square. As stated previously, there are many versions of hypothesis testing. Some of these address hypotheses related to the measure of the strength of the association, unlike Chi-square, which addresses the measure of the existence of the association. The *approximate significance* is actually addressing the hypothesis of the measure of strength rather than the measure of existence.

For Lambda, there is probably no value in using these approximations over Chi-square. For ordinal level data, however, these measures of existence/significance, and those of Spearman's Rho and Pearson's *r*, may have certain advantages over Chi-square. First, the measures of significance for higher-order analyses serve the dual purpose of estimating the existence of a relationship and testing the null hypothesis that the descriptive measure of the strength of the association might hold up even within larger groups. Furthermore, Chi-square only examines whether or not any relationship exists between the two variables; it does not consider the ordering of the variables. The direction of the variables may be a primary interest of the research (see Chapter 10). The approximate significance measures, then, may be preferred over Chi-square because they are a direct measure of the relationship rather than a test of independence only.

Calculation and Interpretation for Ordinal Data

The standard method of examining the approximate significance of ordinal measures of association (Gamma, Somers' d, tau-b and -c) is to use a slightly modified version of the formulas for calculating those measures (see Chapter 9) that utilize a Z score (as in Chapter 6). The general formula for each of these measures can be demonstrated with the formula outlined for Gamma, as shown below:

$$Z = (G - \Gamma)\sqrt{\frac{N_s + N_d}{2N(1 - G^2)}}$$

In this formula, G is a Gamma value calculated in the manner prescribed in Chapter 9. Since this is testing the null hypothesis, we want to argue that the relationship would be zero; therefore, the Γ (the population parameter for G) would be zero. $N_s + N_d$ is the standard numerator for the Gamma formula, and N is the total number of values in the crosstab.

Although these formulas are important, they yield values that are not consistent with those in SPSS output, which is the difference between a Z score significance and an approximate significance. The output from SPSS, then, is shown in **Table 8-8**, and the interpretation of that output is described here instead of the result of the formula above. Although the Approximate T can be used from the SPSS output to establish significance, the better measure is to go directly to the last column (Approx Sig). If the approximate significance found in the output is less than the desired cutoff level (typically, .05 or .01), the null hypothesis can be rejected and it can be said that a relationship exists between these two variables. The values obtained in Table 8-8 would result in a failure to reject the null hypothesis, and a conclusion that this data could be arranged by chance and that no association exists between the two variables.

Spearman's Rho and Pearson's *r*

It is also possible to determine the existence of a relationship between two fully ordered ordinal level variables using *Spearman's Rho* (also known as *Spearman's Correlation*), and it is possible to determine the existence of interval and ratio level data using Pearson's *r*.

Examining the existence of a relationship through statistical significance is perhaps more appropriate at this level because it is more likely that the researcher is attempting to move beyond mere description and toward estimating the values in other groups (i.e., examining criminals in one city but suggesting that the results might apply in other cities). This importance was reflected on by Spearman in 1904: "For, though the correlation between two series of data is an absolute mathematical fact, yet its whole real value lies in our being able to assume a likelihood of further cases taking a similar direction; we want to consider our results as a truly representative sample." It is important, therefore, not only to calculate the co-relation mathematically but also to be able to make some assumptions about how the relationships examined descriptively might hold up when examined in a larger data set or in other places.

TABLE 8-8 SPSS Output for Ordinal Measures of Association and Approximate Significance

Directional Measures

			Value	Asymp. Std. Error[a]	Approx. T[b]	Approx. Sig.
Ordinal by	Somers' *d*	Symmetric	−.061	.056	−1.071	.284
Ordinal		How often do you walk in your neighborhood at night? Dependent	−.044	.041	−1.071	.284
		Is your neighborhood safety changing? Dependent	−.096	.089	−1.071	.284

[a]Not assuming the null hypothesis.
[b]Using the asymptotic standard error assuming the null hypothesis.

Symmetric Measures

		Value	Asymp. Std. Error[a]	Approx. T[b]	Approx. Sig.
Ordinal by	Kendall's tau-b	−.065	.060	−1.071	.284
Ordinal	Kendall's tau-c	−.040	.037	−1.071	.284
	Gamma	−.154	.139	−1.071	.284
N of Valid Cases		256			

[a]Not assuming the null hypothesis.
[b]Using the asymptotic standard error assuming the null hypothesis.

How do you do that?

Measures of Existence for Partially Ordered Ordinal Level Data
1. Open a data set.
   a. Start SPSS.
   b. Select *File*, then *Open*, then *Data*.
   c. Select the file you want to open, then select *Open*.
2. Once the data is visible, select *Analyze*, then *Descriptive Statistics*, then *Crosstabs*.
3. Select the variable you wish to include as your dependent variable and press the ▶ next to the *Row(s)* window.
4. Select the variable you wish to include as your independent variable and press the ▶ next to the *Column(s)* window.
5. Select the *Cells* button and check any of the boxes for information you may want in your crosstab. For a crosstab with full information, select *observed*, *expected*, *row*, *column*, and *total*; then press *Continue*.

6. Select the *Format* button, and check the box marked *Descending.*

7. Select the *Statistics* button at the bottom of the window.

8. Check the box next to "Chi-square", and/or check the box next to one of the "Ordinal" measures discussed in Chapter 9.

9. Select *Continue*, then select *OK.*

10. An output window should appear containing a distribution similar in format to Table 8-8.

Since Spearman's Correlation is simply a Pearson's *r* calculated for the ranks of the values (see Chapter 9), the same formula is often used when significance tests are performed for ordinal and interval level data. A word of caution, however: Generally, in inferential analyses, the underlying joint distribution of the two variables is normal (see the chapters on inferential analyses). By its very nature, a normal distribution typically assumes interval level data at a minimum. It is important, then, when using ordinal level data, to establish whether the data approximates a normal distribution or if it is close to interval level such that the higher-order hypothesis tests may be used. The formula to test the hypothesis of independence between two ordinal level variables is

$$t = r\sqrt{\frac{\text{explained variation}}{\text{total variation}}}$$

which in computational form would be

$$t = r\sqrt{\frac{N-2}{1-r^2}}$$

which is, essentially, a Student's *t*-test (see Chapter 17) with $N-2$ degrees of freedom. For this test, either one- or two-tailed tests may be undertaken (see Chapter 16). A *Z* test of significance for Spearman's Correlation can also be derived using the formula

$$Z = \frac{r_s - 0}{1/\sqrt{N-1}}$$

The data in **Table 8-9** is used in Chapter 9 to show the calculation of the strength of a Pearson's *r*. It can also be used here to show the calculation of the existence/statistical significance of the relationship. This is data on the educational level of criminals and their associated number of crimes committed before imprisonment.

The piece of information needed to calculate the significance is the strength of the association. As shown in Chapter 9, the value of Pearson's *r* for this data is 0.8214, quite a high value. It is necessary to determine, however, whether or not this value was a

TABLE 8-9 **Data for Educational Level and Number of Crimes of Jail Inmates**

X Education	Y Crimes
7	13
4	11
13	9
16	7
10	5
22	3
19	1
ΣX 91	ΣY 49

result of chance, a statistical anomaly of the data. To determine if the Pearson's r is achieved by chance, a test of significance should be performed. The test of significance for Pearson's r using the data above would be calculated as follows:

$$r = -0.8214\sqrt{\frac{7 - 2}{1 - (-0.8214)^2}}$$

$$= -0.8214\sqrt{\frac{7 - 2}{1 - 0.67}}$$

$$= -0.8214\sqrt{\frac{5}{0.33}}$$

$$= -0.8214\sqrt{15.37}$$

$$= -0.8214(3.92)$$

$$= -3.22$$

In this formula, the first value is the obtained Pearson's r from Chapter 9, 7 is the number of jail inmates studied, and the 1 and 2 are constants in the formula. The calculations produce a value of -3.22, which is similar to the obtained value of Chi-square discussed above. As with Chi-square, this value must be checked against the critical value at a particular significance level and for the degrees of freedom. To do this look at Table B-3 in the appendix. For Pearson's r, the degrees of freedom are calculated by $N - 2$. Here the number of degrees of freedom is $7 - 2$, or 5. As shown in the table, the critical value at the .05 level is 2.015, and the critical value at the .01 level is 3.365. When this is compared to the obtained value, -3.22 is greater than -2.015, so the null hypothesis can be rejected at the .05 level; however, -3.22 is less than -3.365, so you must fail to reject the null hypothesis at the .01 level. This shows that even a strong correlation, or Pearson's r, can be nonsignificant.

How do you do that?

Measures of Existence for Fully Ordered Ordinal and Interval Level Data

1. Open a data set.
   a. Start SPSS.
   b. Select *File*, then *Open*, then *Data*.
   c. Select the file you want to open, then select *Open*.
2. Once the data is visible, select *Analyze*, then *Descriptive Statistics*, then *Crosstabs*.
3. Select the variable you wish to include as your dependent variable and press the ▶ next to the *Row(s)* window.
4. Select the variable you wish to include as your independent variable and press the ▶ next to the *Column(s)* window.
5. Select the *Cells* button and check any of the boxes for information you may want in your crosstab. For a crosstab with full information, select *observed*, *expected*, *row*, *column*, and *total*; then press *Continue*.
6. Select the *Format* button, and check the box marked *Descending*.
7. Select the *Statistics* button at the bottom of the window.
8. Check the box next to "Correlations."
9. Select *Continue*, then select *OK*.
10. An output window should appear containing a distribution similar in format to Table 8-10.

Although these calculations can be made by hand, examining the existence/significance of relationships at this level typically involves the use of computers. **Table 8-10** shows the results of a bivariate analysis of the relationship between a person's fear of walking at night and how safe he or she perceives the neighborhood to be. Most of the

TABLE 8-10   SPSS Output for Pearson's *r* and Spearman's Correlation

WALK_NIT (How often do you walk at night)
by
LR_SAFE (How safe is your neighborhood)
Symmetric Measures

		Value	Asymp. Std. Error[a]	Approx. T[b]	Approx. Sig.
Interval by Interval	Pearson's *r*	.125	.071	1.997	.047[c]
Ordinal by Ordinal	Spearman Correlation	.122	.065	1.942	.053[c]
N of Valid Cases		253			

[a] Not assuming the null hypothesis.
[b] Using the asymptotic standard error assuming the null hypothesis.
[c] Based on a normal approximation.

numbers in Table 8-10 are explained in Chapter 9. More important for the topic at hand are the values in the "Approx. Sig." column. These values show that the relationship between walking at night and an attitude of the safety of the city hovers around the .05 level of significance. A significant finding here supports an argument that the measure of the strength value ($-.125$ for Pearson's r and $-.122$ for Spearman's Correlation) is not a product of chance and that the same relationship should hold if examined in other groups. Again, using Pearson's r, we can make statements rejecting the null hypothesis for the *values* of the two variables. Using Spearman's Correlation, however, only allows us to reject the null hypothesis concerning the **ranks** of the values for the two variables.

A final word of caution is needed concerning interpretation of these findings. Failure to reject the null hypothesis can mean there is no relationship between the variables. A failure to reject the null hypothesis using Spearman's Correlation may also mean, however, that the relationship is nonmonotone. Similarly, for Pearson's r, a failure to reject the null hypothesis may mean the relationship is not linear. This is an important issue that is addressed in detail in Chapter 10.

■ An Issue of Significance

In simpler times, the issue of establishing significance was very straightforward. The use of hand calculations and tables allowed researchers to establish a given level of significance that they wished to meet, analyze the data, compare observed and critical values, and make a decision. Cut points for this decision were clear; the table had values for .05, .01, and perhaps, .001.

As with many other things, the use of computers in statistical analysis has made this decision somewhat more complicated. Statistical printouts now provide the exact significance value for the analysis, often to four decimal places. This level of precision can sometimes cause a dilemma in deciding whether or not to reject a null hypothesis. For example, given a significance value of .0551, should you reject the null hypothesis at the .05 level?

The output in Table 8-10 demonstrates the issue of the precision of significance levels. Here, the Pearson's r significance value is .04688. Although this is close to the cutoff of .05, it is still less. Is it far enough below the .05 cutoff, however, that you would feel comfortable rejecting the null hypothesis? Further complicating matters, Spearman's Correlation significance value is .05320. This is technically above the cutoff of .05. Should you conclude that this relationship is nonsignificant, or is it close enough through rounding to reject the null hypothesis? The answer is not clear, and the decision is ultimately left to the researcher. This is an issue that you should consider, however, before conducting research on your own.

■ Conclusion

In this chapter, we have introduced methods of testing the existence of a relationship. The discussion proceeded from methods of determining if any relationship exists to procedures for determining the statistical significance for each level of data. This is generally considered the first step in the process of bivariate analysis. If the relationship between the two variables is significant, it is generally acceptable to continue with other analyses to determine the strength, direction, and nature of the data, which are discussed in Chapters 9 and 10. If the relationship between the two variables is not significant, the researcher must decide whether or not to proceed with other analyses.

◼ Key Terms

approximate significance
association
chance
Chi-square
column percentaged table
degrees of freedom
Delta
Epsilon
existence
expected value

Fisher's Exact Test
model of no association
observed value
ranks
row percentaged table
significance
statistically significant
total percentaged table
Yates correction

◼ Summary of Equations

Expected Value

$$f_e = \frac{\text{row total} \times \text{column total}}{N}$$

Delta

$$\Delta = (f_{o11} - f_{e11}), (f_{o12} - f_{e12}), (f_{o21} - f_{e21}), (f_{o22} - f_{e22}), \ldots$$

Chi-Square

$$\Sigma \frac{(f_o - f_e)^2}{f_e}$$

Degrees of Freedom (Chi-Square)

$$df = (\text{row categories} - 1)(\text{column categories} - 1)$$

Chi-Square (Yates Correction)

$$\chi^2_{\text{Yates}} = \Sigma \frac{(|f_o - f_e| - 0.5)^2}{f_e}$$

Fisher's Exact Test

$$\frac{(a+b)!\,(c+d)!\,(a+c)!\,(b+d)!}{N!\,a!\,b!\,c!\,d!}$$

Z Test for Spearman's Correlation

$$Z = \frac{r_s - 0}{1/\sqrt{N-1}}$$

◼ Exercises

1. For the data and tables that follow (which are the same as those in Exercises 1 and 3 in Chapter 7):
   a. Calculate Epsilon.
   b. Calculate Delta.

c. Calculate the Chi-square.
d. Determine the degrees of freedom.
e. Determine if the relationship is significant by comparing your Chi-square value to the appropriate value in Table B-2 in the appendix.

Victim by Sex

		Sex		
		1	2	Row
Victim		Male	Female	Total
2	Yes	20	37	57
1	No	27	48	75
Column total		47	85	132

Crimes per Month	Income	Crimes per Month	Income
15	$18,000	15	$24,000
5	21,000	15	18,000
15	23,000	15	21,000
10	22,000	15	31,000
10	25,000	20	52,000
5	14,000	7	21,000
20	23,000	7	17,000
20	46,000	20	48,000
5	15,000	20	45,000
5	17,000	10	19,000

f. In the spaces below, construct an ordinal bivariate table for crimes per month (dependent variable) and income (independent variable). For crimes per month, use the categories "10 and less" and "greater than 10." For income level, use the categories "$21,000 and more" and "less than $21,000."

_____ by _____

2. For the following tables, determine the appropriate measure of existence and interpret what it means for the existence of a relationship between the two variables.
3. For the tables in Exercise 2, examine the significance values for Pearson's r and Spearman Correlation and discuss what that would mean for the existence of a relationship if these two variables were fully ordered ordinal or interval or ratio level data.

VICT_Y_N Recoded VICTIM to yes and no * Race of respondent Crosstabulation

| | | | | RACE Race of Respondent | | | |
				1 Black	2 White	3 Other	Total
VICT_Y_N Recoded VICTIM to yes and no	2 No		Count	101	128	2	231
			Expected Count	99.2	129.0	2.8	231.0
			% within VICT_Y_N Recoded VICTIM to yes and no	43.7%	55.4%	0.9%	100.0%
			% within RACE Race of respondent	70.6%	68.8%	50.0%	69.4%
			% of Total	30.3%	38.4%	0.6%	69.4%
	1 Yes		Count	42	58	2	102
			Expected Count	43.8	57.0	1.2	102.0
			% within VICT_Y_N Recoded VICTIM to yes and no	41.2%	56.9%	2.0%	100.0%
			% within RACE Race of respondent	29.4%	31.2%	50.0%	30.6%
			% of Total	12.6%	17.4%	.6%	30.6%
	Total		Count	143	186	4	333
			Expected Count	143.0	186.0	4.0	333.0
			% within VICT_Y_N Recoded VICTIM to yes and no	42.9%	55.9%	1.2%	100.0%
			% within RACE Race of respondent	100.0%	100.0%	100.0%	100.0%
			% of Total	42.9%	55.9%	1.2%	100.0%

Case Processing Summary

| | Cases | | | | | |
| | Valid | | Missing | | Total | |
	N	Percent	N	Percent	N	Percent
VICT_Y_N Recoded VICTIM to yes and no * RACE of respondent	333	96.0%	14	4.0%	347	100.0%

Chi-Square Tests

	Value	df	Asymp. Sig. (2-sided)
Pearson Chi-Square	.840[a]	2	.657
Likelihood Ratio	.785	2	.675
Linear-by-Linear Association	.350	1	.554
N of Valid Cases:	333		

[a]2 cells (33.3%) have expected count less than 5. The minimum expected count is 1.23.

Directional Measures

			Value	Asymp. Std. Error[a]	Approx. T[b]	Approx. Sig.
Nominal by Nominal	Lambda	Symmetric	.000	.008	.000	1.000
		VICT_Y_N Recoded VICTIM to yes and no Dependent	.000	.020	.000	1.000
		RACE Race of respondent Dependent	.000	.000	c	c
	Goodman and Kruskal tau	VICT_Y_N Recoded VICTIM to yes and no Dependent	.003	.006		.658[d]
		RACE Race of respondent Dependent	.000	.002		.872[d]
Ordinal by Ordinal	Somers' d	Symmetric	−.029	.054	−.528	.598
		VICT_Y_N Recoded VICTIM to yes and no Dependent	−.027	.050	−.528	.598
		RACE Race of respondent Dependent	−.031	.059	−.528	.598

[a]Not assuming the null hypothesis.
[b]Using the asymptotic standard error assuming the null hypothesis.
[c]Cannot be computed because the asymptotic standard error equals zero.
[d]Based on chi-square approximation.

Symmetric Measures

		Value	Asymp. Std. Error[a]	Approx. T[b]	Approx. Sig.
Ordinal by Ordinal	Kendall's tau-b	−.029	.055	−.528	.598
	Kendall's tau-c	−.027	.051	−.528	.598
	Gamma	−.062	.118	−.528	.598
	Spearman Correlation	−.029	.055	−.528	.598[c]
Interval by Interval	Pearson's r	−.032	.055	−.591	.555[c]
N of Valid Cases		333			

[a]Not assuming the null hypothesis.
[b]Using the asymptotic standard error assuming the null hypothesis.
[c]Based on normal approximation.

NEIGH_SF: Is your neighborhood safety changing? by COP_OK: In your neighborhood, how well do you think the police perform their duties? Crosstabulation

NEIGH_SF: Is your neighborhood safety changing?			COP_OK: In your neighborhood, how well do you think the police perform their duties?				
			1	2	3 Below Average	4 Not At All	
			Very Well	Average	Average	At All	Total
3	Becoming less safe	Count	21	88	45	8	162
		Expected Count	29.2	87.2	39.6	5.9	162.0
		% within NEIGH_SF: Is your neighborhood safety changing?	13.0%	54.3%	27.8%	4.9%	100.0%
		% within COP_OK: In your neighborhood, how well do you think the police perform their duties?	35.6%	50.0%	56.3%	66.7%	49.5%
		% of Total	6.4%	26.9%	13.8%	2.4%	49.5%
2	Not changing	Count	21	58	33	2	114
		Expected Count	20.6	61.4	27.9	4.2	114.0
		% within NEIGH_SF: Is your neighborhood safety changing?	18.4%	50.9%	28.9%	1.8%	100.0%
		% within COP_OK: In your neighborhood, how well do you think the police perform their duties?	35.6%	33.0%	41.3%	16.7%	34.9%
		% of Total	6.4%	17.7%	10.1%	.6%	34.9%
1	Becoming safer	Count	17	30	2	2	51
		Expected Count	9.2	27.4	12.5	1.9	51.0
		% within NEIGH_SF: Is your neighborhood safety changing?	33.3%	58.8%	3.9%	3.9%	100.0%
		% within COP_OK: In your neighborhood, how well do you think the police perform their duties?	28.8%	17.0%	2.5%	16.7%	15.6%
		% of Total	5.2%	9.2%	.6%	.6%	15.65%

(continues)

NEIGH_SF: Is your neighborhood safety changing?		COP_OK: In your neighborhood, how well do you think the police perform their duties? (*continued*)				
		1 Very Well	2 Average	3 Below Average	4 Not At All	Total
Total	Count	59	176	80	12	327
	Expected Count	59.0	176.0	80.0	12.0	327.0
	% within NEIGH_SF: Is your neighborhood safety changing?	18.0%	53.8%	24.5%	3.7%	100.0%
	% within COP_OK: In your neighborhood, how well do you think the police perform their duties?	100.0%	100.0%	100.0%	100.0%	100.0%
	% of Total	18.0%	53.8%	24.5%	3.7%	100.0%

Case Processing Summary

	Cases					
	Valid		Missing		Total	
	N	Percent	N	Percent	N	Percent
NEIGH_SF Is your neighborhood safety changing? * COP_OK In your neighborhood, how well do you think the police perform their duties?	327	94.2%	20	5.8%	347	100.0%

Directional Measures

			Value	Asymp. Std. Error[a]	Approx. Approx. T[b]	Approx. Sig.
Nominal by Nominal	Lambda	Symmetric	.000	.021	.000	1.000
		NEIGH_SF Is your neighborhood safety changing? Dependent	.000	.039	.000	1.000
		COP_OK In your neighborhood, how well do you think the police perform their duties? Dependent	.000	.000	c	c
	Goodman and Kruskal tau	NEIGH_SF Is your neighborhood safety changing? Dependent	.026	.010		.010[d]
		COP_OK In your neighborhood, how well do you think the police perform their duties? Dependent	.022	.007		.001[d]
Ordinal by Ordinal	Somers' *d*	Symmetric	.172	.048	3.565	.000
		NEIGH_SF Is your neighborhood safety changing? Dependent	.171	.048	3.565	.000
		COP_OK In your neighborhood, how well do you think the police perform their duties? Dependent	.173	.048	3.565	.000

[a]Not assuming the null hypothesis.
[b]Using the asymptotic standard error assuming the null hypothesis.
[c]Cannot be computed because the asymptotic standard error equals zero.
[d]Based on chi-square approximation.

Symmetric Measures

		Value	Asymp. Std. Error[a]	Approx. T[b]	Approx. Sig.
Ordinal by Ordinal	Kendall's tau-b	.172	.048	3.565	.000
	Kendall's tau-c	.158	.044	3.565	.000
	Gamma	.279	.076	3.565	.000
	Spearman Correlation	.192	.053	3.518	.000[c]
Interval by Interval	Pearson's r	.201	.053	3.692	.000[c]
N of Valid Cases:		327			

[a]Not assuming the null hypothesis.
[b]Using the asymptotic standard error assuming the null hypothesis.
[c]Based on normal approximation.

Correlations

		DELINQ: Delinquent Incidents for the Census Tract	BOARDED: Boarded-up Housing Units	OWNOCCUP: Owner-Occupied Housing Units	JUVENILE: Proportion of Juveniles	POPCHANG: Population Change, 1980–1990
DELINQ Delinquent Incidents for the Census Tract	Pearson Correlation	1	.482**	−.004	.339*	−.011
	Sig. (2-tailed)		.001	.981	.023	.941
	N	45	45	45	45	45
BOARDED Boarded-up housing units	Pearson Correlation	.482**	1	−.234	.029	−.222
	Sig. (2-tailed)	.001	—	.122	.848	.143
	N	45	45	45	45	45
OWNOCCUP Owner-occupied housing units	Pearson Correlation	−.004	−.234	1	.752**	.323*
	Sig. (2-tailed)	.981	.122	—	.000	.030
	N	45	45	45	45	45
JUVENILE Proportion of juveniles	Pearson Correlation	.339*	.029	.752**	1	.436**
	Sig. (2-tailed)	.023	.848	.000	—	.003
	N	45	45	45	45	45
POPCHANG Population change, 1980–1990	Pearson Correlation	−.011	−.222	.323*	.436**	1
	Sig. (2-tailed)	.941	.143	.030	.003	—
	N	45	45	45	45	45

**Correlation is significant at the 0.01 level (2-tailed).
*Correlation is significant at the 0.05 level (2-tailed).

References

Conover, W. J. (1967). Some reasons for not using the Yates continuity correction on a 2 × 2 contingency table. *Journal of the American Statistical Association*, 69(346) (June):374–376.

Fisher, R. A. (1922). On the interpretation of χ^2 from contingency tables, and the calculation of *P*. *Journal of the Royal Statistical Society*, 85:87–94.

Grizzle, J. E. (1967). Continuity correction in the χ^2 Test for 2 × 2 tables. *The American Statistician*, 21(4) (October):28–32.

Pearson, E. S. (1947). The choice of a statistical test illustrated on the interpretation of data classed in a 2 × 2 table. *Biometrika*, 37 (January):139–167.

Pearson, K. (1900). On the criterion that a given system of deviations from the probable in the case of a correlated system of variables is such that it can be reasonably supposed to have arisen from random sampling. *The London, Edinburgh, and Dublin Philosophical Magazine and Journal of Science*, Fifth Series, 50, 157–175.

Plackett, R. L. (1964). The continuity correction in 2 × 2 tables. *Biometrika*, 51 (December):327–337.

Spearman, C. (1904). The proof and measurement of association between two things. *American Journal of Psychology*, 15:72–101.

Starmer, C. F., J. E. Grizzle, and P. K. Sen (1967). Comment. *Journal of the American Statistical Association*, 69(346) (June):376–378.

Yates, F. (1934). Contingency tables involving small numbers and the χ^2 test. *Journal of the Royal Statistical Society*, Series B, Suppl. Vol. 1(2):217–235.

For Further Reading

Goodman, L. A., and W. H. Kruskal (1963). Measures of association for cross classifications III: approximate sampling theory. *Journal of the American Statistical Association*, 58(302):310–364.

Notes

1. This process is simplified in a 2 × 2 table. After calculating the expected value for the first cell in the table, other expected values can be determined simply by subtracting that value from the row and column totals. See also the discussion on degrees of freedom.

2. Note that although the existence of a relationship is usually examined with hypothesis testing, not all hypothesis testing examines existence, as some people improperly believe. Hypothesis testing can test any hypothesis. For example, you can test the hypothesis that the value of Lambda or Somers' *d* is nonzero (thereby testing the strength of a relationship as discussed in Chapter 9).

3. There are several normalized Chi-square statistics (values ranging from 0 to 1) which are often used. These include Phi and Cramer's *V*. Both of these attempt to minimize the influence of sample size that adversely affects Chi-square. Phi is the square root of the Chi-square value divided by *N* and is equal to Pearson's Correlation for a 2 × 2 table. For tables larger than 2 × 2, Cramer's *V* is preferred because it can be used to obtain a maximum value of 1 for these tables, whereas

Phi cannot. Cramer's V is calculated as the square root of the Chi-square value divided by $N(k - 1)$, where k is the smaller number of the rows or columns.

4. It should be noted that whereas many see these tests as appropriate for small sample sizes because they make it more difficult to reject the null hypothesis, they both have been criticized as being too conservative, thereby causing researchers to fail to reject the null hypothesis when they should not have, or for being no better at estimating the relationship than using Chi-square alone (Conover, 1967; Grizzle, 1967; Pearson, 1947; Placket, 1964; Starmer, Grizzle, and Sen, 1967).

5. Some conservative statisticians hold that Chi-square should not be used at all if any of the cell frequencies are less than 5. This is a preference rather than a rule, and most people believe that Fisher's Exact Test can correct for such cases.

6. It was Pearson (1900) who first argued that the table of critical values for Chi-square could be limited to 30. In deriving the formulas and setting the method of Chi-square, Pearson stated: "Thus, if we take a very great number of groups our test becomes illusionary. We must confine our attention in calculating P to a finite number of groups, and this is undoubtedly what happens in actual statistics. N (here actually the degrees of freedom, not the number of values) will rarely exceed 30, often not be greater than 12."

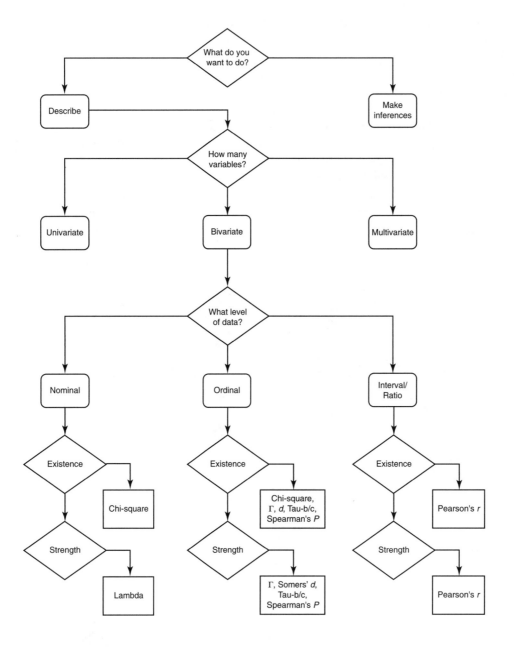

Measures of Strength of a Relationship

CHAPTER

9

In addition to determining the existence of a relationship, it is important to know the **strength of the association**. This dimension, along with the direction and nature of the association, is often termed a **measure of association**. It warrants repeating at this point that the measure of the strength of an association between two variables is based on the assumption that we are dealing with the entire population of the object of study and not a sample taken from the population. These remain, then, *descriptive* statistical procedures rather than inferential ones.

The measure of the strength of an association represents another summary measure of the relationship. A relationship can be described in terms of the univariate measures, as shown in Chapters 4, 5, and 6, or by examining their cross-classification in percentaged tables or crosstabs, as shown in Chapter 7. As Goodman and Kruskal (1954) point out, however, there are two potential problems with publishing results based on such an examination. The first is that there is often so much information that the reader must rely on the author's interpretation of the relationship. If the author's conclusions are erroneous or questionable, it is often difficult for the reader to detect. Similarly, it is possible that the author could choose only the data or comparisons that fit his or her theory. Somers (1962) supports this argument in his statement that percentaged tables can be examined appropriately only if they are 2×2. Anything larger, Somers argues, is difficult to examine without summary measures because there are so many different sets of percentage relationships. Measures of association can alleviate some of those problems. These measures can reduce the mass of tables or crosstabs to a few summary values that can be compared and set against standards. This can greatly increase the value of the research and aid interpretation of the data.

■ What Is Association?

As discussed in earlier chapters, two variables may be compared in a cross-tabulation table (crosstab). Depending on how the categories of the variables are arranged, the two variables may be said to be *statistically significantly related* or not. The next step is to discuss how closely the variables are related—the strength of the association.

Table 9-1 is a crosstab of the data given in **Table 9-2**. In this crosstab, frequencies are shown with the joint probability of occurrence (total percent) in parentheses. If asked to predict the category of D (the dependent variable) for some case drawn at random, knowing nothing about I (the independent variable), which category would you pick? Remembering these questions from Chapter 4, you might pick $D1$ because its marginal total is 177 (the modal category), whereas $D2$ is only 161. In this case, you have a 52.3% chance of being right with $D1$ over a 47.7% chance of being right with $D2$. Alternatively, the probability of getting an error in predicting $D1$ is 47.7, which is calculated by subtracting the probability of being right from 1.

What if you are asked to predict the category of D and are told the category of I from which it will be drawn? If it will be drawn from $I1$, you would probably choose $D2$ because it has the greatest probability of being correct for $I1$. The probability of getting a correct D answer in the $I1$ category is 57.7%, as taken from the following formula:

$$\frac{60}{104} = 0.577$$

Cell $D2$–$I1$ contains 60 of a possible 104 responses for the category of $I1$. This means that cell $D2$–$I1$ contains 57.7% of the values in $I1$ and cell $D1$ contains 42.3% of those values. The probability of getting a correct answer, then, is 57.7%, and the probability of making an error is 42.3%.

Notice that when we knew nothing about I and were attempting to correctly predict the category of D, there was a probability of being wrong 47.7% of the time. Knowing the category of I, however, reduced the number of errors to 42.3%. We can say, then, that there was a reduction of errors associated with knowing information

TABLE 9-1 Table of Joint Frequencies and Probability of Occurrence

	$I1$	$I2$	Total
$D2$	60	101	161
	(0.178)	(0.299)	(0.477)
$D1$	44	133	177
	(0.13)	(0.399)	(0.523)
Total	104	234	338
	(0.308)	(0.692)	(1.00)

TABLE 9-2 SPSS Output for Lambda

**Is crime a serious enough problem that you have considered moving in the past 12 months? *
Recoded VICTIM to yes and no Cross-tabulation**

			Recoded VICTIM to yes and no		
			Yes	No	Total
Is crime a serious enough problem that you have considered moving in the past 12 months?	No	Count	60	101	161
		Expected Count	49.5	111.5	161.0
		% within Is crime a serious enough problem that you have considered moving in the past 12 months?	37.3%	62.7%	100.0%
		% within Recoded VICTIM to yes and no	57.7%	43.2%	47.6%
		% of Total	17.8%	29.9%	47.6%
	Yes	Count	44	133	177
		Expected Count	54.5	122.5	177.0
		% within Is crime a serious enough problem that you have considered moving in the past 12 months?	24.9%	75.1%	100.0%
		% within Recoded VICTIM to yes and no	42.3%	56.8%	52.4%
		% of Total	13.0%	39.3%	52.4%
Total		Count	104	234	338
		Expected Count	104.0	234.0	338.0
		% within Is crime a serious enough problem that you have considered moving in the past 12 months?	30.8%	69.2%	100.0%
		% within Recoded VICTIM to yes and no	100.0%	100.0%	100.0%
		% of Total	30.8%	69.2%	100.0%

(*continues*)

TABLE 9-2   SPSS Output for Lambda *(continued)*

Chi-Square Tests

	Value	df	Asymp. Sig. (2-sided)	Exact Sig. (2-sided)	Exact Sig. (1-sided)
Pearson Chi-Square	6.094[b]	1	.014		
Continuity Correction[a]	5.525	1	.019		
Likelihood Ratio	6.103	1	.013		
Fisher's Exact Test				.018	.009
Linear-by-Linear Association	6.076	1	.014		
N of Valid Cases	338				

[a]Computed only for a 2 × 2 table.
[b]0 cells (.0%) have expected count less than 5. The minimum expected count is 49.54.

Directional Measures

			Value	Asymp. Std. Error[a]	Approx. T[b]	Approx. Sig.
Nominal by Nominal	Lambda	Symmetric	.060	.037	1.575	.115
		Is crime a serious enough problem that you have considered moving in the past 12 months? Dependent	.099	.060	1.575	.115
		Recoded VICTIM to yes and no Dependent	.000	.000	.[c]	.[c]
	Goodman and Kruskal tau	Is crime a serious enough problem that you have considered moving in the past 12 months? Dependent	.018	.014		.014[d]
		Recoded VICTIM to yes and no Dependent	.018	.014		.014[d]

[a]Not assuming the null hypothesis.
[b]Using the asymptotic standard error assuming the null hypothesis.
[c]Cannot be computed because the asymptotic standard error equals zero.
[d]Based on a chi-square approximation.

about *I*. This is the foundation of most of the measures of the strength of an association: reducing errors with knowledge of an independent variable over the number of errors that would occur in predicting the dependent variable without that knowledge. You will come to know this as a **proportional reduction of errors**.

■ Nominal Level Data

At the nominal level, the goal of a measure of association is to predict the mode or category of the dependent variable with knowledge gained from the independent variable. If asked to determine whether or not a person chosen at random was criminal and no other information was available, you would simply have to guess. If given information about the univariate characteristics of the person, however, you could certainly improve the guess. For example, it is well supported in the research that more men than women commit crimes. If you knew, then, that the person was a man or a woman, it would improve your prediction over simply guessing. A measure of association allows improvement even beyond that. For example, if you knew the criminal history of the person, you would probably want to predict that any person chosen with a prior record of criminal behavior was a criminal and any person without a record was not since research has shown that more people are recidivists than people who commit new crimes. What you have done here is to predict the category of criminal or noncriminal (dependent variable) with knowledge of the criminal history of the person (independent variable). That is the goal of most measures of association, especially at the nominal level.

The most popular measure of association at the nominal level is **Lambda** (symbolized by λ). Lambda was developed by Guttman (1941) and expanded on by Goodman and Kruskal in 1954 (in fact, Lambda is often referred to as Goodman and Kruskal's Lambda). Lambda can be defined as an asymmetric measure of association which measures the proportional reduction in errors made in predicting modal values or categories of a dependent variable *Y* when information about an independent variable *X* is used. This definition contains several parts that define Lambda.

First, it is **asymmetric**. This means that it predicts the dependent variable based on information from the independent variable. In other words, instead of simply determining that there is some relationship between the variables, an asymmetric measure indicates how knowledge of a particular variable (generally, the independent variable) would aid in reducing errors in predicting categories of the dependent variable. This is different from a **symmetric** measure, which indicates the strength of an association but does not determine which variable is better able to predict the other. For Lambda, it is possible to be able to predict one variable (*Y*) with information about the other (*X*) but not be able to predict anything about the second variable (*X*) with information about the first (*Y*).

Lambda is also a *proportional reduction of error* (PRE) measure. PRE, which was developed by Goodman and Kruskal (1954), is a ratio of the amount of error predicted when there is no information about the dependent variable and when there is knowledge of the dependent variable gained from the independent variable (as was shown in the section "What Is Association?"). In the example above, it is possible to calculate from several tries how many times a person would be right and wrong in predicting the criminality of a person with no additional knowledge. We could then calculate how many times a person would be right and wrong in predicting the criminality of a person with the additional knowledge of the criminal history of the person. These two values could then be placed in the formula shown below (the same formula as above) to determine the probability of getting a correct answer. For example, with no knowledge at all about the person, all you could do is guess about the criminality of the person. At best, you might be correct 50% of the time. Knowledge about the criminal history of the person, however, significantly increases the probability of being correct. For example, if you knew that 75% of the people in the study who had criminal histories were currently under arrest for a crime, you could use this formula to produce the following result:

$$\frac{50}{75} = 0.666$$

How do you do that?
Measures of Strength of an Association with Crosstabs

Note: The procedures outlined below are for use with a crosstab. If you are analyzing interval or ratio level data without a crosstab, see the box in the section of this chapter on Pearson's *r*.

1. Open a data set.

   a. Start SPSS.

   b. Select *File*, then *Open*, then *Data*.

   c. Select the file you want to open, then select *Open*.

2. Once the data is visible, select *Analyze*, then *Descriptive Statistics*, then *Crosstabs*.

3. Select the variable you wish to include as your dependent variable and press the ▶ next to the *Row(s)* window.

4. Select the variable you wish to include as your independent variable and press the ▶ next to the *Column(s)* window.

5. Select the *Cells* button and check any of the boxes for information you may want in your crosstab. For a crosstab with full information, select *observed*, *expected*, *row*, *column*, and *total*; then press *Continue*.

6. Select the *Format* button, and check the box marked *Descending*.

7. Select the *Statistics* button at the bottom of the window.

8. Check the box next to the appropriate measure of strength, as outlined in this chapter ("Lambda" for nominal level data, "Somers' *d*" for partially ordered ordinal level data, or "Correlations" for fully ordered ordinal level data).

9. Select *Continue*, then select *OK*.

10. An output window should appear containing a distribution similar in format to Table 9-2 through 9-4 and 9-6.

This means that you would improve your probability of a correct prediction by almost 67% with the additional knowledge of the criminal history of the person. This ratio shows what proportion of error has been reduced by knowing the criminal history of the person. Because Lambda is asymmetric and PRE, it can be said to measure how much error is reduced when a value of X is used to predict a value of Y as opposed to not using X in the prediction.

Calculation

When calculating a Lambda, it is labeled λ_{yx}, where y is the dependent variable and x is the independent variable. This denotes which variable is being predicted based on information from the other variable. A symmetric model would have λ without the subscripts. The formula for calculating a Lambda from a crosstab is

$$\lambda_{yx} = \frac{\Sigma m_y - M_y}{N - M_y}$$

where N is the total sample size, M_y the overall mode for the dependent variable, and m_y the mode of the dependent variable for each category of the independent variable. This formula can also be couched in terms of the general formula for a PRE measure. The resulting formula is

$$\lambda_{yx} = \frac{(N - M_y) - (N - \Sigma m_y)}{N - M_y}$$

where the first value in the numerator shows the number of errors that would be made in predicting the dependent variable without any other information, and the second value in the numerator is the number of errors that would be made with the additional information gained from the independent variable. This formula can be simplified as shown below:

$$\lambda_{yx} = \frac{E_1 - E_2}{E_1}$$

where E_1 is the number of errors that would be made in predicting the dependent variable without any other information and E_2 is the number of errors that would be made with the additional information gained from the independent variable. This formula results in a computational formula for Lambda where $E_1 = N - $ largest row total and $E_2 = \Sigma$ (each column total $-$ largest cell frequency in that column), as in

$$\frac{(N - R_t) - [\Sigma(C_t - f_i)]}{N - R_t}$$

In the example in Table 9-2, the calculation of Lambda would be as follows:

$$\lambda = \frac{(338 - 177) - [\Sigma(104 - 60) + (234 - 133)]}{338 - 177}$$

$$\lambda = \frac{(338 - 177) - [\Sigma(44) + (101)]}{338 - 177}$$

$$\lambda = \frac{161 - 145}{161}$$

$$\lambda = 0.09938$$

This means that we would be able to reduce the number of errors in predicting whether a person considered moving in the past six months by about 10% with knowledge of whether he or she had been a victim. Notice also that this is the same value as with *Move* the dependent variable in the Lambda portion of the printout in Table 9-2.

Interpretation

Lambda has a range of values from 0 to 1. A 0 means no strength—the independent variable has no ability to reduce the number of errors in predicting the categories of the dependent variable. A Lambda of 1 means a perfect association—that the additional knowledge of the independent variable allows correct prediction of the dependent variable every time. In the example above, the Lambda was 0.099. This is certainly closer to 0 than to 1, indicating that this is not a very strong association. Lambda will approach 1 if the modal values for each category of the independent variable fall into only one category of the dependent variable: that is, if all the values of the independent variable are in the same column of a crosstab. Lambda can have a value of 0 even when there is an association if the dependent variable is multimodal and the modal values of the dependent variable are in each of the categories of the independent variable (in other words, the largest frequencies in each column lie in the same row of the table). Lambda will also equal 0 if all the values of the independent variable fall in the modal category of the dependent variable. Because we are trying to better our prediction, if all the values of the independent variable are in the same (modal) category of the dependent variable, nothing has been gained—we would have chosen the modal category if we had no information about the independent variable. Finally, Lambda will equal 0 if the sample is highly skewed, even if the variables are related. In instances where Lambda is 0 or there is some question of its ability to properly analyze the data, another of the nominal level measures of association should be used. Both Phi and Cramer's *V* are based on Chi-square and provide a measure of the strength of an association between two variables. Phi is probably better for 2×2 tables because its value is equal to Pearson's Correlation for these tables. Cramer's *V* should be used for larger tables, however, because of the limitations of Phi. Also notice that no negative values are associated with Lambda. This is because there can be no direction associated with a nominal level relationship.

In Table 9-2 there are three values of Lambda for the bivariate relationship of whether a person had been victimized in the preceding year and his or her decision to move. The symmetric value is a combination of the two asymmetric values that follow and is little more than the average of the two. The next two values are the asymmetric measures of the two possible Lambdas, λ_{yx} and λ_{xy}. To facilitate the greatest degree of flexibility and interpretation, SPSS provides asymmetric Lambda values for both variables as dependent.

In this case, the person's consideration of moving is the dependent variable; its categories are in the rows of the table. The Lambda value here is 0.09938, the same as in the preceding calculations. This means that there is almost a 10% reduction in errors in predicting whether a person considered moving with knowledge of whether or not the person had been a victim of crime in the preceding year. Notice also that when attempting to predict victimization with knowledge of consideration of moving, the Lambda is 0. This means that in trying to predict victimization, it is no better knowing whether or not a person had considered moving than not having that information. This makes intuitive sense; there is no reason why information on whether a person had considered moving should have any bearing on whether or not he or she was victimized. Being victimized might influence a person's decision to move, but moving does not influence whether or not a person is victimized. Also note that the symmetric measure is less than the asymmetric with MOVE dependent but more than with VICT_Y_N dependent. This is because this calculation takes both asymmetric measures into account.

A final issue concerning Lambda. It can be used with a table of any size, either square or rectangular. This is not possible with all bivariate measures of association, some of which are appropriate only for square or rectangular tables. For rectangular tables, this creates an additional issue for Lambda. In the example above, Lambda was calculated on a square (2×2) table. This allowed the use of two modes (m_y) to predict two categories of the dependent variable. This is not the case for rectangular tables, however. For example, if there were five categories of the dependent variable and only four categories of the independent variable, Lambda would not be as precise because we can use only four categories to make a prediction rather than the full five. Alternatively, Lambda can be somewhat more precise when the number of categories of the independent variable exceeds that of the dependent variable.

■ Ordinal Level Data

For full use of the ordinal level measures of association, both variables should be ordinal level, dichotomized nominal, or categorized interval level. Ordinal level measures of association are somewhat more powerful than nominal measures because they can indicate both the strength of the association and the direction (see Chapter 10).

With ordinal level data, the goal is to predict the *rank order of pairs of scores* instead of the modal value. The "pairs of scores" are where categories of a dependent variable such as high, medium, or low potential to commit violent crime are compared to categories of an independent variable such as upper-, middle-, or lower-class socioeconomic status. Specifically, the goal is to determine whether or not the rank ordering of cases of one variable is useful in predicting the rank order of cases of the other variable. If knowledge of the rank of cases of one variable predicts the rank of cases of the other variable perfectly, the association is said to be perfect and the measure of association will equal 1. If knowledge of the rank of cases of one variable is of no use in predicting the rank of cases on the other variable, the measure of association will equal 0, showing that nothing is gained by adding information from the independent variable.

There are five possible situations into which the ranks of variables can fall (see **Exhibit 9-1**). *Concordant pairs*, N_s, are pairs that are ranked in the same order on both variables. Using the example above, a person might be in the *low* category on socioeconomic status and in the *low* category in potential to commit violent crime. *Discordant pairs*, N_d, are pairs that are ranked one way on one variable and the opposite way on the other variable. Using the example above, a person might be in the *low* category on

EXHIBIT 9-1   Calculating Ranks and Ties

The concept of ranks, and especially ties, is often difficult to grasp but is essential for understanding ordinal level analyses. Using the survey included in this book, look at how six people answered the questions "Is your neighborhood safety changing?" (HOOD_SAFE) and "How much has your social activity in your neighborhood been affected by fear for your personal safety?" (ACT_FEAR). Their responses determine the type of relationship between the variables.

Respondent	Independent Variable: HOOD_SAFE	Dependent Variable: ACT_FEAR
1	Less safe (3)	Not changing (2)
2	Not changing (2)	Not changing (2)
3	Less safe (3)	Great effect (3)
4	Becoming safer (1)	Not changing (2)
5	Not changing (2)	No effect (1)
6	Becoming safer (1)	Not changing (2)

Here are the possible patterns of these two variables for some of the respondents:

Pair	Second Respondent is _____ on Independent Variable	Second Respondent is _____ on Dependent Variable	Type of Pair
1,5	Lower (3,2)	Lower (2,1)	N_s, concordant
4,5	Higher (1,2)	Lower (2,1)	N_d, discordant
1,2	Lower (3,2)	Tied (2,2)	T_y, tied on dependent variable
1,3	Tied (3,3)	Higher (2,3)	T_x, tied on independent variable
4,6	Tied (1,1)	Tied (2,2)	T_{xy}, tied on both variables

socioeconomic status and in the *high* category in potential to commit violent crime. Pairs can be tied (have the same rank) on the dependent variable but not on the independent variable (T_y). Pairs can also be tied on the independent variable but not on the dependent variable (T_x). Finally, pairs can be tied on both variables (T_{xy}). See Exhibit 9-1 for examples of the ways pairs can be tied.

The difference between the N_s and N_d pairs determines whether the relationship is positive or negative (see Chapter 10). If there are more N_s than N_d pairs, the relationship is positive. If there are more N_d than N_s pairs, the relationship is negative. Values for ordinal level variables range from −1 to 0 to +1. Positive values indicate a positive relationship, negative values indicate an inverse, or negative, relationship. The more N_s

or N_d pairs there are, the closer to 1 the value will be. If there is the same number of N_s and N_d pairs, the measure of association is 0. Also, depending on which measure of the strength of the association is used, ties may heavily influence the outcome.

Several options are available for analyzing ordinal level data. It is most common to use Tau-b and -c, Gamma, or Somers' d with partially ordered, ordinal level variables, and Spearman's Rho with fully ordered ordinal level variables. All of these are (or can be) PRE measures, which means that they measure the amount of error reduced when predicting the rank order of the pairs (positive or negative) as opposed to guessing the rank order or choosing it randomly. The values for each of these measures of association are shown in **Table 9-3**.

Each of the ordinal level measures of association has the same numerator (top value), which measures the difference between the N_s and N_d pairs. The difference between the measures lies in the denominators. Some of the measures exclude more of the possible ties, whereas others do not exclude any. Each of these measures is discussed in the remainder of this chapter.

TABLE 9-3   SPSS Output for Ordinal Level Measures of Association

WALK_NIT How often do you walk in your neighborhood at night? * CRIM_PRB How big a problem is crime? Cross-tabulation

| | | | CRIM_PRB How big a problem is crime? | | | |
			1 No Problem	2 Small Problem	3 Big Problem	Total
WALK_NIT How often do you walk in your neighborhood at night?	3 Often	Count	2	2	0	4
		Expected Count	2.7	1.2	.2	4.0
		% within WALK_NIT How often do you walk in your neighborhood at night?	50.0%	50.0%	.0%	100.0%
		% within CRIM_PRB How big a problem is crime?	1.2%	2.8%	.0%	1.6%
		% of Total	.8%	.8%	.0%	1.6%
	2 Occasionally	Count	20	13	3	36
		Expected Count	23.9	10.4	1.7	36.0
		% within WALK_NIT How often do you walk in your neighborhood at night?	55.6%	36.1%	8.3%	100.0%
		% within CRIM_PRB How big a problem is crime?	12.1%	18.1%	25.0%	14.5%
		% of Total	8.0%	5.2%	1.2%	14.5%

(continues)

TABLE 9-3   SPSS Output for Ordinal Level Measures of Association (*continued*)

WALK_NIT How often do you walk in your neighborhood at night? * CRIM_PRB How big a problem is crime? Cross-tabulation

| | | | CRIM_PRB How big a problem is crime? | | | |
			1 No Problem	2 Small Problem	3 Big Problem	Total
1 Never		Count	143	57	9	209
		Expected Count	138.5	60.4	10.1	209.0
		% within WALK_NIT How often do you walk in your neighborhood at night?	68.4%	27.3%	4.3%	100.0%
		% within CRIM_PRB How big a problem is crime?	86.7%	79.2%	75.0%	83.9%
		% of Total	57.4%	22.9%	3.6%	83.9%
Total		Count	165	72	12	249
		Expected Count	165.0	72.0	12.0	249.0
		% within WALK_NIT How often do you walk in your neighborhood at night?	66.3%	28.9%	4.8%	100.0%
		% within CRIM_PRB How big a problem is crime?	100.0%	100.0%	100.0%	100.0%
		% of Total	66.3%	28.9%	4.8%	100.0%

Chi-Square Tests

	Value	df	Asymp. Sig. (2-sided)
Pearson Chi-Square	3.613[a]	4	.461
Likelihood Ratio	3.539	4	.472
Linear-by-Linear Association	2.462	1	.117
N of Valid Cases	249		

[a]4 cells (44.4%) have expected count less than 5. The minimum expected count is .19.

(*continues*)

TABLE 9-3   SPSS Output for Ordinal Level Measures of Association (*continued*)

Directional Measures

			Value	Asymp. Std. Error[a]	Approx. T[b]	Approx. Sig.
Nominal by Nominal	Lambda	Symmetric	.265	.000	.[c]	.[c]
		WALK_NIT How often do you walk in your neighborhood at night? Dependent	.472	.000	.[c]	.[c]
		CRIM_PRB How big a problem is crime? Dependent	.157	.000	.[c]	.[c]
	Goodman and Kruskal tau	WALK_NIT How often do you walk in your neighborhood at night? Dependent	.010	.013		.268[d]
		CRIM_PRB How big a problem is crime? Dependent	.009	.012		.331[d]
Ordinal by Ordinal	Somers' *d*	Symmetric	.100	.062	1.578	.115
		WALK_NIT How often do you walk in your neighborhood at night? Dependent	.079	.050	1.578	.115
		CRIM_PRB How big a problem is crime? Dependent	.136	.085	1.578	.115

[a]Not assuming the null hypothesis.
[b]Using the asymptotic standard error assuming the null hypothesis.
[c]Cannot be computed because the asymptotic standard error equals zero.
[d]Based on a chi-square approximation.

(*continues*)

TABLE 9-3   SPSS Output for Ordinal Level Measures of Association (*continued*)

Symmetric Measures

		Value	Asymp. Std. Error[a]	Approx. T[b]	Approx. Sig.
Ordinal by Ordinal	Kendall's tau-b	.104	.065	1.578	.115
	Kendall's tau-c	.056	.036	1.578	.115
	Gamma	.263	.149	1.578	.115
	Spearman Correlation	.107	.067	1.686	.093[c]
Interval by Interval	Pearson's *r*	.100	.065	1.574	.117[c]
N of Valid Cases		249			

[a]Not assuming the null hypothesis.
[b]Using the asymptotic standard error assuming the null hypothesis.
[c]Based on normal approximation.

A Bit of History

Tau (known specifically as Kendall's Tau after its developer, Maurice Kendall) was introduced by Kendall in 1938. Although Kendall is credited with the development of Tau, essentially the same calculations existed prior to that. The first person believed to have developed a measure similar to Tau was Fechner (1897), who used the measure for time series analyses. In 1905, G. F. Lipps developed a measure equivalent to Kendall's Tau, along with a measure equivalent to Yule's *Q* (1900). Lipps' work was extended by Gustav Deuchler (1909, 1914), who used the measure to test the null hypothesis of independence of the variables. These early works were written in German, never translated to English, and many were unpublished. It is only through the work of Kruskal (1958) and others who obtained copies of these works that we know of the existence of Tau prior to Kendall. In his 1948 book, Kendall brought Tau to the forefront as a measure of association for ordinal level data.

Tau

Although other statistical procedures were in existence prior to the development of Tau, it is generally considered the first measure of association developed specifically for ordinal level data. There are actually three measures of **Tau** used with ordinal level variables: Tau-a, Tau-b, and Tau-c. Each was developed for specific situations, and each could be said to be an improvement on the other(s).

Calculation

Each Tau can be calculated using a formula with the same numerator. The difference between the various Taus is in the denominator of the equations. The simplest ordinal level measure of association is *Tau-a*. This is the number of concordant pairs minus the number of discordant pairs divided by the total number of unique pairs, as shown in the formula

$$\tau_a = \frac{N_s - N_d}{T}$$

In this formula, T is calculated as

$$\frac{N(N-1)}{2}$$

To calculate Tau, begin in the lower left cell of a table and multiply that value by the sum of all cells above and to the right of it. This will calculate the concordant scores (N_s). This should also be done for any columns that have values above and to the right. For example, in Table 9-3, N_s would be calculated as follows:

$$
\begin{aligned}
N_s &= [(143)(13 + 3 + 2 + 0)] + [(57)(3 + 0)] + [(20)(2 + 0)] + [(13)(0)] \\
&= [(143)(18)] + [(57)(3)] + [(20)(2)] + [(13)(0)] \\
&= 2574 + 171 + 40 + 0 \\
&= 2785
\end{aligned}
$$

Next, begin with the value in the lower right cell of the table and multiply that value by the sum of all cells above and to the left of it. This will calculate the discordant pairs (N_d). Using the data from Table 9-3 would produce the following calculation:

$$
\begin{aligned}
N_d &= [(9)(13 + 20 + 2 + 2)] + [(57)(20 + 2)] + [(3)(2 + 2)] + [(13)(2)] \\
&= [(9)(37)] + [(57)(22)] + [(3)(4)] + [(13)(2)] \\
&= 333 + 1254 + 12 + 26 \\
&= 1625
\end{aligned}
$$

Placing these values in the formula for Tau a will result in the following calculation:

$$\tau_a = \frac{N_s - N_d}{T}$$

Substituting the formula for T would produce the computational formula

$$T = \frac{N(N-1)}{2}$$

$$= \frac{N_s - N_d}{N(N-1)/2}$$

Using the data from Table 9-3 would produce the following calculations:

$$\tau_a = \frac{2785 - 1625}{(249)(249 - 1)/2}$$

$$= \frac{2785 - 1625}{30,876}$$

$$= \frac{1160}{30,876}$$

$$= 0.0376$$

Notice this value is not found in the SPSS printout for Table 9-3. This is because, even though several different types of Taus are included in the printout, Tau-a is not one of them.

Because Tau-a is a PRE measure, it can be said that the additional information obtained from knowing a person's opinion of the crime problem can reduce the number of errors predicted in how often they walk at night by almost 4% (see the interpretation below).

Tau-b is essentially the square root of the two possible Somers' d calculations (see below). It considers ties in each variable (T_y or T_x) but not on both (T_{yx}). Tau-b could therefore be expressed in one of two ways:

$$\tau_b = \sqrt{d_{xy}d_{yx}} \quad \text{or} \quad \tau_b = \frac{N_s - N_d}{\sqrt{(N_s + N_d + T_y)(N_s + N_d + T_x)}}$$

In addition to the calculations required to arrive at the sum of $N_s - N_d$ (as completed above for Tau-a), this formula requires the calculation of ties on the dependent variable (T_y), ties on the independent variable (T_x), and adding the concordant and discordant pairs. T_y is calculated by multiplying the frequencies in each row by the sum of the frequencies to the right of it and then adding those together. The calculation for the data in Table 9-3 is

$$\begin{aligned}
T_y &= [(2)(2+0)] + [(2)(0)] + [(20)(13+3)] + [(13)(3)] \\
&\quad + [(143)(57+9)] + [(57)(9)] \\
&= [(2)(2)] + [(2)(0)] + [(20)(16)] + [(13)(3)] \\
&\quad + [(143)(66)] + [(57)(9)] \\
&= 4 + 0 + 320 + 39 + 9438 + 513 \\
&= 10,314
\end{aligned}$$

The procedure for calculating T_x is the same as that for calculating ties on the dependent variable, except columns are used rather than rows and you work from the top down. For Table 9-3, the calculations would be as follows:

$$
\begin{aligned}
T_x &= [(2)(20 + 143)] + [(20)(143)] + [(2)(13 + 57)] + [(13)(57)] \\
&\quad + [(0)(3 + 9)] + [(3)(9)] \\
&= [(2)(163)] + [(20)(143)] + [(2)(70)] + [(13)(57)] \\
&\quad + [(0)(12)] + [(3)(9)] \\
&= 326 + 2860 + 140 + 741 + 0 + 27 \\
&= 4094
\end{aligned}
$$

Including this value with those obtained for Tau-a would result in the following calculation:

$$
\begin{aligned}
\tau_b &= \frac{N_s - N_d}{\sqrt{(N_s + N_d + T_y)(N_s + N_d + T_x)}} \\
&= \frac{2785 - 1625}{\sqrt{(2785 + 1625 + 10{,}314)(2785 + 1625 + 4094)}} \\
&= \frac{1160}{\sqrt{(14{,}724)(8504)}} \\
&= \frac{1160}{\sqrt{125{,}212{,}896}} \\
&= \frac{1160}{11{,}189.8556} \\
&= 0.103665
\end{aligned}
$$

Using Tau-b rather than Tau-a for the data shown in Table 9-3 will produce a slightly higher than 10% reduction of errors in predicting the frequency of walking at night. This is a substantial increase, which in this case shows how Tau-b is a better method of analyzing this distribution. This is due to the number of ties affecting Tau-a (see below).

Tau-a and -b should be used for square tables. If the table is rectangular, *Tau-c* should be used instead. The calculation of Tau-c is somewhat different from those for the other measures because it must take the unequal shape of the table into account. Tau-c does this by including a value, *m*, which is the smaller of the number of rows or columns in the table. This has the effect of "squaring" the table for the calculation. The formula for Tau-c is

$$
\frac{2m(N_s - N_d)}{N^2(m - 1)}
$$

Table 9-4 is rectangular; it has more columns than rows. Because it is rectangular, this table is not appropriate for use with Tau-a or -b, so Tau-c should be used as a measure of association.

TABLE 9-4   Rectangular Bivariate Table

| | | | In your neighborhood, how well do you think the police perform their duties? | | | | |
			Not At All	Below Average	Average	Very Well	Total
To what extent does the police department need improvement?	No Changes Needed	Count	4	3	0	0	7
		Expected Count	1.2	3.7	1.8	.3	7.0
		% within To what extent does the police department need improvement?	57.1%	42.9%	.0%	.0%	100.0%
		% within In your neighborhood, how well do you think the police perform their duties?	7.3%	1.8%	.0%	.0%	2.3%
		% of Total	1.3%	1.0%	.0%	.0%	2.3%
	Some Changes Needed	Count	48	111	25	0	184
		Expected Count	32.6	97.3	46.3	7.7	184.0
		% within To what extent does the police department need improvement?	26.1%	60.3%	13.6%	.0%	100.0%
		% within In your neighborhood, how well do you think the police perform their duties?	87.3%	67.7%	32.1%	.0%	59.4%
		% of Total	15.5%	35.8%	8.1%	.0%	59.4%
	Many Changes Needed	Count	3	50	53	13	119
		Expected Count	21.1	63.0	29.9	5.0	119.0
		% within To what extent does the police department need improvement?	2.5%	42.0%	44.5%	10.9%	100.0%
		% within In your neighborhood, how well do you think the police perform their duties?	5.5%	30.5%	67.9%	100.0%	38.4%
		% of Total	1.0%	16.1%	17.1%	4.2%	38.4%

(continues)

TABLE 9-4   Rectangular Bivariate Table (*continued*)

| | | In your neighborhood, how well do you think the police perform their duties? | | | | |
		Not At All	Below Average	Average	Very Well	Total
Total	Count	55	164	78	13	310
	Expected Count	55.0	164.0	78.0	13.0	310.0
	% within To what extent does the police department need improvement?	17.7%	52.9%	25.2%	4.2%	100.0%
	% within In your neighborhood, how well do you think the police perform their duties?	100.0%	100.0%	100.0%	100.0%	100.0%
	% of Total	17.7%	52.9%	25.2%	4.2%	100.0%

Chi-Square Tests

	Value	df	Asymp. Sig. (2-sided)
Pearson Chi-Square	83.783[a]	6	.000
Likelihood Ratio	94.117	6	.000
Linear-by-Linear Association	79.094	1	.000
N of Valid Cases	310		

[a]5 cells (41.7%) have expected count less than 5. The minimum expected count is 0.29.

Symmetric Measures

		Value	Asymp. Std. Error[a]	Approx. T[b]	Approx. Sig.
Ordinal by Ordinal	Kendall's tau-b	−.473	.039	−10.926	.000
	Kendall's tau-c	−.396	.036	−10.926	.000
	Gamma	−.778	.047	−10.926	.000
	Spearman Correlation	−.505	.042	−10.270	.000[c]
Interval by Interval	Pearson's r	−.506	.038	−10.294	.000[c]
N of Valid Cases		310			

[a]Not assuming the null hypothesis.
[b]Using the asymptotic standard error assuming the null hypothesis.
[c]Based on normal approximation.

Applying the formula for Tau-c to the data in Table 9-4 would result in the following calculations:

$$\tau_c = \frac{(2)(3)(-25{,}371)}{(310^2)(3-1)}$$

$$= \frac{(6)(-25{,}371)}{(96{,}100)(4)}$$

$$= \frac{-152{,}226}{384{,}400}$$

$$= -0.3960$$

Here 2 is a constant in the formula. Because this is a 4×3 table, the smaller number of columns or rows is 3. The $-25{,}371$ is the result of the calculation (not shown here) of $N_s - N_d$ for the data in Table 9-4. N is 310, which is squared in the formula. Finally, the 3 in the denominator is the same (m) as in the numerator, and 1 is a constant.

The calculations in this case produced a value of -0.3960, which is the same as in Table 9-4. In this case, we are able to reduce the errors in predicting the level of improvements that respondents felt were necessary for the police department by almost 40% by including information about previous contact with the police.

Interpretation

The values of Tau-a, -b, and -c can vary from -1 to 0 to $+1$, depending on whether the association is positive or negative. An association of 0 indicates an equal number of concordant and discordant pairs (no association), whereas a ± 1 indicates that all possible pairs are either concordant or discordant. Anything other than a 0 or a ± 1 indicates there is some association but that it is not perfect.

Tau-a is a symmetric measure that is appropriate for a table of any size. The problem with Tau-a is that it cannot achieve ± 1 if there are any ties, which is usually the case. Tau-a cannot reach ± 1 because the denominator will always be more than the numerator. For this reason, Tau-a has generally been overshadowed by Tau-b or Tau-c.

Tau-b should only be used with square tables and Tau-c with rectangular tables. This is because Tau-b cannot achieve a score of ± 1 on a rectangular table because if the table is square, there can be an equal number of ties on the dependent and independent variable. If the table is rectangular, however, there would be an odd number of possible ties on one or the other variable such that there would be more pairs tied on the variable with the fewest categories than on the other variable; thus, the formula denominator could not equal the numerator. Tau-b will reach ± 1 only if all of the frequencies fall on the diagonal, that is, if all cell frequencies not on the diagonal are 0. Tau-c is used with rectangular tables. Interpretation of Tau-c is the same as that of Tau-b except that it can achieve a value of ± 1 for rectangular tables.

Taus-b and -c have a fairly straightforward interpretation. The formula for Tau-b is simply the number of concordant pairs minus the number of discordant pairs divided by the total number of pairs, including ties on the dependent and independent variable. Starting in the lower left corner and summing above and to the right, in essence, calculates the strength of the positive relationship between the variables (low to high categories for the dependent variable and low to high categories for the independent variable). Starting in the lower right corner and summing above and to the left calculates the strength of the inverse relationship (low to high categories for the dependent variable and high to low categories for the independent variable). Summing all of the values to the right of a dependent variable category includes those values that were excluded as ties in the N_s and N_d calculations; the same can be said for summing all of the values below categories on the independent variable. When these values are included as a ratio, it describes the relationship between the positive relationship and the inverse relationship, including ties.

Probably the most important information here is the ratio of N_s and N_d pairs. If there are more concordant than discordant pairs, Tau will be a positive value, and the more concordant pairs that exist, the closer the value will be to 1. If there are more discordant than concordant pairs, Tau will be a negative value, and the more discordant pairs that exist, the closer the value will be to -1.

For the data in Table 9-3, the value of Tau-a is 0.0376 (not shown in the SPSS printout). This tells us two things. First, there is a positive relationship. As respondents' attitudes moved from seeing crime as no problem to seeing it as a big problem, their frequency of walking at night moved from often to never. The value 0.0376 also tells us that knowing a person's attitude about the crime problem reduced the errors in predicting the frequency of a person walking at night by only about 4%. This is not much of a reduction of errors.

In Table 9-3, the value of Tau-b is 0.10367. This tells us two things: As with Tau-a, it tells us that this is a positive relationship; and it tells us that knowing a person's attitude about the crime problem reduced the errors in predicting the frequency of a person walking at night by about 10%, a more substantial reduction than with Tau-a.

As stated above, although the formulas are different, the interpretation of Tau-c is the same as that for Tau-b. In Table 9-4, the value of Tau-c is –0.39612. Again, this tells us two things. First, the relationship is negative: Those who responded that the police performed their jobs very well in their earlier contacts were more likely to respond that no changes were needed in the police department, and those who were not at all satisfied with their earlier contacts with the police were more likely to respond that many changes are needed. Second, the value -0.39612 is also a fairly strong relationship. This tells us that we can reduce by almost 40% the errors in predicting people's perceptions concerning the need for change in the police department if we know how they perceived their previous contact with the police.

A slightly different interpretation of Tau directly considers the relationship between the two variables. The value of Tau-b in Table 9-3 (0.10367) can be interpreted as the probability that pairs of scores drawn at random will have the same ordering is 10% more likely than drawing pairs of scores ranked oppositely. Similarly, the data for Tau-c from Table 9-4 (-0.39621) can be interpreted as the probability that pairs of scores drawn at random from this data set will be ranked oppositely (discordant pairs) is almost 40% higher than the probability of drawing concordant pairs.

Gamma

The work that expanded the use and usefulness of measures of association probably more than any other single work was a 1954 article by Goodman and Kruskal. One of the measures of association for ordinal level data put forth in their article was **Gamma, which** is (symbolized by Γ). Goodman and Kruskal developed Gamma because they proposed that Tau had no practical interpretation, which at the time was probably correct. As demonstrated in the preceding section, however, practical interpretations have been developed for Tau, primarily by applying the theory Goodman and Kruskal used for Gamma. A primary difference between Gamma and Tau is when many ties are present. Gamma and Tau give the same values when there are no ties in the data. If there are many ties, however, Gamma will give an inflated assessment of the relationship, which may not be desired. There also remain technical and mathematical differences in the structure and interpretation of Gamma and Tau that are not addressed here.

Gamma compares the number of times two variables are ranked the same with the number of times they are ordered differently. It is a PRE measure, but only for untied pairs of scores. That is because it does not consider ties at all; they are excluded from the calculation. In fact, a Gamma value of 1 can be achieved based on only one pair of values if all the rest of the pairs are tied. This is a drawback of Gamma in that it sometimes inflates the value of the relationship and typically yields larger values than Somers' *d* or Tau-b and -c. Gamma is a symmetric measure of association, so it does not specify which variable is independent. Gamma can be used with a table of any size.

Calculation

Gamma is calculated by the ratio of the concordant minus the discordant pairs divided by the concordant pairs plus the discordant pairs:

$$\Gamma = \frac{N_s - N_d}{N_s + N_d}$$

The procedure for calculating Gamma is essentially the same as Tau-a or -b. The same procedures and numbers from Table 9-3 can be used to calculate N_s and N_d. These two values can then be placed in the formula for Gamma to determine its value.

$$\Gamma = \frac{N_s - N_d}{N_s + N_d}$$
$$= \frac{2785 - 1625}{2785 + 1625}$$
$$= \frac{1160}{4410}$$
$$= 0.263038$$

In this case, Gamma is calculated simply by subtracting the N_s from N_d pairs and dividing that result by the sum of N_s and N_d. This produces a proportional reduction of errors of 26% in this case, which is higher than any of the other ordinal level measures of the strength of an association. This is because there are a number of ties in this data, and gamma does not take ties into account. As discussed above, this allows gamma to base the findings on a smaller number of pairs than actually exist in the data, often inflating the actual relationship, as here.

Interpretation

As designed by Goodman and Kruskal, Gamma has a very intuitive interpretation. From the formula we can see it is simply the number of concordant pairs minus the number of discordant pairs divided by the total number of pairs, excluding ties. This is essentially the same interpretation as Tau except that ties are not taken into account. Starting in the lower left corner and summing above and to the right calculates the strength of the positive relationship between the variables (low to high, dependent, and low to high, independent). Starting in the lower right corner and summing above and to the left calculates the strength of the negative relationship (low to high, dependent, and high to low, independent). When these two values are included as a ratio, it describes the relationship between the positive and negative associations. If one is stronger than the other, it will determine the sign of the Gamma. Also, the higher the values, the stronger the positive or negative association and the higher the Gamma value. As with Tau, the values of Gamma range from −1 to 0 to +1.

In this case, the value of Gamma is 0.26308. This indicates two things. The first is that this is a positive relationship. As respondents' attitudes changed from seeing crime as no problem to seeing it as a big problem, their frequency of walking at night changed from often to never. The value 0.2608 also indicates that we have reduced by 26% the errors in predicting the rank ordering of pairs of persons walking at night by knowing their attitudes concerning the crime problem.

In Table 9-4 the value for Gamma is −0.77794, which is almost twice as high as that for any other measure. As discussed above, this shows that the prediction of a person's perception of the amount of improvement needed in this police department could be improved by almost 80% by knowing his or her satisfaction with earlier contact with the police. This value also shows that this is a negative relationship (discussed in Chapter 10). This finding also shows something else; there are probably a large number of ties in this data.

The fact that there are a large number of ties is evident from the ordinal level measures of association. Gamma, which takes no ties into account, is the largest of all the values. Tau has a smaller value because it takes ties into account—in this case, ties that may be reducing the true nature of the concordant relationship. This is something researchers must wrestle with. Larger tables are more likely to have more ties as the data moves toward a continuous scale. This will produce a divergence of the measures of the strength of association such that decisions will have to be made concerning which measure to use.

Gamma will be 0 (or undeterminable) when all the values in the table are in a single row or column of the table. It will also be 0, as shown below, when the values are concentrated on both the diagonals of the table:

5	0	5
0	5	0
5	0	5

Here, the values in one diagonal that will produce a Gamma of $+1$ are completely offset by those that will produce a Gamma of -1. This is the reason that examination of the table is so important. Relying on the summary measure of Gamma would lead to the conclusion that there was no association between the two variables. In reality, there are very strong associations between the variables; in fact, there are two and they offset one another. This phenomenon will also be important later in this chapter and in Chapters 11 and 12 in the discussion of correlation and linearity.

A final note about Gamma. Gamma for a 2×2 table is the same as *Yule's Q*. This value was developed by Yule (1900) as a measure of association using a ratio of the two binomials of the table. Yule named his procedure after Adolphe Quetelet, from whom he apparently took the technique.

Somers' *d*

Somers' *d* is an asymmetric counterpart to Gamma. Somers' *d* was developed by Robert Somers (1962) because he felt the need for an asymmetric counterpart to the symmetric measures of association for ordinal level data. To be asymmetric, ties on the dependent variable must be taken into account, and ties on the independent variable must be excluded. This allows responses on the dependent variable to remain as long as the independent variable is ranked differently. Because Somers' *d* takes into account ties on the independent variable, it is a PRE measure for untied pairs and for pairs tied on the dependent variable but not for ties on the independent variable. Somers' *d* can be used on a table of any size.

Calculation

The calculation for Somers' *d* is similar to Tau-b except that only ties on the dependent variable are included. The formula for Somers' *d* is

$$d_{yx} = \frac{N_s - N_d}{N_s + N_d + T_y}$$

As shown in the calculations for Tau-b, T_y is calculated by multiplying the frequencies in each row by the sum of the frequencies to the right of it and then adding those together. In addition to the calculations used for Gamma, the additional calculation for Somers' *d*, using the data in Table 9-3, is

$$T_y = [(2)(2 + 0)] + [(2)(0)] + [(20)(13 + 3)] + [(13)(3)]$$
$$+ [(143)(57 + 9)] + [(57)(9)]$$
$$= [(2)(2)] + [(2)(0)] + [(20)(16)] + [(13)(3)]$$
$$+ [(143)(66)] + [(57)(9)]$$
$$= 4 + 0 + 320 + 39 + 9438 + 513$$
$$= 10,314$$

Note that the calculation of T_y here is the same as that for Tau-b. Including this value with those obtained for Gamma in the formula for Somers' d results in the following calculation:

$$d_{yx} = \frac{N_s - N_d}{N_s + N_d + T_y}$$

$$= \frac{2785 - 1625}{2785 + 1625 + 10,314}$$

$$= \frac{1160}{14,724}$$

$$= 0.07878$$

This shows that the prediction of a person walking at night can be improved by about 8% by knowing his or her perception of the crime problem. Also notice that this is the same value as shown for Somers' d in the SPSS printout in Table 9-3. Also note that Table 9-3 contains three values for Somers' d. This seems to go against the preceding discussion where the dependent variable was to be taken into account. These three values in SPSS are actually three different calculations of Somers' d. Because SPSS does not know which variable is dependent, it calculates a Somers' d using each variable as the dependent variable. It then calculates a combination of these values for the symmetric measure. Each of the asymmetric measures, however, uses the formula for Somers' d presented above.

Interpretation

Somers' d is interpreted in essentially the same way as Tau and Gamma. Values range from -1 through 0 to $+1$. In cases where all pairs are N_s, Somers' d will be $+1$; in cases where all pairs are N_d, Somers' d will be -1. The interpretation of Somers' d for Tables 9-3 is the same as for the other ordinal level measures. The value 0.07878 shows that there are more N_s pairs than N_d pairs and that the PRE is about 8%.

For square tables, Somers' d will reach 1 only when all the values fall on the diagonal. For a rectangular table such as that in Table 9-4, Somers' d requires only that the values be monotonic. A monotonic relationship exists when the frequencies of the dependent variable move across the table in a stepwise manner, where a single value of the dependent variable does not take more than one value of the independent variable. There can be more than one frequency in a given column, but there cannot be more than one frequency in a given row (see **Figure 9-1**).

On the left in Figure 9-1, the Somers' d value would be 0.91. It does not rise to 1 even though many of the values are on the diagonal because the third row contains two values. Because ties are taken into account on the dependent variable (rows), it increases the value of the denominator, thus decreasing the value of Somers' d. On the right in Figure 9-1, however, there are values that overlap in the independent variable (columns), but there are no values that overlap in the dependent variable. The Somers' d value here is 1.00.

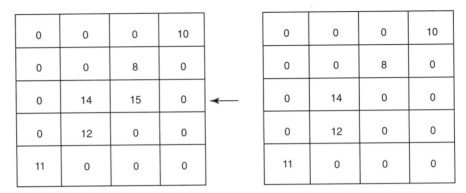

Figure 9-1   Comparison of Somers' *d* Tables

Aside from the mathematical interpretation, the commonsense interpretation of the tables in Figure 9-1 supports the findings of Somers' *d*. Here, ties on the independent variable will lead to ambiguous interpretations of the dependent variable. For example, the third row in the left table of Figure 9-1 has two values for that category of the dependent variable. We should not have two predictors of the independent variable attempting to categorize one category of the dependent variable. In the right table of Figure 9-1, only one category of the independent variable is predicting each category of the dependent variable, even though it is equally good at predicting two categories of the dependent variable. The fact that there are ties on the dependent variable does not mean that the independent variable has no value in making predictions; it means only that the value is reduced and not perfect (1.00). In a real example, it would not be perfect (although it may be acceptable) to know that money and family background predicted that a person would become a thief. In this example, two categories of the independent variable explain one category of the dependent variable. We could maintain a perfect association, however, even if we knew that money predicted a person becoming a thief or a robber. Here, we are still using only one category of the independent variable to predict two dependent variable categories.

Spearman's Rho

Ordinal level data that is fully ordered and interval level data that is not normally distributed—a skewed distribution—is generally analyzed using Spearman's Rho, which is symbolized by r_s. **Spearman's Rho**, often called *Spearman's Correlation*, measures strength, significance (as in Chapter 8), and direction (discussed in Chapter 10) of the differences in ranks of individual cases.

Spearman's Rho was created by Charles Spearman. Although several of his works emphasized the need for a "product moment" (Pearson's *r*) measure for ranked data, he did not begin to develop this measure until 1902 and specifically outlined the statistical procedure in his 1904 article. Spearman's Rho is not a PRE measure, but r_s^2 is a PRE measure. The notion of the square of the value being a PRE measure, and a measure of correlation, was put forth by Spearman in 1904. Here he argued that "[i]n short, not Galton's measure of correlation, but the *square thereof*, indicates the relative influence of the factors in A tending towards any observed correspondence [with B]. . . ."

Spearman's Rho is simply Pearson's *r* (see below) applied to the ranks of data points rather than the points themselves. For ranked data, Spearman's Rho is well accepted. There has been great debate, however, concerning the appropriateness of using Spearman's Rho on ranks rather than Pearson's r on raw data, such as when the data is interval (Pearson's r would be appropriate) but nonnormal (detrimental to Pearson's r). Before the days of computers and hand calculators, it was easier to calculate Spearman's Rho than Pearson's r. In fact, this was the very reason behind Spearman creating this statistical procedure in the first place, to ease calculation of Pearson's r but arrive at essentially the same value. It has become more common, however, for researchers to rank data and to use Spearman's Rho when some of the assumptions of Pearson's r appear to be violated. This seems to be supported by Spearman (1904, p. 81), but it has recently come under serious attack by those who argue that simply ranking the data and using Spearman's Rho (or Pearson's r on the ranked data) does not relieve the researcher of the underlying assumptions of Pearson's r (see Roberts and Kunst, 1990).

Calculation

Spearman's Rho is a symmetric measure of the rankings of scores of the variables. It is calculated using the formula

$$1 - \frac{6 \Sigma D^2}{N(N^2 - 1)}$$

where D is the difference in the rank of the dependent variable minus the rank of the independent variable. Because this value will always be 0, it must be squared to get an actual difference. The denominator, $N(N^2 - 1)$, is a mathematical calculation of the maximum possible difference (D^2) that can be achieved for the particular data set.

Although it looks complicated, this formula is actually very straightforward. Summing D^2 is the primary part of the formula. This is where the ranks of the values are compared. To compare the differences with a standardized value, a ratio is created with the obtained differences in the numerator (ΣD^2) and the maximum number of possible differences in the denominator using the formula

$$\frac{(N(N^2 - 1)}{3}$$

The rest of the formula represents a mathematical derivation to standardize the formula where the sign will be correct and the result will lie between +1 and −1, with 0 showing no association.

It should be noted that SPSS calculates Spearman's Rho differently in different places. In the crosstabs procedure, the formula for Pearson's r is applied to the ranks of the scores. In other places, a specialized Spearman formula is used for the calculations. Also, the treatment often heavily influences calculation of the various Spearman formulas. Because SPSS uses various formulas, the hand calculations of Spearman will not be shown and the SPSS printout will be used for interpretation.

Interpretation

Spearman's Rho will be +1 if the ranks of the variables match perfectly (positive association) and −1 if they are perfectly ranked inversely (negative association). If there is no association between the variables (no pattern in the ranking), Spearman's Rho is 0. Anything other than zero means there is some association.

It is important to note here that Spearman's Rho is a measure of association for the ranks of the data, not the data itself. Any interpretation of the data, therefore, must be specific to the rankings. Also note that Spearman's Rho will often give slightly larger results than those given by other ordinal level measures of association. This is because Spearman's Rho squares the difference, hence giving more weight to extreme differences (discordance) in the ranks. Also, Spearman's Rho will generally be smaller than Pearson's *r* if the data contains many outliers, and it will be larger if there is a monotonic but nonlinear relationship.

An example of the application of Spearman's Rho can be drawn from **Table 9-5**. This table is taken from a study of the productivity of the graduates of doctoral programs in criminal justice and criminology as measured in terms of publications and service to national academic organizations (Walker, 1995). The rank of each school for service and publication is shown. As can be seen from the table, there are several tied scores on the ranks. These ties are reflected in the Spearman's Rho values in part (a) of **Table 9-6**. An artificial correction for ties (arbitrarily assigning different ranks for tied schools) is shown in part (b) of Table 9-6.

Using the data in Table 9-5, the value of Spearman's Rho in Table 9-6 is 0.091 for part (a) and 0.140 for part (b). Both of these show very little strength between the two variables. Squaring these values to achieve a PRE measure accentuates how weak the relationship between the two variables is: In part (a) we can reduce the errors in predicting the rank of the publication of the school by less than 1%; part (b) shows a reduction of errors of only about 2%. The Spearman Correlation is also larger than Pearson's *r* in this case. This shows that the data is ordinal and probably skewed, which is detrimental to accurate measurement with Pearson's *r* (see the discussion of Pearson's *r* below).

TABLE 9-5   Service and Publication Ranks of Doctoral Program Graduates

School	Service Rank	Publication Rank
Michigan State University	1	7
Sam Houston State	2	13
Ohio State University	3	3
Washington State University	3	4
Florida State University	5	2
SUNY	6	1
Indiana University	7	9
University of Maryland	7	12
University of California–Berkeley	10	9
North Carolina University	10	6
University of Pennsylvania	10	5
University of Washington	10	7

Source: Walker (1995).

Table 9-6   SPSS Output for Pearson's *r* and Spearman's Correlation: Symmetric Measures

		Value	Asymp. Std. Error	Approx. T	Approx. Sig.
Ordinal by Ordinal	Spearman Correlation	.091	.282	.290	.778
Interval by Interval	Pearson's *r*	.043	.242	.135	.895
N of Valid Cases		12			
Ordinal by Ordinal	Spearman Correlation	.140	.287	.448	.664
Interval by Interval	Pearson's *r*	.087	.251	.275	.789
N of Valid Cases		12			

The independence of the variables can also be shown when testing the null hypothesis of no relationship between the variables in the population. In this case, the null hypothesis would not be rejected in either case because the values of approximate significance for the Spearman correlation is 0.778 in part (a) and 0.664 in part (b)—both far above the .05 or .01 cutoff for statistical significance.

Limitations

As shown in Table 9-6, Spearman's Rho is influenced by a large number of ties of the ranks. An integral assumption of Spearman's Rho is that there are few or no ties in the data. If there are a large number of ties, Spearman's Rho will be reduced. The number of ties is such a problem, in fact, that some (e.g., Roberts and Kunst, 1990) have proposed doing away with Spearman's Rho and either using the other ordinal level measures discussed above or moving to Pearson's *r*. Stevens (1951) also argued against using Spearman's Rho as anything other than a test of the hypothesis of the ordering of the ranks, not as a measure of the association between the two variables. Stevens argued that to use Spearman's Rho as an approximation for Pearson's *r* is to assume an underlying interval scale and bivariate normal distribution that is too broad for ordinal level data.

One last note about ordinal level measures of the strength of an association. Because you are moving into data that can be ordered and has the possibility of being relatively continuous, the nature of the data (see Chapter 10) becomes more important. A curvilinear relationship (skewed data) can greatly reduce the value of the measure of association and make the interpretation difficult or suspect. If the measure of association produces a value lower than expected, it should signal an examination of the nature of the relationship.

Cold Wave Linked to Temperatures

■ Interval Level Data

If both of the variables are interval or ratio level, it is possible to use the more advanced, interval level statistical analyses to work with them. The real difference between Spearman's Rho and Pearson's *r* is that Spearman's Rho measures the monotonicity of a relationship between two variables, whereas Pearson's *r* requires linearity. A relationship may be monotonic and not linear (see the discussion in Chapter 10).

Although often used incorrectly with nominal and ordinal level data, moving to interval level data facilitates advancement to co-relation (correlation) between the variables under examination. PRE measures, up to and including Spearman's Rho, cannot provide the level of interpretation about the proportion of change in one variable explained by change in another variable; they are capable only of addressing the proportional reduction of errors in making predictions of categories or ranks of the variables. Correlation methods such as those described here move beyond predicting categories to describing the linear relationship between the variables. More important, analyses at this level support the conclusion that an independent variable might be an underlying cause of the variations in the dependent variable (Spearman, 1904).

The goal of interval level bivariate analysis is to make a better prediction of the exact score of a dependent variable than is possible from the mean of the dependent variable. This is undertaken by using information from the independent variable. As discussed in Chapter 4, the mean is a good measure, given no information other than the fact that a data set is interval level, to predict a score in a set of data. If we can improve the prediction of the mean with information about the independent variable, the requirements for a proportional reduction in error have been met.

Suppose we made predictions about where a certain point will fall in a distribution or on a graph. These we will call Y'. By going back and plotting the actual values, we could calculate how well the predictions served as a PRE measure. This is undertaken through the formula for a straight line:

$$Y' = a + b(X)$$

where Y' is the value we want to predict, a the constant starting point for the line (where it crosses the Y axis), b the slope of the line (the angle of the line in a graph), and X the value of the independent variable that corresponds to the dependent variable.

With a number of scores, it would be possible to use these calculations to show how well or how poorly the predicted scores matched the actual scores. The goal of this process is to get the best-fitting straight line that summarizes the linear relationship between the values of the two variables. One method of achieving this goal is to use the least squares method—this is the variance. This method establishes a line in a position where the sum of the squares of distances from each point to the line is the smallest possible value. This minimizes the distance to all pairs of X, Y values and summarizes the relationship between the dependent variable Y and the independent variable X. It will not improve every prediction, but it will improve the average prediction.

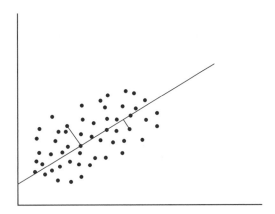

Figure 9-2   Scattergram of Data with Best-Fitting Line Overlay

This process is shown in **Figure 9-2**. Here, the data points are clustered in a relatively uniform pattern. A line can be drawn through these data points such that the variance (the sum of the distance between each point and the line) is the smallest distance. For example, in Figure 9-2, two of the data points have lines drawn from them to the best-fitting line. If this was completed for each point and the distances summed, it would represent the smallest of the summed distances for any line drawn through the values. This is also a graphical description of the variance.

It must be stressed here that this must be a straight line. Curvilinear relationships will always produce lower values, even when a strong relationship exists (as demonstrated in Figure 9-3 and explained further in Chapter 10). For this reason, the nature of the relationship must be carefully monitored and curvilinear relationships transformed or otherwise taken into account (see Chapter 10). Also, achieving a straight line is not always the best answer and must be interpreted in light of the data available. As discussed below, Pearson's *r* may be used with dichotomized data. In these cases, the number of data points (cases of the variable) is important. If you have only one case for each of the variables, plot them, and analyze them with Pearson's *r*, you will get a perfect association. Is it because the variables are perfectly related? No. It is because only two cases will always create a straight line. There are not enough data points to examine the data properly. Stated another way, you have an excellent fit between the two variables because you have so few cases.

Pearson's *r*

The way a best-fitting line is used in bivariate analysis is through Pearson's Product Moment Correlation (or Pearson's *r*). **Pearson's *r*** represents the extent to which the same cases or observations occupy the same relative position on two variables. It can measure the strength of a relationship, examine the existence of the relationship, and determine the direction.

A Bit of History

Correlation is thought to have begun with the work of Bravais in 1846. His work was expanded on by Galton (1886, 1888), who in his example of human anatomy used the term *co-relation* to describe the similar slopes of two regression lines of variables measured on the same scale. In this case, if forearm and head lengths were expressed in units of their probable errors, the slope of the lines (Galton used *r*) would be the same. This "index of co-relation" was an extension of Galton's work on regressions, which he identified as an extension of the regression coefficient to bivariate analysis.

Galton's comments on co-relation were furthered by Karl Pearson (1914–1930), who considered Galton to be the originator of correlation. Pearson even admits to his refinements of Galton's work in his 1896 work, stating: "The investigation of correlation which will now be given does not profess, except at certain stated points, to reach novel results. It endeavors, however, to reach the necessary fundamental formulae with a clear statement of *what assumptions are really made,* and with special reference to what seems legitimate in the case of heredity." In refining Galton's work, Pearson (1896) used the term $\Sigma(xy)$ as an estimation of a "product moment" of heredity. This would ultimately become the trademark for his calculation of correlation, hence the term *Pearson's Product Moment Correlation.*

Pearson's *r* is the most widely used measure of association. It requires normally distributed, interval level data. Pearson's *r* can also be used to examine the relationship between a continuous, interval level variable and a dichotomized variable. It is a technical point, but one that should be made, that the dichotomized variable to be used in this type of analysis, called a *point-biserial correlation*, must *truly* be dichotomized. If there is an underlying continuum between the two values, a point-biserial correlation should not be used. For example, a survey of 1000 people on the death penalty would actually find few that are absolutely for or against the death penalty; thus, there is an underlying continuum of "yes, but" answers that have been forced into yes/no answers. In this case, the variable is not truly dichotomized and Pearson's *r* should not be used. Pearson's *r* can also be used for two dichotomized variables and is called a Phi Coefficient in this instance. One thing to remember when using dichotomized variables is that the direction of the relationship is arbitrary; the characteristic coded as highest will determine the interpretation (see Chapter 10). Pearson's *r* typically precludes the use of bivariate tables because there are usually too many values to put in a table format.

The way Pearson's *r* works is through the least squares line. Pearson's *r* measures the amount of spread around the least squares line and the slope of the line. The amount of spread determines the strength of the association. If all the values are on the line, there will be a perfect association between the variables: A one-unit increase in

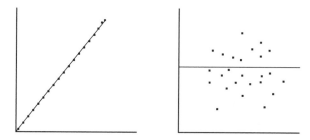

Figure 9-3   Best-Fitting Lines Showing Pearson Correlations

one variable will result in a one-unit increase in the other variable (as shown on the left in **Figure 9-3**). If the values are spread such that the best line that can be drawn runs parallel to the X axis, however, it will indicate a 0 correlation (as shown on the right of Figure 9-3).

An often overlooked requirement in Pearson's r is that both variables must vary or nothing is gained by the procedure. Sometimes, research will be conducted in which one characteristic is a constant (the category or value does not change within the units of analysis). For example, a researcher might conduct research on SAT scores and all students in the research are in the same grade; the grade-level characteristic will be a constant, not a variable. It is not possible, then, to examine whether one variable varies with changes in the other variable because neither variable varies.

Calculation

Calculation of Pearson's r is as follows:

$$r = \frac{N(\Sigma XY) - (\Sigma X)(\Sigma Y)}{\sqrt{[N(\Sigma X^2) - (\Sigma X)^2][N(\Sigma Y^2) - (\Sigma Y)^2]}}$$

How do you do that?
Measures of Strength of an Association for Internal/Ratio Data

1. Open a data set.
   a. Start SPSS.
   b. Select *File*, then *Open*, then *Data*.
   c. Select the file you want to open, then select *Open*.
2. Once the data is visible, select *Analyze*, then *Correlate*, then *Bivariate*.

3. Select the variables you wish to include as your independent and dependent variables and press the ▶ next to the *Variables* window.

4. Check the box in *Correlation Coefficients* next to the statistic appropriate for your data (Spearman for fully ordered ordinal level data, Pearson for interval level data).

5. Do not worry about the *Test of Significance* boxes right now.

6. Make sure the *Flag Significant Correlations* box is checked.

7. Select *Continue*, then select *OK*.

8. An output window should appear containing a distribution similar in format to Table 9-8.

Although it appears imposing, this is actually not a difficult formula. It is simply summing the values of $XY(X \cdot Y)$, X^2, and Y^2. Take as an example the data shown in **Table 9-7** examining the relationship between education level and number of crimes for which a person was convicted. Using that data, Pearson's r would be calculated as follows:

$$r = \frac{(7)(499) - (91)(49)}{\sqrt{[(7)(91^2) - (91)^2][(7)(49^2)(49)^2]}}$$

$$= \frac{(7)(499) - (91)(49)}{\sqrt{[(7)(1435) - 8281][(7)(455) - 2401]}}$$

$$= \frac{3493 - 4459}{\sqrt{[(7)(1435) - 8281][(7)(455) - 2401]}}$$

$$= \frac{-966}{\sqrt{[(7)(1435) - 8281][(7)(455) - 2401]}}$$

$$= \frac{-966}{\sqrt{[10,045 - 8281][3185 - 2401]}}$$

$$= \frac{-966}{\sqrt{(1764)(784)}}$$

$$= \frac{-966}{\sqrt{1,382,976}}$$

$$= \frac{-966}{1176}$$

$$= -0.8214$$

TABLE 9-7 Data for Pearson's r

Education: X	Crimes: Y	XY	X^2	Y^2
7	13	91	49	169
4	11	44	16	121
13	9	117	169	81
16	7	112	256	49
10	5	50	100	25
22	3	66	484	9
19	1	19	361	1
$\Sigma X = 91$	$\Sigma Y = 49$	$\Sigma XY = 499$	$\Sigma X^2 = 1435$	$\Sigma Y^2 = 455$

Here, each of the totals from Table 9-7 has been included in the numerator ($\Sigma XY = 499$, $\Sigma X = 91$, $\Sigma Y = 49$), along with the N of seven people. The values of ΣX, ΣY, and N are also included in the denominator. These calculations produce a Pearson's r of -0.8214, which is very strong (see below).

Correlation Matrixes

It is not common for Pearson's r (or Spearman's Correlation) to be displayed as in Table 9-6. More typically, they are displayed in what is termed a **correlation matrix**. A correlation matrix is shown in **Table 9-8**. There is nothing different in these numbers, nor in their interpretation—only in the way they are displayed. The nature of interval level analyses is that they are often a part of an analysis of many variables. In these cases, it is generally not possible to include separate tables because, as discussed above, the data is often too extensive to include in a table. Furthermore, it would be a waste of space to report all combinations of correlation separately in a display such as Table 9-6, where each value is reported on a different line or different output.

What Table 9-8 shows is the correlation between 14 different variables. Notice that each of these variables is listed across the top of the table and along the side. This allows each variable to be correlated with all other variables. Because each variable is listed twice, if all correlations were included, each correlation would occur twice in the table. For example, the correlation between year in school and grade average (-0.01) would show up in the column under grade average and the row of year in school (not shown in Table 9-8) *and* in the column under year in school and the row of grade average (where it does appear in Table 9-8). Because there is no need for the correlations to be listed twice, a correlation matrix generally takes the form of half a table, with the scores either falling below or above the diagonal. Along the line of the diagonal is a series of values of 1.00. This represents the dividing line between the two sets of correlations. Each of these correlations is 1.00 because it is reporting the correlation between each variable and itself (e.g., the correlation between gender in the gender column and gender in the gender row).

TABLE 9-8   Correlation Matrix of Student Cheating: Intercorrelations between Variables ($N = 330$)

	Gender	Year in School	Grade Average	Moral Beliefs	Shame	Pleasure	Low Self-Control	External Sanctions	Friends' Behavior	High School	Past Year	Caught Cheating	Major	Cheating Intent
Gender	1.00													
Year in School	.09	1.00												
Grade Average	-.21*	-.01	1.00											
Moral Beliefs	-.09	.14*	.12	1.00										
Shame	-.29*	.05	.22*	.47*	1.00									
Pleasure	.07	-.08	-.19*	-.39*	-.44*	1.00								
Low Self-Control	.28*	-.09	.30*	-.22*	-.35*	.32*	1.00							
External Sanctions	-.32*	-.06	.10	.30*	.47*	-.15*	-.29*	1.00						
Friends' Behavior	.03	-.14*	-.26*	-.28*	-.36*	.40*	.25*	-.24*	1.00					
High School	.06	-.21*	-.06	-.31*	-.27*	.19*	.16*	-.24*	.24*	1.00				
Past Year	.05	-.08	-.22*	-.25*	-.26*	.26*	.19*	-.15*	.41*	.17*	1.00			
Caught Cheating	.06	-.09	-.09	-.10	-.14*	.15*	.17*	-.04	.21*	.11	.11	1.00		
Major	-.16*	-.25*	.09	.04	-.00	-.05	-.04	-.03	-.00	-.01	-.09	.05	1.00	
Cheating Intent	.20*	-.09	-.25*	-.51*	-.55*	.48*	.32*	-.30*	.50*	.32*	.52*	.21*	-.03	1.00

$* p < .01$; one-tailed.

Source: Reprinted with permission from S.G. Tibbits, Differences Between Criminal Justice and Non-Criminal Justice Majors in Determinants of Test Cheating Intentions, *Journal of Criminal Justice Education*, Vol. 9, No. 1, pp. 81–94, © 1998, Academy of Criminal Justice Sciences.

Interpretation

Pearson's *r* also has values that fall between ±1, with +1 being a perfect positive relationship and −1 a perfect negative relationship. A Pearson's *r* value of 0 generally means there is no relationship, but caution must be exercised to ensure that the relationship is not curvilinear. Always check the nature of the data to determine if a curvilinear relationship exists (see Chapter 10). Low Pearson's *r* values may mean there is a weak association, but it may mean that the relationship is strong but that because it is curvilinear such that Pearson's *r* is not a true representation of the relationship.

Pearson's *r* is interpreted in the same way whether as a part of a crosstab printout, as in Table 9-6, or in a correlation matrix, as in Table 9-8. The strength of the association is shown by how close the value of Pearson's *r* is to −1.00 or +1.00. For example, the correlation between gender and year in school is 0.09 in Table 9-8. This is a very low association, showing almost no relationship between the variables.

Coefficient of Determination

Pearson's *r* squared (r^2) is known as the **coefficient of determination**. This tells us the proportion of the **variation** in the dependent variable that is accounted for by variation in the independent variable. In the example in Table 9-7, Pearson's *r* was −0.82. Squaring this value makes the coefficient of determination 0.67. When we multiply this value by 100, we get a percentage; this means that 67% of the variation in criminal activity can be explained by variation in years of school completed. This is actually a very strong relationship for social science data.

Interpretation of the coefficient of determination deals with the amount of influence one variable can have on another. Again, think of criminal behavior as a dependent variable. The desire of criminology is to be able to explain criminal behavior: the variation between criminals and noncriminals, or why some people are criminals and others are not. It may never be possible to explain all of the variation between criminals and noncriminals, but what research can show is how criminals and noncriminals differ (vary) in *some* ways. When this occurs, the research has explained *some* of the variation between criminals and noncriminals. In statistical terms, the research has explained some of the variation in the dependent variable—criminal behavior—with information about the independent variable.

A word of caution when comparing correlations, especially the coefficient of determination: The correlation coefficient (r) and the coefficient of determination (r^2) are not ratio level in themselves. An increase of 0.05 in an r^2 of 0.10 is not the same as an increase of 0.05 in an r^2 of 0.80. The reason that correlation coefficients should not be used as ratio level measures is because they become skewed (and therefore nonnormal) the greater the absolute difference of the coefficient from zero. More important for the coefficient of determination, it is a *proportion* of variation. That means it is the proportion of variation explained by adding this variable and the variance yet to be explained. In the example above, 0.05/0.10 = 0.5, while 0.05/0.80 = 0.0625. Explaining an additional 5% of the variation when 80% is already known, therefore, has a much smaller effect than explaining 5% more when only 10% of the variation is known. It is important, then, to be careful when comparing correlation coefficients and coefficients of determination so as not to assume that similar increases or explained variation are equal.

It should also be noted that using r^2 as the coefficient of determination is specific to certain fields, criminal justice being one of them. In other fields, and for other purposes, the correlations themselves often serve as the coefficient of determination. For example, under the statistical theory of common elements, r, not r^2, serves as the coefficient of determination (Ozer, 1985). Furthermore, reliability estimates are correlations that do not require squaring because they are already measures of the proportion of variation between the two variables (Nunnally, 1978).

Correlation and Causation

A phrase heard often in statistics courses is that "correlation does not equal **causation.**" This is because many people conclude incorrectly that if two variables are correlated, one must be the cause of another. This is not necessarily true. Research may bring us *closer* to determining the cause, but it is not close enough to be able to draw a causal nexus. It can be argued, however, that correlation is a *necessary* but not *sufficient* condition to establish a causal relationship. It is difficult to imagine (although possible, at least under laboratory conditions) a situation where a variable would be the cause of another but where the two variables would not show a correlation. Alternatively, though, an empirical association (correlation) between two variables is only a starting point for examining causality. A strong and significant correlation definitely supports an argument of causality and should certainly lead to further exploration of the possibility of causation. Other methodological tools, such as temporal ordering or eliminating rival causal factors, must be used before any notions of causality can be established (see Chapter 1 and Labovitz and Hagedorn, 1971).

Limitations

The most serious limitation of Pearson's r is that it is highly susceptible to nonnormal data. Curvilinear relationships may show a low strength (and perhaps nonsignificance), even though there may be definite patterns in the data. As discussed in Chapter 4, curvilinearity shows up in the univariate analysis of the data. If the data is skewed, truncated, or even symmetrical but with many outliers, it may not be normally distributed. A more detailed examination of nature and linearity is discussed in Chapter 10. If the data violates the assumption of normality, nonparametric measures of association, such as the Mann–Whitney or Kolmogorov–Smirnov tests, are generally more appropriate.

Pearson's r is also seriously influenced by sample size. Weak correlations may be found to be significant when the sample size is large. Additionally, small sample sizes, as in Table 9-7, may produce unstable correlations. In that example, one additional data value could have changed the correlation dramatically.

■ Conclusion: Selecting the Most Appropriate Measure of Strength

In this chapter, we introduced methods of testing the strength of a relationship. The discussion builds on that of Chapter 8 to determine if the relationship found to be statistically significant has any ability to predict values of the dependent variable. This is generally considered the second step in the process of bivariate analysis. This step is necessary because measures of significance do not provide all needed information. If we obtain a statistically significant χ^2, we can say that the two variables are not independent of each other. It does not, however, provide information on the strength of that relationship. We know statistical relationships that are almost nonexistent can show up as highly statistically significant χ^2 results if the sample size is large. This can delude a researcher into a false sense of accomplishment. Measures of the strength of the association, however, determine the ability to predict (the mode, median, or mean) and add validity to findings.

One issue that remains is what measure to use. Some situations will dictate which measure is most appropriate. For example, nominal level data requires the use of Lambda. In other instances, however, several measures may seem appropriate, particularly for ordinal level data. In these cases, the researcher must examine the data closely and attempt to determine which measure would best represent the data. You are cautioned, however, not to fall into the trap of using the measure with the greatest value. As Spearman (1904) put it: "For such auxiliary methods are very numerous and their results, owing to accidents, will diverge to some extent from one another; so that the unwary, 'self-suggested' experimenter may often be led unconsciously—but none the less unfairly—to pick out the one most favorable for his particular point, and thereby confer upon his work an unequivocality to which it is by no means entitled." Whether unconsciously or consciously, choosing a measure because it provides the most favorable results is dangerous. It is best to select the method before analyses are conducted and to choose the method based on articulable criteria.

Regardless of the outcome here, it is usually advisable to examine the direction and nature of the data. This will further define the relationship between the two variables and may explain any lower than expected measures of the strength of the association. This examination is the topic of Chapter 10.

■ Key Terms

asymmetric	Pearson's r
causation	proportional reduction of errors
coefficient of determination	Somers' d
correlation matrix	Spearman's Rho
degree of association	strength of association
Gamma	symmetric
Lambda	Tau
measure of association	variation

■ Summary of Equations

Lambda

$$\lambda = \frac{(N - R_t) - (\Sigma(C_t - f_i))}{(N - R_t)}$$

Tau-a

$$\tau_a = \frac{N_s - N_d}{T}$$

T in Tau-a Formula

$$\tau_b = \frac{N(N - 1)}{2}$$

Tau-b

$$\tau_b = \frac{N_s - N_d}{\sqrt{(N_s + N_d + T_y)(N_s + N_d + T_x)}}$$

Tau-c

$$\tau_c = \frac{2m(N_s - N_d)}{N^2(m - 1)}$$

Gamma

$$\Gamma = \frac{N_s - N_d}{N_s + N_d}$$

Somers' d

$$d_{yx} = \frac{N_s - N_d}{N_s + N_d + T_y}$$

Spearman's Rho

$$\rho = 1 - \frac{6\Sigma D^2}{N(N^2 - 1)}$$

Pearson's Product Moment Correlation

$$r = \frac{N(\Sigma XY) - (\Sigma X)(\Sigma Y)}{\sqrt{[N(\Sigma X^2) - (\Sigma X)^2][N(\Sigma Y^2) - (\Sigma Y)^2]}}$$

■ Exercises

For the tables that follow (which are the same as those used in the exercises for Chapter 8):

1. Determine the level of measurement for each variable.
2. Determine the appropriate measures of the strength of the association and discuss why it was chosen.
3. Conduct the appropriate calculations for each of the measures of association.
4. Discuss the strength of the relationship and which variable is causing the variation.

VICT_Y_N Recoded VICTIM to yes and no * Race of respondent Cross-tabulation

| | | | | RACE Race of Respondent | | | |
				1 Black	2 White	3 Other	Total
VICT_Y_N Recoded VICTIM to yes and no	2 No	Count		101	128	2	231
		Expected Count		99.2	129.0	2.8	231.0
		% within VICT_Y_N Recoded VICTIM to yes and no		43.7%	55.4%	0.9%	100.0%
		% within RACE Race of respondent		70.6%	68.8%	50.0%	69.4%
		% of Total		30.3%	38.4%	0.6%	69.4%
	1 Yes	Count		42	58	2	102
		Expected Count		43.8	57.0	1.2	102.0
		% within VICT_Y_N Recoded VICTIM to yes and no		41.2%	56.9%	2.0%	100.0%
		% within RACE Race of respondent		29.4%	31.2%	50.0%	30.6%
		% of Total		12.6%	17.4%	.6%	30.6%
Total		Count		143	186	4	333
		Expected Count		143.0	186.0	4.0	333.0
		% within VICT_Y_N Recoded VICTIM to yes and no		42.9%	55.9%	1.2%	100.0%
		% within RACE Race of respondent		100.0%	100.0%	100.0%	100.0%
		% of Total		42.9%	55.9%	1.2%	100.0%

Case Processing Summary

| | Cases | | | | | |
| | Valid | | Missing | | Total | |
	N	Percent	N	Percent	N	Percent
VICT_Y_N Recoded VICTIM to yes and no * RACE of respondent	333	96.0%	14	4.0%	347	100.0%

Chi-Square Tests

	Value	df	Asymp. Sig. (2-sided)
Pearson Chi-Square	.840[a]	2	.657
Likelihood Ratio	.785	2	.675
Linear-by-Linear Association	.350	1	.554
N of Valid Cases:	333		

[a]2 cells (33.3%) have expected count less than 5. The minimum expected count is 1.23.

Directional Measures

			Value	Asymp. Std. Error[a]	Approx. T[b]	Approx. Sig.
Nominal by Nominal	Lambda	Symmetric	.000	.008	.000	1.000
		VICT_Y_N Recoded VICTIM to yes and no Dependent	.000	.020	.000	1.000
		RACE Race of respondent Dependent	.000	.000	c	c
	Goodman and Kruskal tau	VICT_Y_N Recoded VICTIM to yes and no Dependent	.003	.006		.658[d]
		RACE Race of respondent Dependent	.000	.002		.872[d]
Ordinal by Ordinal	Somers' d	Symmetric	−.029	.054	−.528	.598
		VICT_Y_N Recoded VICTIM to yes and no Dependent	−.027	.050	−.528	.598
		RACE Race of respondent Dependent	−.031	.059	−.528	.598

[a]Not assuming the null hypothesis.
[b]Using the asymptotic standard error assuming the null hypothesis.
[c]Cannot be computed because the asymptotic standard error equals zero.
[d]Based on chi-square approximation.

Symmetric Measures

		Value	Asymp. Std. Error[a]	Approx. T[b]	Approx. Sig.
Ordinal by Ordinal	Kendall's tau-b	−.029	.055	−.528	.598
	Kendall's tau-c	−.027	.051	−.528	.598
	Gamma	−.062	.118	−.528	.598
	Spearman Correlation	−.029	.055	−.528	.598[c]
Interval by Interval	Pearson's r	−.032	.055	−.591	.555[c]
N of Valid Cases		333			

[a]Not assuming the null hypothesis.
[b]Using the asymptotic standard error assuming the null hypothesis.
[c]Based on normal approximation.

NEIGH_SF: Is your neighborhood safety changing? by COP_OK: In your neighborhood, how well do you think the police perform their duties? Cross-tabulation

NEIGH_SF: Is your neighborhood safety changing?			COP_OK: In your neighborhood, how well do you think the police perform their duties?				
			1 Very Well	2 Average	3 Below Average	4 Not At All	Total
3 Becoming less safe		Count	21	88	45	8	162
		Expected Count	29.2	87.2	39.6	5.9	162.0
		% within NEIGH_SF: Is your neighborhood safety changing?	13.0%	54.3%	27.8%	4.9%	100.0%
		% within COP_OK: In your neighborhood, how well do you think the police perform their duties?	35.6%	50.0%	56.3%	66.7%	49.5%
		% of Total	6.4%	26.9%	13.8%	2.4%	49.5%
2 Not changing		Count	21	58	33	2	114
		Expected Count	20.6	61.4	27.9	4.2	114.0
		% within NEIGH_SF: Is your neighborhood safety changing?	18.4%	50.9%	28.9%	1.8%	100.0%
		% within COP_OK: In your neighborhood, how well do you think the police perform their duties?	35.6%	33.0%	41.3%	16.7%	34.9%
		% of Total	6.4%	17.7%	10.1%	.6%	34.9%

(continues)

NEIGH_SF: Is your neighborhood safety changing? by COP_OK: In your neighborhood, how well do you think the police perform their duties? Cross-tabulation (*continued*)

NEIGH_SF: Is your neighborhood safety changing?		COP_OK: In your neighborhood, how well do you think the police perform their duties?				
		1 Very Well	2 Average	3 Below Average	4 Not At All	Total
1 Becoming safer	Count	17	30	2	2	51
	Expected Count	9.2	27.4	12.5	1.9	51.0
	% within NEIGH_SF: Is your neighborhood safety changing?	33.3%	58.8%	3.9%	3.9%	100.0%
	% within COP_OK: In your neighborhood, how well do you think the police perform their duties?	28.8%	17.0%	2.5%	16.7%	15.6%
	% of Total	5.2%	9.2%	.6%	.6%	15.65%
Total	Count	59	176	80	12	327
	Expected Count	59.0	176.0	80.0	12.0	327.0
	% within NEIGH_SF: Is your neighborhood safety changing?	18.0%	53.8%	24.5%	3.7%	100.0%
	% within COP_OK: In your neighborhood, how well do you think the police perform their duties?	100.0%	100.0%	100.0%	100.0%	100.0%
	% of Total	18.0%	53.8%	24.5%	3.7%	100.0%

Case Processing Summary

	Cases					
	Valid		Missing		Total	
	N	Percent	N	Percent	N	Percent
NEIGH_SF Is your neighborhood safety changing? * COP_OK In your neighborhood, how well do you think the police perform their duties?	327	94.2%	20	5.8%	347	100.0%

Directional Measures

			Value	Asymp. Std. Error[a]	Approx. T[b]	Approx. Sig.
Nominal by Nominal	Lambda	Symmetric	.000	.021	.000	1.000
		NEIGH_SF Is your neighborhood safety changing? Dependent	.000	.039	.000	1.000
		COP_OK In your neighborhood, how well do you think the police perform their duties? Dependent	.000	.000	c	c
	Goodman and Kruskal tau	NEIGH_SF Is your neighborhood safety changing? Dependent	.026	.010		.010[d]
		COP_OK In your neighborhood, how well do you think the police perform their duties? Dependent	.022	.007		.001[d]
Ordinal by Ordinal	Somers' d	Symmetric	.172	.048	3.565	.000
		NEIGH_SF Is your neighborhood safety changing? Dependent	.171	.048	3.565	.000
		COP_OK In your neighborhood, how well do you think the police perform their duties? Dependent	.173	.048	3.565	.000

[a]Not assuming the null hypothesis.
[b]Using the asymptotic standard error assuming the null hypothesis.
[c]Cannot be computed because the asymptotic standard error equals zero.
[d]Based on chi-square approximation.

Symmetric Measures

		Value	Asymp. Std. Error[a]	Approx. T[b]	Approx. Sig.
Ordinal by Ordinal	Kendall's tau-b	.172	.048	3.565	.000
	Kendall's tau-c	.158	.044	3.565	.000
	Gamma	.279	.076	3.565	.000
	Spearman Correlation	.192	.053	3.518	.000[c]
Interval by Interval	Pearson's r	.201	.053	3.692	.000[c]
N of Valid Cases:		327			

[a]Not assuming the null hypothesis.
[b]Using the asymptotic standard error assuming the null hypothesis.
[c]Based on normal approximation.

5. For each of the combinations of variables in the correlation matrix that follows, discuss the value of Pearson's r in terms of the strength of the relationship.

Correlations

		DELINQ: Delinquent Incidents for the Census Tract	BOARDED: Boarded-up Housing Units	OWNOCCUP: Owner-Occupied Housing Units	JUVENILE: Proportion of Juveniles	POPCHANG: Population Change, 1980–1990
DELINQ Delinquent Incidents for the Census Tract	Pearson Correlation	1	.482**	−.004	.339*	−.011
	Sig. (2-tailed)		.001	.981	.023	.941
	N	45	45	45	45	45
BOARDED Boarded-up housing units	Pearson Correlation	.482**	1	−.234	.029	−.222
	Sig. (2-tailed)	.001	—	.122	.848	.143
	N	45	45	45	45	45
OWNOCCUP Owner-occupied housing units	Pearson Correlation	−.004	−.234	1	.752**	.323*
	Sig. (2-tailed)	.981	.122	—	.000	.030
	N	45	45	45	45	45
JUVENILE Proportion of juveniles	Pearson Correlation	.339*	.029	.752**	1	.436**
	Sig. (2-tailed)	.023	.848	.000	—	.003
	N	45	45	45	45	45
POPCHANG Population change, 1980–1990	Pearson Correlation	−.011	−.222	.323*	.436**	1
	Sig. (2-tailed)	.941	.143	.030	.003	—
	N	45	45	45	45	45

**Correlation is significant at the 0.01 level (2-tailed).

*Correlation is significant at the 0.05 level (2-tailed).

References

Bravais, A. (1846). Analyse Mathématique sur les Probabilités des Erreurs de Situation d'un Point. *Mémoires Presentés par Divers Savants a l'Académie Royale des Sciences de l'Institut de France*, 9:255–332.

Deuchler, G. (1909). "Beiträge zur Erforschung der Reaktionsformen." *Psychologische Studien*, 4:353–430.

Fechner, G. T. (1897). *Kollektivmasslehre*. G. F. Lipps, ed. Leipzig: Wilhelm Engelmann.

Finley, J. P. (1884). Tornado predictions. *The American Meteorological Journal*. 85–88.

Galton, F. (1886). Family likeness in stature. *Proceedings of the Royal Society of London*, 40:42–73.

Galton, F. (1888). Co-relations and their measurement, chiefly from anthropometric data. *Proceedings of the Royal Society of London*, 45:135–145.

Goodman, L. A., and W. H. Kruskal. (1954). Measures of association for cross-classification. *Journal of the American Statistical Association*. 49:732–764.

Guttman, L. (1941). An outline of the statistical theory of prediction. In Horst, P. (ed.) *The Prediction of Personal Adjustment*. Bulletin 48, Social Science Research Council, New York: 253–318.

Kendall, M. G. (1938). A new measure of rank correlation. *Biometrika*, 30:81–93.

Kendall, M. G. (1948) (second edition 1955). *Rank Correlation Methods.* New York: Hafner.

Kruskal, W. H. (1958). Ordinal level measures of association. *American Statistical Association Journal*, 53:814–861.

Labovitz, S., and R. Hagedorn. (1971). *Introduction to Social Research.* New York: McGraw-Hill.

Lipps, G. F. (1905). Die bestimmung der abhängigkeit zwischen den merkmalen eines gegenstades. *Berichte über die Vorhandlungen der Königlich Sachsischen Gesellschaft der Wissenschaften zu Leipzig, Mathematisch-Physische Klasse*, 57:1–32.

Nunnally, J. (1978). *Psychometric Theory.* New York: McGraw-Hill.

Ozer, D. J. (1985). Correlations and the coefficient of determination. *Psychological Bulletin*, 97:305–315.

Peirce, C. S. (1884). The numerical measure of the success of predictions. *Science*, 4:453–454.

Pearson, K. (1896). Contributions to the Mathematical Theory of Evolution. III. Regression, Heredity and Panmixia. *Philosophical Transactions of the Royal Society of London.* 191(A):253–318.

Pearson, K. (1914–1930). *The Life, Letters and Labours of Francis Galton.* Cambridge: Cambridge University Press.

Roberts, D. M., and R. E. Kunst. (1990). A case against the continuing use of the spearman formula for rank-ordered correlation. *Psychological Reports*, 66:339–349.

Somers, R. H. (December 1962). A new asymmetric measure of association for ordinal variables. *American Sociological Review*, 27:799–811.

Spearman, C. (1904). The proof and measurement of association between two things. *Journal of American Psychology*, 15:72–101.

Stevens, S. S. (1951). Mathematics, measurement and psychophysics. In Stevens, S. S. (ed.) *Handbook of Experimental Psychology*, New York: Wiley.

Walker, J. T. (1995). Setting the Stage: Productivity of Doctoral Program Graduates. Presented at the Annual Meeting of the Southwestern Association of Criminal Justice; Houston, TX.

Yule, G. U. (1900). On the association of attributes in statistics: With illustrations from the material from the childhood society, etc. *Philosophical Transactions of the Royal Society of London*, Ser A, 194:257–319.

◼ Note

1. A symmetric Lambda can be calculated by combining the two asymmetric measures in a separate calculation.

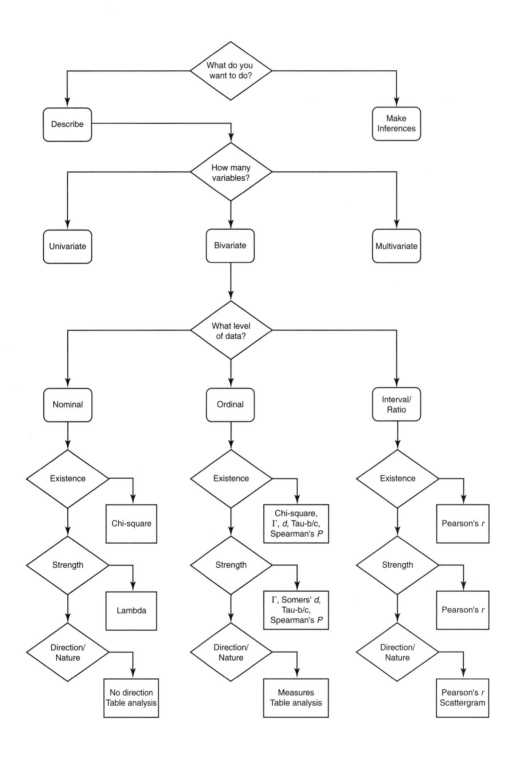

Measures of Direction and Nature of a Relationship

CHAPTER

10

Obviously, knowing the existence (significance) and strength of an association is important. Two additional pieces of information are also important to a full understanding of a bivariate relationship of ordinal, interval, and ratio level data. First, it is important to know the **direction** of the relationship. We introduced this concept in our discussion of concordant and discordant pairs in Chapter 9. Here you will learn how to recognize when high scores on one variable are associated with high scores on the other or when high scores are associated with low scores. The final piece of information concerns the **nature** of the relationship. This builds on the concepts discussed in Chapter 6 concerning the form and can indicate when the data is arranged in such a manner as to make the bivariate analyses erroneous.

■ Direction of the Association

Where both variables are at least ordinal level, the direction of the association can be determined. It is not possible to measure the direction of nominal level data directly because the ordering of the categories is arbitrary. It is possible to make statements about the general direction of the values in a nominal bivariate table, but conclusions about the direction of the relationship between the two variables exceeds what the data can produce.

Direction is established by the ratio of N_s to N_d pairs in ordinal level data and the way the scores are arranged in interval and ratio level data. If high values or ranks on one variable are associated with high values or ranks in the other variable, we have a **positive relationship**. If the high values of one variable are associated with low values of the other variable, we have a **negative relationship**.

The direction of a relationship can be established in one of two ways. A less precise method is to examine the distribution of values in a table or scatterplot. A more common and more accurate method is to use one of the ordinal or interval level measures of association discussed in Chapter 9, which also indicate the direction of the relationship.

Establishing Direction for Ordinal Level Data

Generally, examining the distribution of scores in a table is useful only for ordinal level variables. Establishing direction in this manner for nominal level data is suspect at best, and interval and ratio level data typically are not placed in tables because of the complexity and number of categories. So when considering the use of a table to examine direction, make sure the data is at least ordinal level and that there are not too many interval or ratio level categories.

The procedure for determining the direction of the relationship in a table larger than 2 × 2 begins by underlining the highest column percentage values in each row (see **Table 10-1**). This is essentially the same as calculating an Epsilon, and is an indicator of where the bulk of cross-tabulated values is in the distribution. Once those values are underlined, it is a simple matter of determining how those highest values are arranged in the table. This procedure requires that the table has been set up with the variables and categories arranged properly, as shown in Chapter 7.

To determine the direction of the value arrangement, draw a line through each of the highest values. If the line drawn through the underlined values generally moves from lower left to upper right (as is the case in Table 10-1), it is a positive association. If the line drawn through the underlined values generally moves from upper left to lower right, it is a negative association. If the line drawn through the values runs directly across the table, there is no association or the association is nonlinear or nonmonotonic (as would probably be supported by Chi-square and the measures of strength). The reason for this is in the ability of the independent variable to predict values of the dependent variable. Using the example of a PRE, if all of the highest values are in the same row, nothing has been gained with the additional information. In predicting criminality, as discussed in Chapter 9, if you were to guess whether 10 people were criminal or not and they were all criminal, you would be correct only when you predicted them to be criminal. If additional information was added, such as income, it would not make any difference and you would not reduce your errors in prediction with the additional information.

As discussed in Chapter 9, the stronger the association between two variables, the higher the percentage of the values that will lie directly on the diagonal. We also noted, however, that there are almost never perfect associations between variables, although the example in Table 10-1 shows a very strong positive relationship. Herein lies a problem for determining direction by examining the table. Most of the time, the highest values in each column will not lie on the diagonal. These values will probably vary above and below the diagonal, often alternating between lying above and below it. It is a judgment call, then, concerning the general direction. If most of the values are in a positive direction, the relationship is probably positive; alternatively, if most of the values are in a negative direction, the relationship is probably negative. If the relationship is curvilinear, as discussed below, the direction may change radically within the table. When this occurs, the only way to predict the direction accurately is with one of the measures of association that provides an indication of direction.

Typically, 2 × 2 tables contain only nominal level data, which does not permit a measure of direction. Ordinal or higher-level data typically contains more than two categories of data. Also, it is very difficult to interpret a 2 × 2 table, even with ordinal level data, because of the inter- and intracategory differences between the variables. On the limited occasions that there is a need to determine direction using a 2 × 2 table, the direction can be determined by placement of the values. First, determine the location of the largest frequency in the table (the mode). If the largest frequency is in the lower left-hand corner, the relationship is positive, indicating low values of the dependent variable and low values of the independent variable. If the largest frequency is in the upper

TABLE 10-1 Crosstab of Perception of Fear of Walking in the Daytime by Fear for Children to Play Alone in the Neighborhood Showing Direction of the Data

How much has fear of crime affected your decision to walk during the daytime? * How much has fear of crime affected supervision of your children? Cross-tabulation

			How much has fear of crime affected supervision of your children?			
			1 No Effect	2 Small Effect	3 Great Effect	Total
How much has fear of crime affected your decision to walk during the daytime?	3 Great Effect	Count	20	17	43	80
		Expected Count	39.4	19.1	21.5	80.0
		% within How much has fear of crime affected your decision to walk during the daytime?	25.0%	21.3%	53.8%	100.0%
		% within How much has fear of crime affected supervision of your children?	14.9%	26.2%	58.9%	29.4%
		% of Total	7.4%	6.3%	15.8%	29.4%
	2 Small Effect	Count	49	25	18	92
		Expected Count	45.3	22.0	24.7	92.0
		% within How much has fear of crime affected your decision to walk during the daytime?	53.3%	27.2%	19.6%	100.0%
		% within How much has fear of crime affected supervision of your children?	36.6%	38.5%	24.7%	33.8%
		% of Total	18.0%	9.2%	6.6%	33.8%
	1 No Effect	Count	65	23	12	100
		Expected Count	49.3	23.9	26.8	100.0
		% within How much has fear of crime affected your decision to walk during the daytime?	65.0%	23.0%	12.0%	100.0%
		% within How much has fear of crime affected supervision of your children?	48.5%	35.4%	16.4%	36.8%
		% of Total	23.9%	8.5%	4.4%	36.8%

(continues)

highest column % for this row

TABLE 10-1   Crosstab of Perception of Fear of Walking in the Daytime by Fear for Children to Play Alone in the Neighborhood Showing Direction of the Data (*continued*)

How much has fear of crime affected your decision to walk during the daytime? * How much has fear of crime affected supervision of your children? Cross-tabulation

| | | How much has fear of crime affected supervision of your children? | | | |
		1 No Effect	2 Small Effect	3 Great Effect	Total
Total	Count	134	65	73	272
	Expected Count	134.0	65.0	73.0	272.0
	% within How much has fear of crime affected your decision to walk during the daytime?	49.3%	23.9%	26.8%	100.0%
	% within How much has fear of crime affected supervision of your children?	100.0%	100.0%	100.0%	100.0%
	% of Total	49.3%	23.9%	26.8%	100.0%

right-hand corner, the relationship is still positive: high values of the dependent variable associated with high values of the independent variable. If the largest frequency is in the lower right-hand corner, the relationship is negative, indicating low values on the dependent variable and high values on the independent variable. If the largest frequency is in the upper left-hand corner, the relationship is still negative: high values of the dependent variable associated with low values of the independent variable.

Regardless of whether the data is ordinal, interval, or ratio level, perhaps the best way to examine direction is through one of the measures of association discussed in Chapter 9. All of the ordinal and higher measures of association are able to examine direction as part of their interpretation. As shown in **Table 10-2**, the ordinal level measures of association used with the data in Table 10-1 all show that there is a positive association, and a fairly strong one at that. This is indicated because these measures all have positive values. If these measures had negative values, the relationship would have been considered negative. This method has the additional advantage of allowing a determination of the direction of the data as part of the other bivariate analyses rather than having to underline portions of the table.

Establishing Direction for Interval and Ratio Level Data

Because interval and ratio level variables are not typically analyzed using bivariate tables, the determination of direction is actually often easier. Aside from using a scattergram where the data is plotted and a visual determination of direction is made, the only practical way to establish the direction of a relationship for interval and ratio level data is by using a measure of the strength of the association. As described above, Pearson's r would be used to determine the direction, with positive values of r denoting positive relationships and negative values, negative relationships. In **Table 10-3** the

TABLE 10-2 Bivariate Analyses of Fear of Walking in the Neighborhood During the Day by Fear for Children's Safety Within the Neighborhood

Symmetric Measures

		Value	Asymp. Std. Error[a]	Approx. T[b]	Approx. Sig.
Ordinal by Ordinal	Kendall's tau-b	.336	.050	6.592	.000
	Kendall's tau-c	.326	.049	6.592	.000
	Gamma	.496	.068	6.592	.000
	Spearman correlation	.373	.055	6.611	.000[c]
Interval by Interval	Pearson's *r*	.385	.055	6.853	.000[c]
N of Valid Cases		272			

[a]Not assuming the null hypothesis.
[b]Using the asymptotic standard error assuming the null hypothesis.
[c]Based on a normal approximation.

Directional Measures

			Value	Asymp. Std. Error[a]	Approx. T[b]	Approx. Sig.
Ordinal by Ordinal	Somers' *d*	Symmetric	.336	.050	6.592	.000
		How much has fear of crime affected your decision to walk during the daytime? Dependent	.345	.052	6.592	.000
		How much has fear of crime affected supervision of your children? Dependent	.327	.049	6.592	.000

[a]Not assuming the null hypothesis.
[b]Using the asymptotic standard error assuming the null hypothesis.

relationships between median rental value (MED RENT) and female heads of household (FEM HOUSE) and between median rental value (MED RENT) and delinquency (DELINQUENT) are negative, whereas the relationship between female heads of household (FEM HOUSE) and delinquency (DELINQUENT) is positive.

A final note concerning direction. It is often easy to confuse the concepts of direction, asymmetry, and one-tailed versus two-tailed tests. This is understandable because

TABLE 10-3 Correlation Matrix of Delinquency and Social Characteristics

	FEM HOUSE	MED RENT	DELINQUENT
FEM HOUSE	1.00		
MED RENT	−0.1351	1.00	
DELINQUENT	.6008**	−0.3891**	1.00

*$p < .05$ **$p < .01$ (2-Tailed)

these concepts show similar characteristics of data or represent similar statistical procedures. They are quantitatively different, however. The direction of a relationship concerns the data itself. One cannot alter the direction of the data without altering the data itself. It is, simply, how the scores of one variable are arranged in terms of the scores of another variable for the same case. Symmetric or asymmetric measures of association examine the strength of the relationship between two variables in a certain manner. Asymmetric measures examine, typically, the proportional reduction of errors that can be made when the categories or values of the independent variable are used to predict the category or value of the dependent variable. This is a decision made by the researcher in terms of which variable will be dependent and whether an asymmetric procedure will be used. For each pair of variables, two asymmetric measures (one for each variable as dependent) and one symmetric measure can be calculated. Which one of these to choose is up to the researcher. Finally, there is the issue of one-tailed versus two-tailed *t*-tests. Again, this has little to do with the data and more to do with what the researcher wants to examine. *One-tailed tests* look for specific change or a specific association between two (or more) variables, whereas *two-tailed tests* look for statistical significance but do not specify the change desired. For example, in research concerning weight change, a one-tailed test would examine only weight loss, but a two-tailed test would not care whether subjects gained or lost weight, only that they did not stay the same. In this way, one-tailed tests are similar to asymmetric measures except they examine the significance of a relationship rather than the strength.

■ Nature of the Association

The final characteristic of a bivariate analysis is the nature of the association. Essentially, the nature of an association is an examination of the degree of **linearity** or **monotonicity** of the relationship. This characteristic does not necessarily add to information concerning the bivariate relationship, but it does provide a great deal of information about how the measures of association can be influenced by the relationship between the variables.

Establishing the Nature of the Distribution for Nominal and Ordinal Level Data

The nature of the association for nominal and ordinal level variables can build on the steps taken to determine the direction of an association using a bivariate table. Whereas column percentages were used to determine direction, to determine the nature of an association, the highest row percentage values in each row of the table should be underlined, as shown in **Table 10-4**. Also differing from the method of determining direction, it is generally helpful to underline any row percentage values in a row that are close to the highest value. Although not a bright-line rule, generally, any value within 1% of the highest value should also be underlined.

The pattern of underlined values will indicate the nature of the association. If the underlined values are arranged generally along the diagonal, the association probably approaches linearity or monotonicity. If the values curve, form an arc or a U (parabola), or seem randomly distributed, the nature of the relationship is nonlinear or nonmonotonic.

TABLE 10-4   Crosstab of Perception of Fear of Walking in the Daytime by Fear for Children to Play Alone in the Neighborhood Showing the Nature of the Data

How much has fear of crime affected your decision to walk during the daytime? * How much has fear of crime affected supervision of your children? Cross-tabulation

			How much has fear of crime affected supervision of your children?			
			1 **No Effect**	**2** **Small Effect**	**3** **Great Effect**	**Total**
How much has fear of crime affected your decision to walk during the daytime?	3 Great Effect	Count	20	17	43	80
		Expected Count	39.4	19.1	21.5	80.0
		% within How much has fear of crime affected your decision to walk during the daytime?	25.0%	21.3%	53.8%	100.0%
		% within How much has fear of crime affected supervision of your children?	14.9%	26.2%	58.9%	29.4%
		% of Total	7.4%	6.3%	15.8%	29.4%
	2 Small Effect	Count	49	25	18	92
		Expected Count	45.3	22.0	24.7	92.0
		% within How much has fear of crime affected your decision to walk during the daytime?	53.3%	27.2%	19.6%	100.0%
		% within How much has fear of crime affected supervision of your children?	36.6%	38.5%	24.7%	33.8%
		% of Total	18.0%	9.2%	6.6%	33.8%
	1 No Effect	Count	65	23	12	100
		Expected Count	49.3	23.9	26.8	100.0
		% within How much has fear of crime affected your decision to walk during the daytime?	65.0%	23.0%	12.0%	100.0%
		% within How much has fear of crime affected supervision of your children?	48.5%	35.4%	16.4%	36.8%
		% of Total	23.9%	8.5%	4.4%	36.8%

(continues)

TABLE 10-4   Crosstab of Perception of Fear of Walking in the Daytime by Fear for Children to Play Alone in the Neighborhood Showing the Nature of the Data (*continued*)

How much has fear of crime affected your decision to walk during the daytime? * How much has fear of crime affected supervision of your children? Cross-tabulation

| | | How much has fear of crime affected supervision of your children? | | | |
		1 No Effect	2 Small Effect	3 Great Effect	Total
Total	Count	134	65	73	272
	Expected Count	134.0	65.0	73.0	272.0
	% within How much has fear of crime affected your decision to walk during the daytime?	49.3%	23.9%	26.8%	100.0%
	% within How much has fear of crime affected supervision of your children?	100.0%	100.0%	100.0%	100.0%
	% of Total	49.3%	23.9%	26.8%	100.0%

For nominal and ordinal level data, it is generally more accepted to speak of the nature in terms of monotonicity rather than linearity. Monotonicity addresses the general nature (and direction) of a relationship. It can show whether a variable remained the same or shifted in a certain direction with change in a second variable. A **monotone-increasing relationship** is represented by a positive direction and a generally straight line in the table (for an example, review Tables 9-4 and 10-4). A **monotone-decreasing relationship** is represented by a negative direction and a generally straight line in the table (a stair step in the table in a negative direction). A relationship with multiple modes or one displaying serious departures from a straight line in a table is generally considered nonmonotonic. Monotonicity is especially important when examining rectangular tables. It is not possible to establish perfect linearity in these tables because of the way the cells are arranged. Monotonicity can be established, however, as indicated by a stair step of values. Although it is difficult to tell in only a 3 × 3 table (as is often the case with these types of analyses), the data in Table 10-4 seem to be fairly monotonic (monotone increasing). In Table 10-4, the values increase in a fairly straight, stair step from the bottom left of the table to the top right. The stair step is what indicates monotonicity.

Monotonicity addresses the issue of how the categories or ranks of the two variables stand in relation to each other rather than the linear nature of the variables. Attempting to address the linear nature of the variables is beyond what nominal or ordinal data can support. Linear relationships are always monotonic, but there are other relationships, many more closely matching nominal and ordinal level data, that are monotonic but not linear. To say that a relationship is linear implies that there are sufficiently constant intervals in the data to address a shift in values of a dependent variable given a certain amount of shift in the independent variable. Nominal and ordinal level data does not have sufficiently constant intervals to be able to make these assertions.

A measure of the nature of an association can also be obtained with univariate analyses. The degrees of skewness and kurtosis are very helpful in addressing the nature of an association. Recall that skewness or kurtosis values greater than ±1 indicate nonnormal (skewed or kurtose) data. A nonnormal table (distribution) will show up as a random pattern when the row values are underlined. Patterns that show as nonmonotonic in a table are probably skewed, kurtose, or both. This can be displayed as anywhere from a slight departure from monotonicity to a table where the row underlines look as if they were drawn at random.

Monotonicity and linearity for nominal and ordinal level data are important for bivariate analyses because some bivariate measures of association are heavily influenced by skewed or kurtose distributions. For example, Lambda is heavily influenced by skewness in the dependent variable and should be interpreted with caution for those variables that are highly skewed.

Establishing the Nature of the Distribution for Interval and Ratio Level Data

Establishing the nature of an association for interval and ratio level data is probably more crucial to the analysis and interpretation than with nominal and ordinal level data. The reason is that the primary bivariate measure of association for interval and ratio level data is Pearson's r, which explicitly assumes a linear relationship between the two variables.

At the most basic level, the nature of an interval level association can be examined using a scattergram of the data and overlaying a best-fitting line, as discussed in Chapter 3. Where this becomes important for examining the nature of an association is in what the line looks like and where it is on the scattergram. Remember that a regression line extending from lower left to upper right or upper left to lower right at a 45° angle is indicative of a perfect correlation (as shown in **Figure 10-1**). Also remember that a line extending parallel to the X axis is indicative of no association. These lines are not always straight, however. A best-fitting line might have to take a number of curves and bends to be the best fitting for a distribution. As shown on the right in **Figure 10-2**, a best fitting-line might even have to make an arc to fit the data. In reality, the best-fitting line would not bend or curve, it would simply shift to horizontal to create the best fit. This would result in a correlation (measure of strength) of zero.

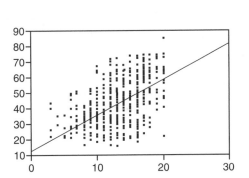

Figure 10-1   Scattergram of a Strong Positive Correlation

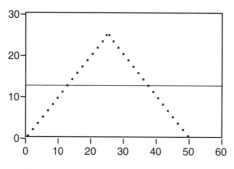
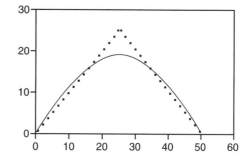

Figure 10-2   Scattergrams of Curvilinear Data

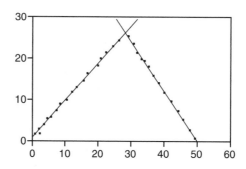

Figure 10-3   Positive and Negative Best-Fitting Lines for Curvilinear Data

A regression line made to fit the scattergram shown in Figure 10-2 would show no association, as shown on the left in Figure 10-2. The correlation or regression would probably also show no association between the two variables. This is because the straight line that would best fit the data would have to be horizontal. As shown in **Figure 10-3**, though, this data set could actually be represented by two perfect association lines: one positive and one negative. These offset to produce a zero association, but the two variables are actually highly correlated, just in a curvilinear nature.

These scattergrams demonstrate why it is important to know the nature of interval and ratio level data. Simply relying on Pearson's *r* in this case would produce an erroneous interpretation. Pearson's *r* would no doubt be 0 and the interpretation would be no association. Understanding the nature of the data, however, allows an interpretation that more accurately represents the true association of the data. Knowing that the data was curvilinear would also facilitate more accurate analyses because it would alert the researcher that the data should be transformed. A simple exponential transformation of the data shown in Figure 10-3 would probably produce a positive linear association that would approach a Pearson's *r* of 1.00.

■ Conclusions

In this chapter, we introduced methods of testing the direction and nature of a relationship. The discussion builds on that in earlier chapters and completes the analysis of two variables. Knowing the existence, strength, direction, and nature of the relationship between two variables provides a fairly full understanding of their interaction. Like the univariate analyses discussed in earlier chapters, the analyses performed here can complete a research project or can form a foundation for further analyses in terms of the multivariate (more than two variables) analyses discussed in the chapters to follow. Understanding individual bivariate relationships is essential to a thorough understanding of the interaction of two or more variables. It is to the discussion of multivariate analysis that we turn next.

■ Key Terms

best-fitting line
direction
linearity
monotone-decreasing
monotone-increasing

monotonicity
nature
negative relationship
positive relationship

■ Exercises

For the data and tables that follow (which are the same as those used in the exercises for Chapter 9):
1. Determine the direction of the association from the table.
2. Determine the direction of the association from the univariate measures.
3. Determine the nature of the association from the table.
4. Determine the nature of the association from the univariate measures.
5. Examine and discuss the nature of the association for each of the data sets.

VICT_Y_N Recoded VICTIM to yes and no * Race of respondent Cross-tabulation

			RACE Race of Respondent			
			1 Black	2 White	3 Other	Total
VICT_Y_N Recoded VICTIM to yes and no	2 No	Count	101	128	2	231
		Expected Count	99.2	129.0	2.8	231.0
		% within VICT_Y_N Recoded VICTIM to yes and no	43.7%	55.4%	0.9%	100.0%
		% within RACE Race of respondent	70.6%	68.8%	50.0%	69.4%
		% of Total	30.3%	38.4%	0.6%	69.4%
	1 Yes	Count	42	58	2	102
		Expected Count	43.8	57.0	1.2	102.0
		% within VICT_Y_N Recoded VICTIM to yes and no	41.2%	56.9%	2.0%	100.0%
		% within RACE Race of respondent	29.4%	31.2%	50.0%	30.6%
		% of Total	12.6%	17.4%	.6%	30.6%

			RACE Race of Respondent			
			1 Black	2 White	3 Other	Total
Total		Count	143	186	4	333
		Expected Count	143.0	186.0	4.0	333.0
		% within VICT_Y_N Recoded VICTIM to yes and no	42.9%	55.9%	1.2%	100.0%
		% within RACE Race of respondent	100.0%	100.0%	100.0%	100.0%
		% of Total	42.9%	55.9%	1.2%	100.0%

NEIGH_SF: Is your neighborhood safety changing? by COP_OK: In your neighborhood, how well do you think the police perform their duties? Cross-tabulation

			COP_OK: In your neighborhood, how well do you think the police perform their duties?				
NEIGH_SF: Is your neighborhood safety changing?			1 Very Well	2 Average	3 Below Average	4 Not At All	Total
3	Becoming less safe	Count	21	88	45	8	162
		Expected Count	29.2	87.2	39.6	5.9	162.0
		% within NEIGH_SF: Is your neighborhood safety changing?	13.0%	54.3%	27.8%	4.9%	100.0%
		% within COP_OK: In your neighborhood, how well do you think the police perform their duties?	35.6%	50.0%	56.3%	66.7%	49.5%
		% of Total	6.4%	26.9%	13.8%	2.4%	49.5%
2	Not changing	Count	21	58	33	2	114
		Expected Count	20.6	61.4	27.9	4.2	114.0
		% within NEIGH_SF: Is your neighborhood safety changing?	18.4%	50.9%	28.9%	1.8%	100.0%
		% within COP_OK: In your neighborhood, how well do you think the police perform their duties?	35.6%	33.0%	41.3%	16.7%	34.9%
		% of Total	6.4%	17.7%	10.1%	.6%	34.9%

(continues)

NEIGH_SF: Is your neighborhood safety changing? by COP_OK: In your neighborhood, how well do you think the police perform their duties? (*continued*)

NEIGH_SF: Is your neighborhood safety changing?		COP_OK: In your neighborhood, how well do you think the police perform their duties?				
		1 Very Well	2 Average	3 Below Average	4 Not At All	Total
1 Becoming safer	Count	17	30	2	2	51
	Expected Count	9.2	27.4	12.5	1.9	51.0
	% within NEIGH_SF: Is your neighborhood safety changing?	33.3%	58.8%	3.9%	3.9%	100.0%
	% within COP_OK: In your neighborhood, how well do you think the police perform their duties?	28.8%	17.0%	2.5%	16.7%	15.6%
	% of Total	5.2%	9.2%	.6%	.6%	15.65%
Total	Count	59	176	80	12	327
	Expected Count	59.0	176.0	80.0	12.0	327.0
	% within NEIGH_SF: Is your neighborhood safety changing?	18.0%	53.8%	24.5%	3.7%	100.0%
	% within COP_OK: In your neighborhood, how well do you think the police perform their duties?	100.0%	100.0%	100.0%	100.0%	100.0%
	% of Total	18.0%	53.8%	24.5%	3.7%	100.0%

Statistics

	Is your neighborhood safety changing?	In your neighborhood, how well do you think the police perform their duties?	Race of respondent	Recoded VICTIM to yes and no
Number valid	336	335	333	347
Number missing	11	12	14	0
Mean	2.34	2.13	1.58	1.69
Median	2.00	2.00	2.00	2.00
Mode	3	2	2	2
Std. deviation	0.731	0.756	0.518	0.461
Variance	0.534	0.571	0.268	0.213
Skewness	−0.618	0.332	−0.073	−0.848
Std. error of skewness	0.133	0.133	0.134	0.131
Kurtosis	−0.905	−0.129	−1.393	−1.288
Std. error of kurtosis	0.265	0.266	0.266	0.261
Range	2	3	2	1

Correlations

		DELINQ: Delinquent Incidents for the Census Tract	BOARDED: Boarded-up Housing Units	OWNOCCUP: Owner-Occupied Housing Units	JUVENILE: Proportion of Juveniles	POPCHANG: Population Change, 1980–1990
DELINQ Delinquent Incidents for the Census Tract	Pearson Correlation	1	.482**	−.004	.339*	−.011
	Sig. (2-tailed)		.001	.981	.023	.941
	N	45	45	45	45	45
BOARDED Boarded-up housing units	Pearson Correlation	.482**	1	−.234	.029	−.222
	Sig. (2-tailed)	.001	—	.122	.848	.143
	N	45	45	45	45	45
OWNOCCUP Owner-occupied housing units	Pearson Correlation	−.004	−.234	1	.752**	.323*
	Sig. (2-tailed)	.981	.122	—	.000	.030
	N	45	45	45	45	45
JUVENILE Proportion of juveniles	Pearson Correlation	.339*	.029	.752**	1	.436**
	Sig. (2-tailed)	.023	.848	.000	—	.003
	N	45	45	45	45	45
POPCHANG Population change, 1980–1990	Pearson Correlation	−.011	−.222	.323*	.436**	1
	Sig. (2-tailed)	.941	.143	.030	.003	—
	N	45	45	45	45	45

**Correlation is significant at the 0.01 level (2-tailed).
*Correlation is significant at the 0.05 level (2-tailed).

Statistics

	Delinquent Incidents for the Census Tract	Boarded-up Housing Units	Owner-Occupied Housing Units	Proportion of Juveniles	Population Change, 1980–1990
Number valid	45	45	45	45	45
Number missing	0	0	0	0	0
Mean	39.02	17.31	792.16	688.29	5.3563
Median	28.00	3.00	727.00	643.00	−5.7374
Mode	9[a]	0	1046	476	−60.00
Std. deviation	40.194	29.602	752.134	506.344	51.27772
Variance	1615.568	876.265	565,705.453	256,383.846	2629.40449
Skewness	1.465	2.874	1.868	1.321	2.084
Std. error of skewness	0.354	0.354	0.354	0.354	0.354
Kurtosis	1.814	10.104	5.010	3.625	4.380
Std. error of kurtosis	0.695	0.695	0.695	0.695	0.695
Range	162	154	3810	2613	250.37

[a]Multiple modes exist. The smallest value is shown.

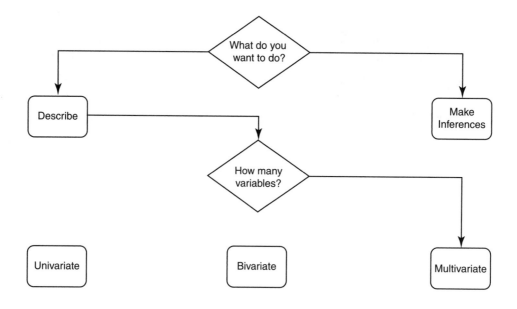

Introduction to Multivariate Statistics

When Two Variables Just Aren't Enough

In the previous few chapters, we dealt exclusively with two variables in research. As you well know, however, the world is often much too complex to be able to examine only two variables at a time. We must be able to examine several, and often many, variables at a time to get a true picture of something as complex as human behavior or crime. This is where **multivariate analyses**, analyses of more than two variables, come into play. The statistical procedures discussed in this section of the book, as well as many others, give researchers the ability to examine a dependent variable using models that can include all of the independent variables that appear to be associated with the dependent variable. With this new ability, however, come new problems for the analysis procedures and new issues that must be dealt with theoretically and methodologically.

In this chapter, we introduce theoretical and methodological issues for addressing multivariate analyses, and provide an overview of how they may be used in the research process. Although many of the issues addressed in this chapter are more methodological than statistical, they are important in multivariate analysis because they can strongly influence the analysis or interpretation of findings.

Interaction Among Variables

Analyzing the interaction among variables is the specialty of multivariate analysis techniques. Some of the more popular statistical procedures for dealing with interaction are discussed in the following chapters: ordinary least squares (OLS) regression, logistic regression, and regressions designed especially for limited dependent variables, factor analysis, and structural equation modeling (SEM). Although there are other, higher-level statistical analyses (hierarchical linear modeling and various time series analyses) that can be used to examine the relationship between multiple variables, the discussion here is limited to the various regression techniques and structural equation modeling, as these are the most popular in criminological research.

In its most general sense, interaction can be any change in one variable that produces, or is associated with, change in another variable. This is the essence of what was discussed in Chapters 8–10. When most researchers speak of interaction among variables, however, they typically are speaking of the influence that two or more variables have on the dependent variable. This is the combined effect, cause, or association of all of the independent variables on the dependent variable.

If two variables explain some variation in a dependent variable, they can be said to interact with one another and with the dependent variable. These two variables have at least an **additive effect.** That is, they each explain a part of the variation in the dependent variable, much as you would explain an entire pie chart simply by explaining what each part is and how much it contributes to the total. This interaction could be discovered through the bivariate analyses discussed in Chapter 10. For example, in **Table 11-1**, both of the independent variables (boarded-up homes and owner occupancy) explain part of the variation in the dependent variable (delinquency). Here, boarded up homes explain 23% of the variation in delinquency (0.48^2), and owner occupancy explains about 5% (0.24^2). Together, these two variables can explain about 28% of the variation in delinquency.

Interactions can explain more of the variation in the dependent variable than the simple sum of their separate effects, however. Variables should come together in a **synergy effect**, where the sum of their contribution is greater than their individual effects (Cohen, 1978). This is most typically how interaction of variables is discussed in research. An example here might be patrol characteristics and an officer's probability of making an arrest for a violent crime. We can say with some certainty that the area of town to which an officer is assigned has some influence on the probability that he or she will make an arrest; higher crime, more violence, and similar factors vary positively with violent crime arrests. It can also be said with some certainty that an officer's shift has some influence on his or her probability of making an arrest for a violent crime; night shifts typically have a higher incidence of such crimes. If these are combined, however, they may explain much more of the variation in arrests for violent crimes than possible using the variables separately. In this case, midnight shifts in a "bad" part of town may be highly correlated with violent crime arrests. It is this synergetic effect that multivariate analysis seeks to examine. This interaction effect occurred with the data in **Figure 11-1**. Although it is not shown, an SPSS printout of the regression analysis of these variables indicates the R^2 increases from 0.28 to 0.31 (a significant increase) when a variable representing the interaction between these two variables is added. This shows the synergetic effect of the interaction of the two variables beyond their additive value.

It is also important to understand the various types of interactions. For example, with three independent variables—IV_1, IV_2, and IV_3—there are three *zero-order correlations* that can be derived from the effect of these variables on the dependent variable.

TABLE 11-1   Correlation Matrix Showing Additive Effect of Independent Variables

	Delinquent	Boarded-up Homes	Owner-Occupied Homes
Delinquent	1.00		
Boarded-up homes	0.48**	1.00	
Owner-occupied homes	−0.24	0.00	1.00

**$p < .01$

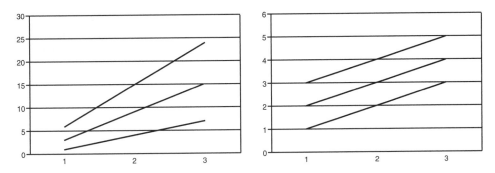

Figure 11-1 Interaction of Variables

These are the bivariate measures of association discussed in Chapters 8–10: the association between each of the independent variables and the dependent variable. These are

$$IV_1 \rightarrow D \qquad IV_2 \rightarrow D \qquad IV_3 \rightarrow D$$

There are also three *first-order interactions* that can be derived from these three variables (IV_{12}, IV_{23}, and IV_{13}):

$$IV_1 \cdot IV_2 \rightarrow D \qquad IV_2 \cdot IV_3 \rightarrow D \qquad IV_1 \cdot IV_3 \rightarrow D$$

Finally, there is a *second-order interaction* that can be formed from these three variables (IV_{123}):

$$IV_{123} \rightarrow D$$

All four of these interactions are important to the model, and all should be explored in any statistical analysis that is undertaken.

The interaction of variables can be shown graphically, although they typically are not graphed. Examining interaction graphically involves plotting the change in each independent variable that is associated with change in the dependent variable and also change in the independent variables' relation to each other. If there is an association between the independent variables and the dependent variable but no interaction among the independent variables, the lines will show some change; that is, they will not be parallel to the X axis. They will be parallel to each other, however, showing that each contributes only its individual amount to explaining variation in the dependent variable (on the right in Figure 11-1). If, however, the three variables interact (have a synergetic effect) on the dependent variable, the bivariate lines will be unequal. The reason is that the independent variables are varying in relation to changes in the category of the dependent variable and also varying in relation to changes in the category of the other independent variables (as shown on the left in Figure 11-1). There is more information that can be obtained from these graphs, such as whether the interaction is ordinal or not; but these types of graphs are not particularly useful to an explanation of the relationship, so printouts of the analysis and discussion of the findings usually takes the place of such graphs.

Beyond using graphs, there are several ways to determine the value of an interaction, or whether or not the interaction contributes to the variation in the dependent variable. All of the procedures in Chapter 12, along with others, are useful. Regression analysis is particularly useful in this respect. As discussed in Chapter 12, the value of R^2 expresses the contribution, or added contribution, of a variable in explaining variation in the dependent variable. Suppose, for example, you entered the three variables described above (IV_1, IV_2, and then IV_3) into a regression model and you determined that each contributes significantly to explaining variation in the dependent variable. By adding the interaction variables (IV_{12}, IV_{13}, and IV_{23}) to the model, you can now determine if each of these makes a contribution to the model. If there is a statistically significant increase in the R^2, the interaction variable is useful; otherwise, it may need to be dropped.[1] It is also necessary to test the second-order interaction variable (IV_{123}). As with the first-order interaction variables, if there is a statistically significant increase in the R^2, the interaction variable is useful; otherwise, it may need to be dropped.

A final note: Be careful of interactions in your research. As discussed in Chapter 2, there are many ways that confounding variables can enter into research. If those confounding variables are interacting with an independent variable, the dependent variable, or both, you may get a result that is unexpected or erroneous. For example, by looking at the individual properties of both hydrogen and oxygen, you would not think they could produce water when combined. Although interaction among variables is a part of research, and even desired when it can be identified and accounted for properly, unintended interaction can be a problem.

■ Causation

Interaction among multiple variables promotes another characteristic of multivariate analysis: causation. In Chapter 9, we warned against making inferences of causality concerning bivariate association. It is even more tempting, but no more appropriate, to make causal inferences when a number of variables are used in a multivariate model. Multivariate analysis, however, does lend itself to the possibility of drawing causal inferences between variables if the causality is supported methodologically or theoretically.

Causal ordering is the reason many researchers undertake multivariate analysis, although it is possible to establish causality in a bivariate analysis. Establishing causality requires certain steps, including establishing association among the variables, determining temporal ordering, and eliminating confounding variables (discussed in the next section).

Association

As discussed earlier in the book, association among variables is an important element of research and analysis. There must be some relationship between the variables, or nothing is gained from the research. The first step in attempting to establish causality is determining if an association exists. Although not infallible, a general rule is that if two variables are not associated, one cannot be the cause of the other. The procedures used in earlier chapters to examine a bivariate relationship, and the univariate measures upon which those bivariate analyses are based, then, should be the first step in any attempt to establish causality.

There are two components to association that should be stressed. These are the magnitude and consistency of the association. The magnitude of an association is the existence and strength of the association, as established in Chapters 8 and 9. *Consistency* (often called *reliability*) concerns the fact that an association remains constant from one study to the next and under a variety of conditions. If an association has sufficient magnitude (it has a substantial and statistically significant association) and it is consistent over time, it lends credence to the argument that some causal nexus may exist.

Association alone is rarely the goal of research, however; there are many associations in criminology and criminal justice. For example, it has been known for more than 75 years that there is an association between crime and certain urban characteristics, such as high population turnover, physically deteriorated neighborhoods, and a proportionally small number of homeowners. In this example, there is certainly some contribution that is made by the urban characteristics in explaining variation in criminal behavior. The associations have endured too long and over too many different cities to think otherwise. There may even be some partial causation that can be established for these variables based on their enduring association. The elusive holy grail, however, has been to find a causal relationship between those characteristics and crime. Establishing true causality, however, is complex and requires a number of extra steps in the examination.

Temporal Ordering

The second major step in establishing causality is examining the **temporal ordering** of the variables. Temporal ordering is the relationship in time between two variables. Temporal ordering examines the variables at time 1 and any change that has occurred at time 2. The change, difference, or relationship between time 1 and time 2 is the temporal ordering. Temporal ordering comes into play in causal modeling because the independent, causal variable must either occur first or change prior to the dependent variable changing. Remember that research typically examines change in the independent variable that is producing or related to change in the dependent variable. If the independent variable does not change and the dependent variable does, it makes it difficult to establish a relationship based on that change. Additionally, if the dependent variable changes before the independent variable changes, it would be difficult to state that the change in the independent variable produced the change in the dependent variable. If the independent variable changes first, however, and then the dependent variable changes, there is at least some probable cause to further examine the relationship to determine whether the variables just happened to change in this order or whether the change in one produced change in the other.

Typically, four methods are used to establish temporal ordering: observation, logic, theory, and data analysis. None of these in themselves will establish causality, even if they do establish temporal ordering. This is an additional step that can be used, however, in causation analysis. Also, these four methods may be used in combination to strengthen a temporal ordering or causation argument.

The first method of establishing temporal ordering is *observation*. This process is simply examining the independent and dependent variables over time to determine if any change is occurring and when change in each occurred. If there is change in the variables and the independent variable consistently changes prior to the dependent variable, temporal ordering may be established. The easiest way to establish temporal ordering through observation is with experiments. In an experiment in a controlled environment, a researcher can manipulate (change) the independent variable and watch for any change in the dependent variable. This also makes it easier to measure the amount of change in the independent variable associated with change in the dependent variable. Unfortunately for criminal justice, experiments in a controlled environment are almost impossible. Criminal justice research occurs in the messiness of everyday life and human actions. It is sometimes possible to approximate experimental conditions, but criminal justice researchers virtually never have the control of the environment needed to manipulate independent variables.

Given the inability to control the environment and change in the independent variable, criminal justice research often turns to a second method of establishing temporal ordering: *logic.* Observation alone makes it difficult to establish temporal ordering in criminal justice research because data is often gathered after events in both variables have occurred. It is usually very difficult, therefore, to determine which variable changed first (temporally) or which variable may produce change in the other through observation alone. For example, one of the primary criminological theories is control theory. One element of this theory is that breaks in parental bonds or attachments may "free" a youth to commit crime. It is easy to show, however, that criminal behavior could cause a breakdown in the parental relationship. What is the temporal ordering in this instance? Here, observation will probably not be useful because data will probably be gathered on youths who have already committed crimes; thus both events have occurred. In such cases, logic can help sort out the temporal ordering by rejecting the reverse temporal ordering. A simple example of establishing temporal ordering through logic is the relationship between arrest and prison. A person must first be arrested before going to prison. Furthermore, the reverse cannot be true—that a person can go to prison without being arrested (OK, perhaps on a technical parole violation; but work with us here). By rejecting one possible temporal ordering, the other possible temporal ordering may be established. The example above between parental attachment and crime is more difficult to establish using logic. In this case, a solid theoretical model and appropriate statistical analysis would be helpful.

The third method of establishing temporal ordering is through *theory.* As discussed in Chapter 1, all research should have a carefully crafted theoretical model that flows from the primary question to the research questions to the concepts to the variables and onto the data. Using a theoretical and methodological model involving clearly spelled out assumptions, the temporal ordering of variables can be hypothesized. Statistical analysis confirming the theoretical model supports an argument for the hypothesized temporal ordering; that is, if the theoretical model is supported, it can be reasoned that the hypothesized temporal ordering would also be supported.

The final method of determining temporal ordering is *data analysis.* This is accomplished through a technique called *cross-lagging.* In cross-lagging, two variables are compared by first taking the initial value of one variable and pairing it with the value of another variable at a later time. The sequence is then reversed. The ordering that shows the highest association (the association that has the highest strength value, as discussed in Chapter 9) is assumed to be the situation showing the proper temporal/causal model. The same can also be done using bivariate analysis and an asymmetric measure of association; however, less support can be derived for a bivariate measure of association because it is typically a cross-section technique, in which data for both variables is gathered at the same time rather than over a period of time.

Elimination of Confounding Variables

The final step in dealing with causality is eliminating *confounding variables*, often called *rival causal factors* when used in this context. As discussed in Chapter 2, confounding variables are often present in research. Unknown confounding variables can seriously influence the explanation of a causal relationship, both theoretically and in the interpretation of the analysis, because you do not have all the variables that are important to the model identified for examination. Confounding variables can be dealt with either theoretically or through analysis. Eliminating rival causal factors is important because the theory, explanation, and interpretation of a causal relationship require not only that the independent variable changes first but also that it causes changes in the dependent variable. If there is another variable that is causing the change, even if the independent variable is causing change in the confounding variable, which is then causing change in the dependent variable, the relationship is not adequately explained and the model is not complete. For example, it is a common hypothesis in criminology and criminal justice research that broken homes may lead to delinquency, but this model is far too simplistic to be of any value. Certainly, broken homes do not lead to delinquency; there are many single-parent homes that have wonderful, successful children. To justify that broken homes lead to delinquency, a more complete model must be developed that includes characteristics of broken homes that may play some role in increased delinquency, such as a breakdown of parental attachment, lack of supervision, or other factors. Until these rival causal factors are either included in the model because they are truly independent variables or they are eliminated from the model through theory and analysis, the model is not properly specified.

In practice, eliminating rival causal factors is much more difficult than in the confines of a book. Fully accounting for and controlling confounding variables is rarely possible in the social sciences because only in experiments where it is possible to strictly control the process is it possible to account for all confounding variables. In the world of human behavior, rival causal factors abound, and it is doubtful that criminal justice researchers will ever be able to specify a complete model accounting for all the factors in an examination of criminal behavior.

Attempts to include or eliminate all rival causal factors have lead to a current trend in criminal justice and criminology research of using a procedure called *structural equation modeling*. This is an advanced technique using path analysis and confirmatory factor analysis to examine multiple relationships among many variables. Common statistical programs for this analysis are LISREL and AMOS. The complexity of this type of analysis is shown in **Figure 11-2**, which is a structural equation model of the effect on delinquent behavior of delinquent peers, peer reactions to a person's behavior, and a person's beliefs. This model shows how complex multivariate analysis can be when attempting to account for a number of rival causal factors.

The steps listed in this portion of the chapter are important in establishing any causal relationship among variables. Even if all of these measures are accomplished, however, it is no guarantee that causation can be established. These are simply means to support an argument of causation, not the ends themselves.

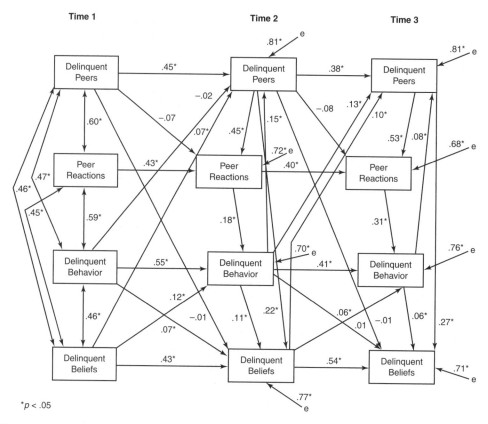

Figure 11-2 LISREL model of relationship of delinquent pers, peer reactions, delinquent beliefs, and delinquent behavior.

Source: Reprinted with permission from Thornberry et al., © 1994, American Society of Criminology.

■ Additional Concepts in Multivariate Analysis

Discussed briefly here are several additional concepts that are important in multivariate analysis. Two of these concepts, robustness and error, are applicable to bivariate analysis but are discussed more often with multivariate analysis. A third concept, parsimony, is directly applicable to multivariate analysis.

Robustness

Robustness of a statistical procedure is important when one wishes to use analyses appropriate for interval and ratio level data. As discussed in Chapter 10, interval level analyses require the data to be interval level and normally distributed. In certain cases, though, there is the exception that dichotomized nominal data or ordinal data of a sufficiently large scale can be used with interval level analyses. In these cases, the data should be as close to a normal distribution as possible. How close the data is to interval level and how close the data is to a normal distribution are often subjectively determined—one person might say that the data is appropriate for particular analyses, whereas another would say that the data is not appropriate. Robustness helps to determine how much deviation from normality the data can be before the statistical procedure becomes unusable.

The word *robust* has been used since at least the Middle Ages to refer to someone who was strong, yet often crude (Stigler, 1973). Since the 1800s, mathematicians have wrestled with the degree of violation of assumptions that a particular statistical procedure can withstand and still remain effective. G. E. P. Box ultimately applied the term *robust* to statistics in his 1953 work on use of the variance for nonnormal data. Since then, the term has been used to refer to a statistical procedure that is able to withstand violations of its basic assumptions. Robustness is the ability of a statistical procedure to provide accurate results despite the fact that its assumptions have been violated. Typically, this translates to an interval level statistical procedure that is able to work with less than truly interval level data or with data that is not normally distributed.

Often associated with a robust statistical procedure is the concept of **resistance**. A statistical procedure is said to be resistant if its value is not heavily influenced by outliers or changes in the data. As discussed in Chapter 4, outliers often heavily affect certain statistics, such as the mean and statistics based on the mean; even a few outliers can produce a mean that is not consistent with the data being analyzed. Alternatively, the median is not influenced by outliers because it considers only the distribution and not individual values. The median, then, is a resistant estimator (in the case of the measure of central tendency), whereas the mean is not. Resistant estimators are not heavily influenced by changes in the data, either by large changes in a small number of the data points, such as a few outliers, or by small changes in a large number of cases, such as rounding values.

The issue of robustness is key to multivariate analysis because the value of a particular analysis procedure may depend on how robust it is. For example, ordinary least squares, a type of regression analysis, is generally not considered to be robust or resistant because it is based on the mean and the variance. Using ordinary least squares, then, would not be advisable if the data was less than truly interval or if it was not normally distributed.

The robustness of most statistical procedures has been well documented. If you want to use a procedure and do not know how robust it is, a Monte Carlo procedure can be conducted to determine its robustness. This procedure allows many samples to be drawn or created from a single data set. Monte Carlo procedures can be used to slowly violate the assumptions of normality, level of measurement, and other factors. A robust measure will not change dramatically between the different samples, whereas a measure that is not robust will be strongly influenced.

Error

Related to the issue of robustness is that of **error**. If you have had a research methods course, you have learned that error is present in all research and that there are various types of error: measurement error, observation error, random error, systematic error, and others. Error is such a part of multivariate analysis that it is specifically accounted for in some statistical analyses. For example, the formula used for regression is

$$Y = a + b(x) + e$$

The *e* in this formula represents random error in the model. This *e* can simply be thought of as what we do not know. For instance, if we were examining the effects of social bonds to the incidence of juvenile offending, we would want to examine both familial and peer relationships. In research it is common not to be able to capture variables for both these concepts. As such, if one were omitted, we have only information on family relationships and not on peer relationships, or vice versa. Here, *e* captures the missing information we did not have in the analytic model. This persistent error is an issue that must be addressed in any multivariate analysis. Also, each of the associations in Figure 11-2 have attached error values (the *e* in the model). This is also an attempt to estimate how close or how far off the model may be from a true representation of the relationship.

In his book, Schumacher (1973, p. 209) outlines a humorous scenario that typifies the bane of statistics and the raison d'être (reason for being) for statisticians. Here, he pictures God making the Earth and thinking the following:

If I make everything predictable, these human beings, whom I have endowed with pretty good brains, will undoubtedly learn to predict everything, and they will thereupon have no motive to do anything at all, because they will recognize that the future is totally determined and cannot be influenced by any human action. On the other hand, if I make everything unpredictable, they will gradually discover that there is no rational basis for any decision whatsoever and, as in the first case, they will thereupon have no motive to do anything at all. Neither scheme would make sense. I must therefore create a mixture of the two. Let some things be predictable and let other things be unpredictable. They will then, among other things, have the very important task of finding out which is which.

Error must be dealt with and accounted for as much as possible in theoretical and methodological planning and in the statistical analysis of any research project. The issue of error will be explained in more detail in virtually all of the remaining chapters.

Parsimony

Parsimony is essentially the opposite of the interaction of variables in a model. As discussed above, attempts to obtain a complete model often lead researchers to include every variable imaginable. Although this may result in the highest amount of explained variation in the dependent variable, it may make the model more complex than necessary. Many of the variables included may not add much to the model, may not add anything to the model, or may actually reduce the variation explained. In these cases, it may be better not to have the variables in the model at all. Balancing the need to include all necessary variables, the need to explain as much of the variation in the dependent variable as possible, and the need to have the most straightforward model is the issue of parsimony.

Put simply, parsimony is the attempt to identify the smallest number of the most important influences on a dependent variable. Especially in something as complex as studying criminal (human) behavior, there are almost always more influences on a model than can possibly be measured. A 14th-century philosopher, William of Ockham, developed what has come to be known as *Occam's Razor*, a rule that states basically that the simplest of competing theories is usually the best. As applied to statistics, it means that if a variable can be removed from a model without affecting it seriously, the variable should be removed so the model may be as simple as possible. Shaving off these additional variables is a process that researchers should consider in any theoretical or statistical model they develop. It is important to remember, though, that the principle of parsimony does not mean that the most simple model should always be accepted, only that if two models are equal and competing, the simplest should be chosen or that if a variable can be removed without seriously affecting the model, it should be removed and the model made simpler.

■ Conclusion

In this chapter, we introduced some concepts and issues that are important for multivariate analysis. Although these do not directly affect the interpretation of analyses, they will be very important if and when you begin to do your own research. Causation is the holy grail that all researchers look for in statistical analysis and research. A key concept in causation is causal and temporal ordering of variables. Because the field of criminal justice and criminology studies the complex system of human behavior, researchers must balance how many variables are important to a model and the possible interaction of those variables with the need for parsimony in the model. Finally, with multivariate analysis, especially at the interval level, issues of robustness and resistance are important. Although not addressed directly in the chapters that follow, you will see numerous places where these concepts are important.

■ Key Terms

additive effect
causal ordering
causation
error
interaction
multivariate analysis

parsimony
resistance
robustness
synergy effect
temporal ordering

■ Summary of Equations

Regression Equation/Slope-Intercept Formula

$$Y = a + b(x) + e$$

■ Exercises

Find journal articles that address the issues brought up in this chapter. You may be able to find a single article that addresses several of these issues; or you may be required to use several articles to address all of these concepts. Be sure to include articles that address:

    a. Variable interaction

    b. Temporal and causal ordering

    c. Robustness of a statistical procedure

    d. Error

    e. Parsimony

◼ References ◼

Box, G. E. P. (1953). Non-normality and tests of variances. *Biometrika*, 40:318–335.

Cohen, J. (1978). Partialed products are interactions; partialed vectors are curve components. *Psychological Bulletin*, 85:858–866.

Pedhazur, E. J. (1973). *Multiple Regression in Behavioral Research*. Ft. Worth, TX: Holt, Rinehart and Winston.

Schumacher, E. F. (1973). *Small Is Beautiful: A Study of Economics as if People Mattered*. London: Blond and Briggs.

Stigler, S. M. (1973). Simon Newcomb, Percy Daniell and the history of robust estimation, 1885–1920. *Journal of the American Statistical Association*, 68(344):872–879.

Thornberry, T. P., A. J. Lizotte, M. D. Krohn, M. Farnworth, and S. J. Jang (1994). Delinquent peers, beliefs and delinquent behavior: a longitudinal test of interaction theory. *Criminology*, 32(1):47–83.

■ Note

1. Some authors (see, e.g., Pedhazur, 1973) argue that interaction effects should not be included in regression analyses except when used in conjunction with time-series experiments. Although this convention is often violated, Pedhazur's arguments should be considered.

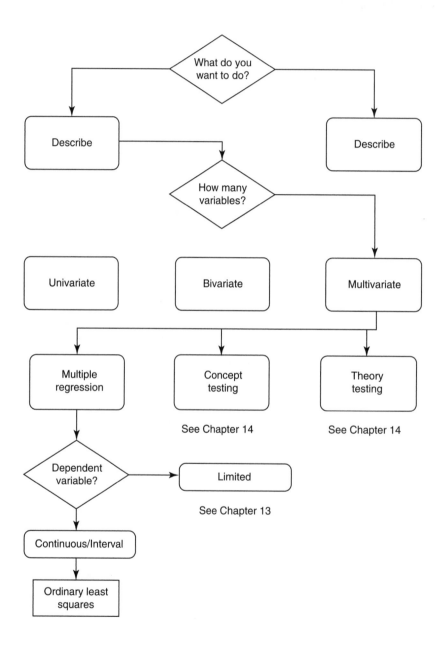

Multiple Regression I: Ordinary Least Squares Regression

Ludwig Boltzmann, who spent much of his life studying statistical mechanics, died in 1906 by his own hand. Paul Ehrenfest, carrying on the work, died similarly in 1933. Now it is our turn to study statistical mechanics. Perhaps it will be wise to approach the subject cautiously.

—David L. Goodstein, *States of Matter* (1985)

It is certainly common to conduct research using bivariate, even univariate, methods of analysis. For those doing "real research," though, it is much more common to use a multivariate analysis procedure. Some of the concepts important to multivariate analysis were introduced in Chapter 11. In the present chapter we expand on that discussion and introduce you to the most common multivariate statistical procedure: multiple regression. Although there are many other multivariate analyses that could be employed, some form of multiple regression is often the standard of research today. Of the many different regression techniques, the first and still most used regression analysis is ordinary least squares (OLS) regression. OLS regression will be where we begin our discussion of multivariate analysis.

■ Regression

OLS regression is probably the most popular multivariate analysis procedure in criminal justice and criminology research. Generally, when researchers say they conducted a regression, they are implying that they conducted an OLS regression. Although the terms *OLS regression* and *regression* are used interchangeably, they are not the same. This will become an important distinction in Chapter 13 when we discuss other regression techniques. OLS regression is not necessarily the simplest statistical procedure, but it is often taught first when introducing students to multivariate analysis because it is simply an extension of Pearson's Product Moment Correlation (*r*) to multivariate analysis that you learned in Chapter 9.

Like Pearson's *r*, regression is a measure of association for interval level data (measuring existence, strength, and direction). In practice, it may be bivariate or multivariate (a bivariate regression is the same as a Pearson's Correlation). Also like Pearson's *r*, the goal of regression is to determine the best-fitting straight line that summarizes the linear relationship between values of two or more variables. Regression takes advantage of the least squares analysis of a line where the **sum of the squares** of the distance from each point to the line is the smallest possible value. This method minimizes the distance to all pairs of X, Y values and summarizes the relationship between the dependent variable (Y) and the independent variables (X_1, X_2, . . . X_i)—this determines the strength and direction of the relationship. Finally, it should be reinforced that, like Pearson's r, regression will not always improve individual predictions; it will only improve the average prediction.

As a means of statistical analysis, **multiple regression** shows the linear relational effects on a dependent variable of an independent variable, controlling for the effects of other independent variables, plus the combined effects of all the independent variables: that is, predicting scores on a dependent variable using the combined predictive scores of several independent variables. The basic goal is to produce a linear combination of independent variables that will correlate as highly as possible with the dependent variable. This linear combination can then be used to predict values of the dependent variable, and the importance of each independent variable can be assessed. In this, we are actually utilizing the first derivative of an equation. Associated with calculus and physics, the **first derivative** (dy/dx or $\delta y/\delta x$) of an equation for a dependent variable is the rate of change in a dependent variable for a change in the independent variable.

A Bit of History

Regression is said to have been developed by Galton in 1885. Galton established the crude beginnings of his work on regression in his 1869 work on heredity. By 1877, Galton had developed his statistical procedure further and termed it *an empirical law of reversion*. By 1885, Galton had changed the name of the procedure to *regression* based on his research, which showed that the mean size of sweet peas tended to *regress* toward mediocrity and that their variances were equal (as discussed in his 1885 address). Regression then went through several iterations and changes until being put into its current form by Yule in 1897.

Although OLS regression as discussed in this chapter utilizes formulas and mathematical calculations, the formulas often get quite complicated because of the nature of **multivariate analysis**. So, although the formulas will be presented, not a lot of time will be spent discussing them. The formula for a simple regression is one of the most common of all multivariate analyses[1]:

$$Y = a + bX + e$$

where Y is the **predicted value** of the independent variable, a the intercept, b the slope, X the value of the independent variable in question, and e the error associated with the model. You will recognize this formula as the one identified with Pearson's r; students of algebra will recognize this formula as the slope-intercept equation for a straight line. This basic formula can be expanded for multiple regression (regression for multivariate analysis) by increasing the number of independent variables in the equation as follows:

$$Y = a + b_1X_1 + b_2X_2 + b_3X_3 + \cdots + b_nX_n + e$$

Here, each b represents a coefficient associated with an independent variable, and each X represents a value of an independent variable in the multiple regression model.

■ Assumptions

Most students who begin to learn multiple regression find that the procedures are very straightforward and fairly easy to learn. Students also learn, however, that a number of assumptions must be met when using multiple regression analysis. Examining the data in terms of these assumptions and ensuring that they are met is one of the most difficult parts of a regression analysis. Below we discuss briefly the assumptions of regression analysis and how they influence analysis and interpretation. The assumptions here are taken primarily from Berry (1993) and Studenmund (1997); students are encouraged to consult these sources for an even deeper understanding of the assumptions of OLS regression.

The first assumption is that the variables in the model are interval level. Regression uses the mean and variance as the roots of its analysis. Data that is not suitable to be examined using the mean and variance should not be used in a regression analysis. The one exception (for independent variables only) is dichotomized nominal level data or data that can be dichotomized or **dummy coded**. While some researchers will use ordinal level data such as Likert scales (1 = strongly agree, 2 = agree, 3 = neutral, 4 = disagree, 5 = strongly disagree) as explanatory variables, it is recommended that ordinal level variables be dummy coded. Dummy coding can be used to make categorical data approximate interval level because only two points are used, thus allowing a straight line between the two variables. The easiest way to use categorical (dichotomized) data in a regression analysis is if the variable has only two categories. As discussed previously, when analyzing a dichotomous, or dummy coded, variable, generally we code the value we are most interested in as 1 and the value we are less interested in as 0. For instance, if we were studying the influence of race on judicial sentences, we might explore whites versus nonwhites. In this case, we would be most interested to see if there was some disparity in sentencing, thus whites would receive a score of 0 and nonwhites would receive a score of 1. Dummy coding can be performed with more than two categories, but it is time intensive and the outcome of breaking many categories down into all possible combinations and using them in a regression analysis is suspect at best. The other issue of dummy coding is that it should not be used if one of the categories contains less than 20% of the cases. This is because the assumption of a normal distribution would be violated too greatly. Dummy coding independent variables will be discussed more fully later in this chapter.

The second assumption of regression is that the variables are linearly associated in the coefficients and the error terms. As discussed with Pearson's Correlation, a data set that is curvilinear will produce low regression coefficients even if there is a strong association among the variables. As also discussed in Chapter 3, there are ways to detect nonlinearity, the most common being an examination of bivariate scatter plots. Scatter plots of the residuals plotted against the predicted values should show no recognizable pattern. As with normality, if a scatter plot shows a definite pattern, there are methods of correcting it. Most of the same transformations (logging or squaring) that correct nonnormality will also work for nonlinearity. Be careful, though, that a transformation designed to correct one problem does not cause problems in the other (a correction for nonnormality that makes an otherwise linear relationship nonlinear). Also, the transformations must be supported by theory and methodology.

The next assumption of regression analysis goes back to the discussion of complete but parsimonious models in Chapter 11. This is that there is no **specification error** in the model, which means that every independent variable critical to explaining the dependent variable is included in the model, but those that are not critical are excluded. There are four conditions under which this assumption is violated: (1) independent *variables are included* that should have been excluded; (2) independent *variables are excluded* that should have been included; (3) the proper variables are included but in the wrong form (e.g., a curvilinear relationship that needs to be transformed); or (4) highly multicollinear variables are included in the model (see below).

Each of these **misspecifications** requires different methods of detecting the specification error. The caution here is that misspecification is a theoretical rather than a statistical issue; therefore, misspecification must be dealt with at a theoretical and methodological level rather than at a statistical level. For inclusion of erroneous variables, the suspected variable can be removed from the regression model. If there is little change in the **b coefficient** or R^2, the variable might be extraneous and is a candidate to be removed. If a variable is an integral part of the theoretical model, however, it may have to be left in the model even if there is some specification error. Always remember: Theory drives the analysis, not the other way around. Dealing with variables that should be in the model and are not is much more difficult. It becomes an issue of theory, as discussed in Chapters 1 and 11, and requires that you rethink your research more than rethink the analysis. It is possible to determine if a variable may have been left out of a model. If IV_1 and X_2 are highly correlated and X_2 is left out of the model, the b's (see interpretation below) of both variables will be shown in the b of IV_1 to the degree of the correlation between IV_1 and X_2, and the s_b will be low. This is not always easy to detect, however, and should only be a signal from the analysis that there may be a problem. Similarly, a low R^2 value may indicate that some critical variable has been excluded from the model. What that variable is, however, is more an issue of theory than the analysis. The third misspecification was addressed in the second assumption, above. Finally, **multicollinearity** may be detected by examining the bivariate correlation of all variables and examining the **standard error** (SE) of the regression model. If IV_1 and IV_2 are highly correlated, the s_b and the SE values may be large.

The fourth assumption of OLS regression concerns the **error term**. As discussed in earlier chapters, error is a part of any research. When using regression, it is important that every effort be taken to eliminate measurement error. There are two types of measurement error that are important for regression analysis. The first is *random* or **unsystematic error**. This usually arises from coding mistakes and sometimes from using an instrument inappropriate for measurement (you are not really measuring what you think you are measuring, as discussed in Chapter 1). This heavily influences the reliability of the findings of the regression analysis. An example of this is a variable measuring age. When age is entered correctly, the most we could ever expect would be a little over 100. If you have a value of over 100, someone could have accidentally added an extra number when entering the data. Thus you might get a value of 165 when the value is supposed to be 16. This is random error.

The second type of error is *nonrandom* or **systematic error**. This is often the result of not eliminating rival causal factors, as discussed in Chapter 11. This source of error heavily influences the validity of the findings. The consequences of excessive error are quite dramatic. At the statistical level, the most common results are an inflation of the variance and unpredictable results (the *b* coefficients are unstable).

Also related to error in the model is the fifth assumption, which is that residuals should have a mean of zero. Although the coefficients will remain unbiased if this assumption is violated, the intercept in the equation will be biased. This is an assumption that is easy to check but is complex in its assumption. A greater discussion of this assumption will no doubt occur in any class focused on multivariate modeling.

The sixth assumption is also related to error, which is avoiding **heteroskedasticity**. Heteroskedasticity occurs when the **variance of the error term** is not constant over the entire length of the regression line. Rather, we want a constant variance across the length of the regression line; this is referred to as *homoskedasticity*.

The seventh assumption of OLS regression is there should be no **autocorrelation** in the model. Autocorrelation, both temporal (over time) and spatial, implies that the errors are correlated across cases. Autocorrelation can lead to biased coefficients and underestimates of the standard errors associated with *b* coefficients.

The eighth assumption of OLS regression is that the independent variables have to be uncorrelated with the error term. This is typically not an issue if sound methodology was followed in choosing and developing the variables. The final assumption deals with normality. This is not normality among variables as we discussed earlier (which is a common misperception about multiple regression). Normality in this case is examined in conjunction with the error term. The error terms must be normally distributed about the regression line. If they are not, the significance tests are completely invalidated and the model means nothing. Ways of dealing with measurement errors include conducting reliability tests prior to construction of the regression model and using more than one indicator or combing several items into a scale.

This final assumption of OLS regression pertains to the error term as well and is one of the most common and most difficult to deal with. This assumption is that there is neither perfect nor excessive multicollinearity among any two independent variables or group of independent variables. For proper regression analysis, the independent variables should be correlated with the dependent variable but not with each other. Multicollinearity occurs in all research to some extent; there must be some multicollinearity between independent variables in the social sciences because, otherwise, it is unlikely that they would be correlated with the dependent variable. Regression is robust enough to withstand some multicollinearity. Researchers must therefore balance enough multicollinearity to make the model work but not so much as to destroy the model. While we will discuss multicollinearity in greater detail later in the chapter, understand for now that if there is perfect multicollinearity, the regression analysis (in any statistical program) will not run. On the other hand, severe multicollinearity does have some remedies.

A full list of the assumptions of Ordinary Least Squares Regression is presented in **Table 12-1**.

TABLE 12-1   Assumptions of OLS Regression

Assumption	
1	The variables are interval level
2	The variables are linearly associated in the coefficients and the error terms
3	There is no specification error in the model
4	The elimination of most measurement error (random and systematic)
5	Residuals should have a mean of zero (0)
6	No heteroskedasticity in the model
7	No autocorrelation in the model
8	The independent variables are not correlated with the error term
9	There is no perfect multicollinearity between two (2) independent variables

If any of these assumptions of OLS regression are violated, the value we get for the regression coefficients will not be a correct estimate of the effect of an independent variable on the dependent variable. When all of the assumptions are met, several things will happen. First, regression coefficients will be unbiased: The mean of *b* coefficients will be centered on the true population of *b*. Second, coefficients achieve minimum variance by centering and clustering tightly about the mean of *b*. When we have a tight distribution, we have smaller standard errors, which are more likely to produce statistical significance. Third, we gain consistency. As we increase our sample size, it will converge even more about the true value of the regression coefficient. Fourth, our distribution of coefficients will be normally distributed, which allows us to carry out significance tests. Thus, it is important that we meet all of the assumptions of OLS regression before attempting to interpret any output from statistical programs.

■ Analysis and Interpretation

In this section of the chapter, we discuss a general procedure for conducting a regression analysis. Although this is a very simplified version, it should suffice to introduce you to the process. A thorough understanding of regression analysis will require a much deeper study of the material. The data for this example comes from sentencing data in Arkansas. It focuses on the number of months in prison a guilty defendant receives. We will use several independent variables to discern what influences judicial decision making has on the length of prison sentences.

The first step in conducting a regression analysis is to develop the variables. This sounds simplistic, but as discussed above in the assumptions and in Chapter 11, it is important because this is the step that may produce misspecification error or reduce multicollinearity. It is also important at this point to determine if the data is appropriate for use in a regression. As discussed previously, the data should be interval level or dichotomized nominal level data. If there are variables that do not meet these requirements, they may need to be dropped and data gathered that is at the proper level of measurement—or use a statistical procedure other than regression.

The next step is to gather and code the data. Again, this preliminary step is important because attention to detail here is critical to ensure there is no random or systematic error in the coding. In the case of the data used in this example (and in most social science/criminal justice data), the data was collected by a governmental agency. The data had to be cleaned so that undesirable cases were omitted and variables could be analyzed by SPSS. The next step is to conduct univariate statistical analysis. The object here is to examine the data and determine if it is appropriate for regression analysis or whether assumptions have been violated. Key characteristics to examine include the presence of outliers, the mean and standard deviation, and the skewness and kurtosis. Each of these examine whether the variables are normally distributed.

The next step is to conduct a bivariate analysis of the data. The primary statistical procedure in this step is a correlation analysis of the variables to examine the data for high levels of multicollinearity. This step can also provide an initial glimpse of what to expect from the regression.

The next step is to run the regression. This is somewhat more involved than it would seem. Although many people do it, running a regression is not as simple as dumping the variables in a hat and waving a statistical wand over it. The variables should be entered in accordance with the theory, and each step of the regression model should be analyzed carefully to determine the effect each variable and combination of variables has on the dependent variable.

There are two philosophies regarding this step. The first philosophy states that the variables to be included in the regression should be crafted according to theory; then the results of the regression should be analyzed and presented as they are. In the example here, the variables were placed in the model in accordance with current literature on sentencing research. This is the essence of theory testing, where a model is created and tested and the results presented regardless of the outcome. The second philosophy is that a model should be tweaked to explain as much of the variation in the dependent variable as possible. If many of the variables were insignificant, the researcher would probably want to rerun the regression using only variables that made a significant contribution. Additionally, the order of the variables might even be changed so the one explaining most of the variation would be entered first. This is generally considered model building rather than theory testing.

At this point, a brief discussion on the issue of stepwise regression models is in order. Stepwise regression has been the topic of heated debates for a number of years. A **stepwise regression** is a procedure by which the researcher determines what variables will be considered in the regression model (hopefully, based on theory); then the computer is allowed to determine which variables will stay in the model, which variables will be removed, and the order in which those in the model will be entered. This is accomplished by calculating all possible combinations of the variables and examining the relative contribution (strength and significance) of each variable. Those with the greatest mathematical contribution are entered in the model first. Other variables with significant contributions are entered in the order of their contribution and the variables that do not have significant contributions are dropped from the model.

The controversy over this method surrounds how the variables get chosen. Those in support of stepwise regression argue that the computer has been programmed with the parameters that the researcher would use in creating the regression model and that the computer can make the calculations faster and more accurately than the researcher. These people contend that a stepwise regression should produce the same model that a researcher would arrive at building a regression model manually (using model building rather than theory testing). Those against stepwise regression argue that variables should be entered only as they appear in the theoretical model and that allowing the computer to choose the ranking of variables is not sound research. They contend that the theoretical model should be tested as originally proposed. The researcher should then report the results of the test of the theoretical model: good or bad. These people argue that using a stepwise regression is contrary to theory testing and is nothing more than a "hopper analysis," where all possible variables are gathered and the computer is used to sort out which variables might have some association with the dependent variable. Ultimately, it will be up to you to determine whether or not you will use stepwise regression in your analysis.

The example that follows uses Arkansas sentencing data. The dependent variable is the number of months in prison a defendant received. This does not include those offenders who receive any sentence other than time in prison (jail, probation, electronic monitoring, etc.). The independent variables include the offender's criminal history score, age, sex, race, if the offender was on probation or parole (supervision status), the total number of counts an offender was charged with, the offense seriousness, whether or not the offense was a felony, whether or not the offense was violent or was drug related, whether or not a warrant was issued for arrest, the type of trial (whether the defendant plead guilty or went to trial), and whether or not the case occurred in the Mississippi delta region of the state or elsewhere (a proxy variable for the wealth of the particular area of the state). Each of these variables has been shown or argued theoretically to contribute to judicial decision making on sentencing. These variables were also included to show examples of how different types of variables (e.g., dichotomized variables) perform in OLS regression analysis.

How do you do that? OLS Regression

1. Open a data set.
   a. Start SPSS.
   b. Select *File*, then *Open*, then *Data*.
   c. Select the file you want to open, then select *Open*.
2. Once the data is visible, select *Analyze*, then *Regression*.
3. A list of different types of regression techniques will appear. Select *Linear*. There are other types of regression, some of which are discussed in Chapter 13. For OLS regression, simply select *Linear*.
4. Select the dependent variable you wish to include in your analysis and press the ▶ next to the window marked *Dependent*.
5. Select the independent variables you wish to include in your analysis and press the ▶ next to the window marked *Independent*.
6. Make sure the word in the box for *Method* reads *Enter*.
7. For now, do not worry about the options boxes at the bottom of the window; simply press *OK*.
8. An output window should appear containing output similar to Table 12-1.

The output for OLS regression is generally more number oriented than anything you have seen to this point. This is because of the number of variables we used in this example. The SPSS output for the OLS regression example is presented in **Table 12-2**. The first

TABLE 12-2 OLS Regression Results with SPSS Output

Model Summary

Model	R	R Square	Adjusted R Square	Std. Error of the Estimate
1	.489[a]	.239	.238	131.565

a. Predictors: (Constant), Age, Total Number of Counts, Supervision Status, Violent Offense, Delta vs. Other Counties, Sex, Warrant Status, Type of Trial, Criminal History Total, Race, Drug Offense, Felony, Offense Seriousness

ANOVA[b]

Model		Sum of Squares	df	Mean Square	F	Sig.
1	Regression	1.4E + 008	13	10876218.24	628.339	.000[a]
	Residual	4.5E + 008	26053	17309.471		
	Total	5.9E + 008	26066			

a. Predictors: (Constant), Age, Total Number of Counts, Supervision Status, Violent Offense, Delta vs. Other Counties, Sex, Warrant Status, Type of Trial, Criminal History Total, Race, Drug Offense, Felony, Offense Seriousness

b. Dependent Variable: Sentence in Months

Coefficients[a]

Model		Unstandardized Coefficients		Standardized Coefficients	t	Sig.
		B	Std. Error	Beta		
1	(Constant)	233.701	5.519		42.342	.000
	Offense Seriousness	−38.228	.696	−.430	−54.902	.000
	Felony	−33.478	2.625	−.094	−12.754	.000
	Criminal History Total	3.156	1.022	.017	3.089	.002
	Total Number of Counts	23.024	.671	.190	34.327	.000
	Supervision Status	−5.541	3.760	−.008	−1.474	.141
	Type of Trial	18.661	2.550	.040	7.319	.000
	Violent Offense	52.109	2.699	.120	19.306	.000
	Warrant Status	2.009	3.269	.003	.615	.539
	Drug Offense	−19.241	2.226	−.057	−8.642	.000
	Delta vs. Other Counties	−7.826	3.505	−.012	−2.233	.026
	Race	4.166	1.719	.014	2.423	.015
	Sex	−17.023	2.433	−.038	−6.996	.000
	Age	.674	.088	.042	7.692	.000

a. Dependent Variable: Sentence in Months

TABLE 12-3   Presentation Table of SPSS OLS Regression Output

	b	Std Error	β
Offense Seriousness	-38.228*	0.696	-0.429
Felony (1 = Yes)	-33.478*	2.625	-0.093
Criminal History Total	3.156*	1.022	0.017
Total Number of Counts	23.023*	0.671	0.19
Supervision Status (1 = Parole/Probation)	-5.541	3.76	-0.008
Type of Trial (1 = Trial)	18.661*	2.549	0.04
Violent Offense (1 = Violent Offense)	52.108*	2.699	0.12
Warrant Status	2.008	3.269	0.003
Drug Offense (1 = Drug Offense)	-19.241*	2.226	-0.057
Delta vs. Other Counties (1 = Delta Counts)	-7.825*	3.505	-0.012
Race (1 = Non-white)	4.166*	1.719	0.014
Sex (1 = Female)	-17.023*	2.433	-0.038
Age	0.674*	0.088	0.042
R^2	0.239*		
F	628.34		

* $p < 0.05$

two parts of Table 12-2 pertain to the overall model, telling us if the model was significant and how much variance was explained in predicting the dependent variable by knowing all of the independent variables in the model. The last part presents coefficients on the effects of individual independent variables on the dependent variable. Although three tables of output are generated, they are fairly easy to interpret.

In most research articles, the information from OLS regression is compiled into a single table for ease of interpretation. **Table 12-3** gives an example of a presentation table associated with the SPSS output in Table 12-2. Notice the differences between the SPSS output and this table. This table is actually much simpler to examine when evaluating the results. It is presented here as a means of showing the contrast between the output that is received from a statistical package and what one typically sees in research articles. A presentation table allows the results to be simplified and interpreted more easily. The primary difference between these two table styles is the use of asterisks (*). An **asterisk** implies that something was statistically significant in the model. This allows the column of significance to be omitted, saving space for other, more useful bits of information.

Steps in OLS Regression Analysis

There are five primary pieces of information that should be evaluated in any multiple regression output. These represent the existence of a relationship for the entire model, the strength of the relationship for the entire model, the existence of a relationship for individual independent variables, the strength of the relationship for

TABLE 12-4   Steps for Interpreting OLS Regression

Step	Key Information Needed	Statistic
1	What is the probability that the results of the model could have happened by chance?	$p < 0.05$ (F)
2	How good are the results? How closely do the points cluster about the mean?	R^2
3	Are the unstandardized regression coefficients (b) statistically significant?	$p < 0.05$ (t)
4	Interpret the unstandardized effects of the independent variables.	b or B
5	Interpret the standardized effects of the independent variables and rank by importance.	β

individual independent variables, and the relative importance of the independent variables. These pieces of information are summarized in **Table 12-4** along with the coefficients or values from OLS regression output that correspond to these five items. Each of the items is explained in the sections that follow, using the example of Arkansas sentencing data.

1. *What is the probability that the results of the model could have happened by chance?*

The first piece of information is whether or not the model is significant (whether the combined effect of all the independent variables on the dependent variable is statistically significant). This analysis is the same as discussed in Chapter 8. As with a bivariate analysis, if the relationship is not statistically significant, there is no real reason to continue the analysis. There is a minimal difference between the process here and that for bivariate analysis. In OLS regression, there are several variables in the model. Especially if you are doing theory testing, where you put all variables related to the theory in the model, there may be a situation where one or more of the variables is making the entire model not statistically significant. If that happens, it may be possible to remove those variables and make the model statistically significant. Of course, this may violate what you are trying to accomplish with theory building, but it may be very useful if you are willing to tweak your model some to determine what variables may be associated with the dependent variable.

You can tell if the model is statistically significant by examining the significance associated with the F-test. This is found in the *Sig* column of Table 12-2 immediately to the right of the F column in the table labeled *ANOVA*. Like other analyses, if the significance is less than 0.05, the model is significant. In the presentation table, this can be discerned if there is one or more asterisks by the R^2 value. In the example model above, the probability associated with F is 0.000. This is less than 0.05 and indicates that the model is statistically significant. This is also indicated on the presentation table by the asterisk. Because the model is significant, we can proceed on to step 2 in evaluating the OLS regression results.

2. How good are the results? How closely do the points cluster about the mean?

The second piece of information addresses the strength of the overall model. This indicates how powerful the model is (i.e., how much variance does the independent variable produce in the dependent variable). If the model is significant, the next logical question is how much the model explains in predicting the dependent variable. You can ascertain this by looking at the **multiple R** (R in the SPSS output) or the multiple correlation coefficient (R^2, or R Square in the SPSS output). The value of R^2 is the proportion of variation in the dependent variable associated with variation in the independent variables. For a bivariate regression, this is equal to Pearson's r; for a multiple regression, this is the correlation between the dependent variable and all of the independent variables. It tells how good a predictor X (or X's) is of Y with a particular regression line. The **adjusted R^2** is sometimes preferable to the regular R^2. The adjusted R^2 is corrected for the number of cases. A small number of cases relative to the number of variables in the regression can bias the estimate of R^2 upward. An adjusted R^2 takes the number of variables into account and adjusts the R^2 for that number: in effect, controlling for the number of variables entered into the model.

In the example, the R^2 is equal to 0.239. When multiplied by 100, you get the percentage of variance explained by the entire model. In this case, we have explained almost 24% of the variance in predicting how much time a defendant will be sentenced to prison. Although this is relatively small, for criminal justice research this is actually fairly substantial.

3. Are the unstandardized regression coefficients (b) statistically significant?

The third key measure to evaluate for OLS regression is whether or not the individual variables in the model are significant. If you know the model is significant and how much variance it explains, you will want to know more about each variable. The first step in determining the importance of an individual variable is to explore the significance associated with t. This can be found in the table labeled *Coefficients* in the SPSS output in Table 12-2 or look for the asterisk next to specific variables in a presentation table such as Table 12-3. In SPSS output, look in the column labeled *Sig.* As always, you want to determine if the variable is statistically significant. In this case, you are actually looking at a bivariate relationship, just as you did in Chapter 8. The only difference is that you will be using t here instead of Chi-square or one of the other measures. Consistent with earlier chapters, you want this value to be less than 0.05.

Evaluating the significance associated with t in Table 12-2 and the asterisks associated with b coefficients in Table 12-3, we see that all of the variables in the model are statistically significant, with the exceptions of whether there was a warrant issued for an offender and whether the offender was on probation or parole. This is therefore a good model of independent variables for predicting the number of months in prison a defendant received. For the variables that were not significant, this should be noted. Aside from this, there is no further interpretation for variables that are not statistically significant.

4. Interpret the unstandardized effects of the independent variables.

The fourth examination is the strength of the individual contributions of the independent variables on the dependent variable. This represents an analysis of the **unstandardized effects** of independent variables on the dependent variable. The unstandardized effects of independent variables are represented in Table 12-2 in the *Coefficients* table as either a *b* or a *B*; and *Std. Error* is the standard error of that coefficient. The unstandardized coefficient represents several parts of an OLS regression. First, it is an adjustment of the scale of the independent variable to the scale of the dependent variable. This also converts the labels/units of the independent variable into the labels/units of the dependent variable. Third, it minimizes the sum of squared errors. Finally, it provides an estimate of how the value of the dependent variable changes with the unit of change in an independent variable. This is an important value to use when explaining results to practitioners and informing policy, as it is understandable and easy to interpret.

The unstandardized coefficient is interpreted generally as: "We can expect an (increase or decrease) of (the value of the *b* coefficient) in the (dependent variable) for a 1-unit change in the (independent variable)." For example, we might say, "We can expect an increase of 2.56 in the number of crimes for every 1-unit change in drug use." The increase or decrease is determined by the sign of the *b* coefficient. If the sign is negative, it is a decrease; if the sign is positive, it is an increase.

For the independent variable measuring the number of counts using the wording denoted above, we can expect an increase of 23 months sentenced to prison for each additional count for which an offender is charged ($B = 23.024$ in Table 12-2). In this case, a higher number of counts levied against a defendant resulted in over 23 months more for each additional count. This shows a certain culpability on the defendant's behalf when he or she is charged with more than one offense. The courts in Arkansas obviously take this into account when sentencing the defendant to prison.

The other three independent variables are much more straightforward to interpret. We can expect a decrease of 38.23 months in prison for a 1-unit change in the offense seriousness ($B = -32.228$ in Table 12-2). If a defendant was charged with three counts, that defendant would receive on average 46 more months in prison than an offender who was charged with only one offense. Age has relatively small unstandardized effects on the number of months of prison received ($B = 0.674$ in Table 12-2). In terms of criminal history, we can expect an increase of 3 months in the prison for a 1-unit change in the defendant's criminal history score ($B = 3.156$ in Table 12-2).

When evaluating the unstandardized effects, it is best to distinguish between continuous and dichotomous independent variables. Dealing with dichotomous independent variables is somewhat easier, as there is only one of two values an independent variable can take. It is important to remember how the variable was coded, though. This is one way in which a presentation table is better than a statistical table. In the example, the majority of variables are dichotomous in nature. As far as being tried in either the Mississippi delta region or other parts of the state, individuals in the poorer delta received almost 8 months less in prison ($B = -7.82$ in Table 12-2). Offenders who chose to go to trial rather than plead guilty received a little over 18 more months in prison ($B = 18.661$ in Table 12-2). This is often referred to as the "trial tax." If you choose to go to trial rather than plead out, most research has found that the number of months received in prison increases dramatically. If the defendant was charged with a violent offense rather than a property offense, the defendant received 52.1 months more in prison ($B = 52.109$ in Table 12-2). Defendants charged with drug crimes received 19 months less in prison ($B = -19.241$ in Table 12-2) than defendants charged with other types of crimes. All of the rest of the dichotomous variables can be interpreted in a similar fashion. Try a few on your own.

Although the unstandardized coefficient is important, it cannot be used to judge the relative importance of an independent variable. This is because the size of the B coefficient depends on how the independent variable was measured. To see which variable is most important in the analysis relative to the other variables, another coefficient is necessary, a standardized coefficient.

5. Interpret the standardized effects of the independent variables and rank by importance.

The final step is to examine the **standardized effects** of independent variables on the dependent variable. This process puts all values into a standard measure and examines which variables make a greater contribution to the model. There are two types of effects that should be examined in an analysis of regression results: the unstandardized coefficient (B, discussed above) and the standardized coefficient (β). The **beta** (β) *value*, sometimes referred to as the beta weight, is the standardized correlation coefficient.[2] This standardization allows beta to measure the contribution of variables irrespective of the scale on which they were measured. Standardized scores are created when the data is converted to Z scores. Standardization also sets the mean to zero and the standard deviation to 1. This allows the beta values to be directly comparable between variables. The beta is the effect of a change of one standard deviation in an independent variable on the standard deviation of the dependent variable. In reality, beta represents the relative contribution a particular variable makes to explaining variation in the dependent variable when other variables in the model are controlled for (kind of like a bivariate correlation without the effect of the other variables). The standardization of coefficients allows you to rank order the variables that are most important to the model.

The interpretation of the standardized coefficients is relatively simple, although there is great controversy over how much information can be determined from the beta values. One group of statisticians argue that the beta values can be used to address the amount of variation in the total model accounted for by a particular beta. Another group of statisticians argue that beta values are only good for examining the rank-order of values in the model. Both interpretations will be presented here. Beta (β) values are found in the *Beta* column in both Tables 12-2 and 12-3.

To examine the amount of variation in the model assumed by a particular variable, multiply the value in the *Beta* column by 100 to convert to a percentage. After doing this, the variable with the most influence is the offense seriousness category, predicting almost 43% of the variance in the model. This indicates that judges strongly relied on how serious an offense was in determining how long to send an offender to prison. From there on, the relative importance of the other variables drops significantly. The total number of counts is the next most important variable, with 19% of the variance. The rest of the variables range from explaining only 0.3 to 12% of the variance in the model. These are relatively low percentages of the variance. Note that it does not matter if the value is a negative. The size of the beta weight is the most important thing. The bigger the better is the rule for beta weights.

For an interpretation that only takes rank order of the beta into account, first take the absolute value of the beta; this gets rid of any negative value. Then rank-order the beta values from largest to smallest. The rule of thumb is that the larger the beta the better. What we are doing when rank-ordering the independent variables is determining which values are more important to the model. A Beta value of 0.1 or less represents a weak effect of a variable on the model. A Beta value of 0.2 to 0.4 represents a moderate effect on the overall model. A Beta value of greater than 0.4 represents a strong effect on the analytic model. To interpret the standardized effects of the independent variables, take the values listed in the columns for the significant variables, and rank them according to size. As discussed above, offense seriousness is the most powerful variable in the model ($|\beta| = 0.43$). Note it is very strong as the value is greater than 0.4. The next closest variable is the total number of counts ($|\beta| = 0.19$). This variable has a moderate effect on the model. With the exception of if the defendant was charged with a violent offense, the remainder of the variables has Beta values less than 0.1, which indicates weak effects on the model. This rank-ordering allows us to say that in Arkansas, offense seriousness is the key variable for predicting sentence length. The other variables, although having an influence, have a weak effect on the overall model.

Other OLS Regression Information

There are several other pieces of information in OLS regression output. These are either not terribly important to interpretation of the results, or are a part of a more advanced discussion, and will be given little attention here. The *t* is the obtained *t*-value for the beta. This is similar to the obtained value for Chi-square in Chapter 8. The mean square is nothing more than the average sum of squares. The degrees of freedom (*df*) is the number of independent variables minus one. The standard error of the estimate is the same as the root mean square of errors. This is the standard deviation of the errors in the model. The *F* is used in regression as a method of testing a null hypothesis that includes more than one variable. The **F-test** determines whether the fit of an equation is significantly reduced by keeping the equation confined to the null hypothesis. Residuals are another item that can be analyzed in the output above. The veracity of the model can be determined by examining the **residuals** of the regression equation. The residuals represent the difference between the value of the dependent variable and the value predicted by the regression model. If analysis of the plots or residuals shows any problems with the regression model, it may be necessary to transform one of more of the variables.

Limitations of OLS Regression

For all its power and popularity, regression is not without its limitations. As discussed previously, regression is generally not considered a terribly robust statistical procedure. Its heavy reliance on the sum of squares, the mean, and the variance make it susceptible to problems when using nonnormal, nonlinear, or less than interval level data. Regression is also seriously influenced by sample size, such that weak correlations are often significant with large samples or populations. Finally, the dependent variables often used in criminal justice research often do not conform to the requirement of OLS regression for continuous, unbounded dependent variables. These are in essence limited dependent variables. Regression techniques that analyze these types of dependent variables are discussed in greater depth in Chapter 13.

■ Independent Variables with Lower Levels of Measurement and Nonlinear Relationships

As discussed previously, OLS regression requires certain assumptions to be met. This is not always possible using social science data. Sometimes we are most interested in the effect of independent variables that have a lower level of measurement, such as

Likert scale values from a survey. Other times we might be interested in the relationship of dependent and independent variables that are not linearly related. This section examines some solutions to independent variables that do not meet the assumptions of OLS regression.

Dummy Variables

Oftentimes we want to evaluate independent variables that are either nominal or ordinal in nature in a multiple regression analysis. This occurs when we want to evaluate different groups (Catholics, Protestants, Jews, etc.), or people's attitudes or beliefs (on a scale from strongly agree to strongly disagree). The need to analyze these kinds of variables occurs frequently across criminal justice and criminological research. One example is fear of crime research. In this line of research, independent variables are comprised largely of statements that ask survey participants whether they agree or disagree (Are you afraid to walk in your neighborhood alone at night?). These scale independent variables are then evaluated in relation to an individual's fear of crime. Although the use of nominal or ordinal variables in OLS regression is generally forbidden, researchers can elect to utilize **dummy variables**. The use of dummy variables can allow the introduction of nominal or ordinal level variables into OLS regression without violating the assumptions.

Dummy variables are used to represent broader variables or factors. As Hardy (1993:20) states, for a dummy variable, the "predicted value of Y_i changes B_k units each time membership in the specified category is switched on or off, because a 'unit' change in a dummy variable (from 0 to 1 or 1 to 0) indicates membership or nonmembership in the designated category." Dummy variables can range from the simplistic (sex: male = 0 and female = 1) to the complex (religion: Protestant = 0, Catholic = 1, Jew = 2, and other = 3).

Dummy variables are created by treating each category or characteristic of a variable as an individual variable. Like dichotomous variables, scores are arbitrarily given for each and every category or characteristic (usually 0 and 1) to determine the presence or absence of those particular categories or characteristics. Thus, the dummy variables can now be treated as dichotomous nominal variables and can be used in statistical techniques that require interval/ratio level data in their assumptions. These variables would be entered into the model while leaving one of them out as the reference category.

All dummy variables generated from a nominal or ordinal level variable are not included in the regression analysis. Because one dummy variable is omitted, it does not mean we lose that information. Rather, all of the other dummy variables are interpreted in relation to the excluded dummy variable/category or characteristic. If all of the dummy variables were included, there would be no way to interpret the results. Likewise, the inclusion of all dummy variables results in perfect multicollinearity and the loss of all ability to interpret the dummy variables.

The example provided in **Table 12-5**, shows a nominal level independent variable testing whether the area of employment in which a person works will determine if the person is likely to engage in a white-collar crime or a street-level crime. The different categories of this variable will be used to create dummy variables. Notice there are six categories associated with the employment variable. Thus, we can expect to generate five (5) new variables while omitting the sixth category, which becomes the reference group.

TABLE 12-5   Employment Sample Independent Variable to Dummy Code

Variable	Categories	Coding
Employment	Construction	1
	Farming	2
	Professional	3
	Service	4
	Government	5
	Other	6

TABLE 12-6   Dummy Coded Variables for the Independent Variable for Employment

Type of Cases	D1	D2	D3	D4	D5
Construction	1	0	0	0	0
Farming	0	1	0	0	0
Professional	0	0	1	0	0
Service	0	0	0	1	0
Government	0	0	0	0	1
Other	0	0	0	0	0

Table 12-6 provides the dummy coded variables created from the independent variable employment listed above. The new dummy variables are denoted as D1 through D5. In D1 construction is compared to all other types of employment (denoted as 1 if the individual was a construction worker and 0 if not). In D2 farming is compared to all other types of employment (denoted as 1 if the individual was a farmer and 0 if not). This pattern follows for the other variables.

Remember, creating dummy variables is a starting point. Each of these new dummy variables (D1–D5) has to be evaluated using univariate and bivariate statistics and then entered into the regression equation if OLS assumptions are not violated. When dummy variables are added to regression analyses, the coefficients that are generated provide the distance between the mean of the net category (reference group) and the mean of an explicit group. In essence, we interpret the outcome of a dummy variable in relation to the comparison group/dummy variable excluded from the regression analysis. In the preceding example, we omitted the "other employment" category, making it the reference category; so all interpretation is based on the difference between "other employment" and each of the other categories. A final note on using dummy variables: When using dummy variables, it is imperative to omit the original variable from which the dummy variables were created. This will prevent multicollinearity in the model.

Interaction Terms

Sometimes research utilizes complex independent variables that may be a measure of two or more separate independent variables. An example of this is the concept of socioeconomic status. What is socioeconomic status? How would you measure it? In many instances, criminologists use income as a measure of socioeconomic status (SES), but that is not the only connotation of the concept of SES. In its most abstract conceptualization, SES is a measure of affluence, and simply measuring it using income is not enough. Thus, some researchers will include education, job title, etc., as interaction variables in an effort to better measure the combined effect of SES on criminality.

Interaction terms generally imply creating a combined measure of two of more independent variables. For instance, you might want to combine sex and race. If each of these two variables has only two outcomes, white, non-white and male, female, this would result in four potential outcomes (white males, white females, non-white males, and non-white females) for the new variable (see **Table 12-7**). As shown in Table 12-7, it is simple to create interaction terms in the form of one variable. Although it is relatively easy to code these two variables into a combined variable, there are two problems in the interpretation.

First, the new variable has four categories. It must now be dummy coded to be utilized in the regression analysis, resulting in three new variables in the analytical model (remember to leave out one dummy variable as your comparison group or your analysis could indicate perfect multicollinearity).

Further, if there are two variables with more than four combined outcomes, such as interval/ratio level measures, it becomes tedious and difficult to interpret. If you have interval or ratio level data, this process can be simplified somewhat by creating categories of the most important groups you are interested in in relation to another variable. For instance, if we want to combine age and race, the combined group we will be most interested in is non-whites who are in the 18–30 year old age group because this group is represented more in official crime statistics. Race would be coded as previously indicated, but age might be turned into three categories, 18–30, 31–45, and 46 and above. In this case we would have six categories in the interaction variable. Again, this would have to be dummy coded, resulting in five added variables in the analytical model.

If you feel that the creation of interaction terms could mask important changes in independent variables, it might be better to use a different statistical technique. There are several analyses to evaluate the combined effect of two or more interval/ratio level independent variables on a dependent variable. These techniques are examined in Chapter 14 in relation to factor analysis and structural equation modeling (SEM).

TABLE 12-7 Interaction Variable for Combined Race and Sex

Individual Variables		Interaction Variable	
Race	White = 0	Race/Sex	White-Male = 1
	Non-White = 1		White-Female = 2
			Non-White-Male = 3
Sex	Male = 0		Non-White-Female =4
	Female = 1		

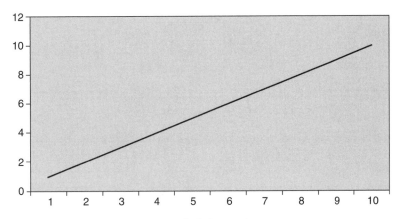

Figure 12-1 Linear Relationship of Variables in OLS Regression

Nonlinear Relationships and Transformations

OLS regression is designed to evaluate linear relationships among a dependent variable and several independent variables. A graphical representation of the relationship of a linear independent and dependent variable in OLS regression is provided in **Figure 12-1**. This linearity allows researchers to interpret a positive or negative change in the dependent variable in relation to some change in an independent variable on a straight line. Linear associations are not the only types of relationships that exist, however.

Sometimes variables and variable interactions will not be linear in association. When this occurs, relationships can be misrepresented in regression analysis (typically no statistical significance is indicated when a relationship does exist). In criminal justice research, there are two common types of nonlinear relationships: parabolic functions and logarithmic functions. If parabolic or logarithmic relationship is present, a variable may have to be "transformed" so the relationship between the dependent variable and an independent variable can be correctly evaluated in regression analyses. **Transformations** involve the mathematical manipulation of data so a relationship between two variables may be "linearized" for purposes of analysis. This could involve addition, subtraction, multiplication, and/or division of the data.

Though transforming data is rather simplistic, there is a debate over why a transformation is necessary. Generally, a nonlinear relationship between nonlinear variables can be suggested by an examination of lower level statistics (univariate and bivariate). According to some, this is not a rationale for immediately transforming a variable. These researchers maintain there has to be a theoretical reason for the transformation. Examples of theoretical rationales for transforming variables are presented in the next section. For now, though, know there should be a theoretical basis for transforming a variable, not simply so the variable can remain in the model. The second issue is the interpretation of transformed data. This will be discussed with respect to specific transformations.

Parabolic Functions

In geometry and/or trigonometry classes, you have probably been exposed to a parabola, or a **parabolic function**. An example of a parabola is provided in Figure 12-2. A parabola has a distinct shape, either a U or an upside down U. If we were to put a

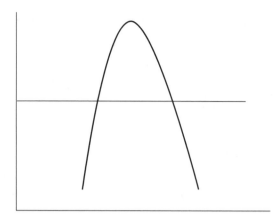

Figure 12-2 Parabolic Relationship of Two Variables

best fitting line through the center of a parabola, much like what occurs in OLS regression, it would not be a good fit, as shown in **Figure 12-2**. In regression, a relationship like this could easily mask the influence of independent variables on the dependent variable. This is a problem.

One instance where this occurs in criminal justice and criminological research is in the age-crime curve. One of the most consistent findings across criminology is the fact that most criminals will begin engaging in crime in their teens, accelerate their involvement through their early to mid 20's, and then desist by their late 20's and early 30's. This aging out process is illustrated in **Figure 12-3**.

Thus the relationship between age and criminality is a parabolic function. Notice that the graph is not a full parabola as seen above; this abrupt stop in the graph indicates an individual's desistance from crime in his or her late 20's.

To transform a variable to account for a parabolic relationship, the independent variable must be squared. The equation for a regression model with a parabolically transformed independent variable is

$$y = a + b_1 x_1 + b_2 x_1^2 + b_n x_n + e$$

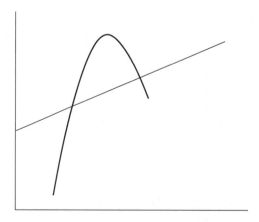

Figure 12-3 The Age–Crime Curve

where y is the dependent variable, a is the slope, $b_1 x_1$ is the original untransformed variable and b-coefficient, $b_2 x_1^2$ is the transformed variable, $b_n x_n$ is any other independent variables and b-coefficients, and e is the error associated with our model. The sign of the b-coefficient is what determines if the U shape of the parabola is upturned or downturned.

In SPSS it is relatively simple to transform any variable using the compute function. **Table 12-8** shows age and age transformed.

Notice that the utility of squared transformations is that it increases the distance between values, or the length of the interval (256-961 versus 16-31). This is what creates the linearity between the independent and dependent variable.

Though it is relatively easy to transform variables, interpretation of what comes out of the regression analysis is more difficult. What does the b-coefficient mean and how would it be interpreted? We cannot interpret a transformed variable's b-coefficient in the same manner as discussed for OLS regression because a parabola is not linear but rather additive as the slope is constantly changing. In the case of a squared independent variable, the b-coefficient generated is a change in y for a change in x given by the slope of the line tangent to the curve. This opens the door to more than one coefficient.

Interpreting a parabolic coefficient must be completed by hand as most statistical packages will not do it for you. First, it is necessary to determine which form of the variable should be used. This is accomplished by evaluating the statistical significance of the transformed variable. If the variable is statistically significant, then it is appropriate to use the transformed variable. Remember, though, to examine the effects of the transformed variable, not the regular version of the variable. If the transformed variable is not statistically significant, use the original, untransformed variable and use the normal procedure for interpreting the regression results.

Next, because it is not advisable to interpret the overall influence of the transformed variable, certain points along the parabola must be examined. Important points on a parabola are obviously at the very tip of the parabola (Why at that age does participation in crime peak?), point of maximum increase, and points of minimum

TABLE 12-8   The Variables for Age and Age Squared

Age	Age-Squared
31	961
26	676
25	625
24	576
24	576
23	529
22	484
18	324
16	256

decrease (Why at those ages does crime increase and decrease more than at other age points?). To evaluate the maximum points, use the following equation:

$$x_1 = -b_1/2b_2$$

where x_1 is a value of the independent variable, b_1 is the b-coefficient associated with the untransformed independent variable, and $2b_2$ is 2 times the b-coefficient associated with the transformed independent variable. By inserting the b-coefficients from the regression output, it is possible to determine both the maximums. To determine the other critical values, or ages in this case, use the following equation:

$$dy/dx = b_1 + 2b_2x$$

where dy/dx is a change in x for a change in y (derivative), b_1 is the b-coefficient associated with the untransformed independent variable, $2b_2$ is 2 times the b-coefficient associated with the transformed independent variable. With respect to important values, simply add in the values in place of the x and then solve for dy/dx (or a change in y for a change in x). This will allow you to see how dy/dx changes with respect to a certain value of an independent variable.

To demonstrate this procedure, we have run a regular OLS regression and have added a squared version of age, based on our knowledge of the age-crime curve, to predict the number of delinquent acts completed by a sample of high school students. **Table 12-9** shows the results of this regression analysis.

TABLE 12-9   OLS Regression Results of Squared Variable

Coefficients[a]

Model		Unstandardized Coefficients		Standardized Coefficients	t	Sig.
		B	Std. Error	Beta	B	Std. Error
1	(Constant)	8.172	5.076		1.610	.009
	Age of juvenile	2.80	.655	-.444	-.428	.049
	Age Squared	.08	.021	.401	.386	.000

[a]Dependent Variable: Number of delinquent activities

Notice we included both age and age squared to determine which one is the more effective in predicting the number of reported delinquent incidents. Because age squared is statistically significant, we would use it. To find the maximum point effect of age on self-reported delinquency, we must use the b-coefficients of both age variables. Recalling the formula for maximum impact points on delinquency, insert the b-coefficients into the equation.

$$x_1 = -2.8/(2 \times 0.08)$$

$$x_1 = -17.5$$

After taking the absolute value of -17.5, we find that 17.5 years is the age where the most delinquency was reported by respondents, or where the peak of the parabola occurs.

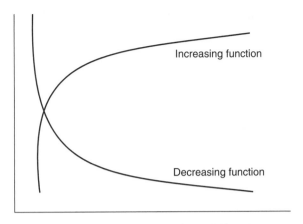

Figure 12-4 Logarithmic Function of Two Variables, Positive and Negative

We might also want to know how much delinquency is associated with the average age of the group of respondents. This is accomplished by using the second equation above. In this case, the mean of the group was 16 years of age. Adding the b-coefficients into the equation we get:

$$dy/dx = 2.8 + 2(0.08)x$$

$$dy/dx = 2.8 + 2(0.08)(16)$$

$$dy/dx = 5.36$$

This suggests that on average, respondents who were 16 years old reported being involved in over 5 delinquent events.

Logarithmic Functions

Another common form of data found in the social sciences is a logarithmic function. An example of this kind of function is provided in **Figure 12-4**.

In both types of logarithmic graphs, notice how there is a quick increase or decrease followed by a leveling off period once some threshold has been achieved. As with a parabolic function, a straight line plotted through either of these graphs would provide very little information.

One of the best examples of these types of functions is the relation between crime and the number of police officers on the streets. It is thought that more officers on the streets would cause crime rates to decrease. Research indicates the number of officers added has an influence up to a point; after that, it does not matter how many officers are added, crime rates will remain stable, as shown in the decreasing curve in Figure 12-4.

To transform a variable to account for this kind of relationship, we must take the logarithm of the independent variable. The equation for a regression model with a logarithm transformed independent variable is

$$y = a + b_1x_1 + b_2\ln x_2 + b_nx_n + e$$

where y is the dependent variable, a is the slope, b_1x_1 is the first variable and b-coefficient, $b_2\ln x_2$ is the logged variable and b-coefficient, b_nx_n is any other independent variables and b-coefficients, and e is the error associated with the model. The b-coefficient is what determines if the function is increasing or decreasing.

TABLE 12-10   **The Variables for Number of Police Officers and the Logarithm of the Number of Officers**

Officers	Officers Logged
529	6.27
576	6.35
961	6.86
324	5.78
256	5.54
484	6.18
625	6.43
676	6.51
576	6.35

In SPSS, it is relatively simple to transform any variable using the compute function. **Table 12-10** shows a variable of the number of police officers in a city and the logarithm of the number of officers in a city.

Transforming independent variables is relatively easy. Notice that the real utility of logged transformations is that they compress and reduce the level of skew in a variable. This makes logging an ideal method of dealing with severely skewed variables. Remember, you need to have a theoretical reason for making this transformation. Do not transform the variable just so you can add it to your regression model.

A logged independent variable has a more straightforward interpretation than a parabolic function. The b-coefficient provided in SPSS output can be interpreted as the rate of change in y for a unit change in the logarithm of x. This interpretation is impractical, though. The easy fix to this is to simply divide the b-coefficient for the logged variable by 100. This will give you the expected change in y units for a one percentage change in x. Then, the b-coefficient can be interpreted as normal.

■ Multicollinearity

A major part of regression analysis is to determine if the model has any severe collinearity issues. This should be undertaken as an assumption of regression and should be tested prior to conducting regression analysis; however, there are some techniques of examining multicollinearity that are undertaken after the regression model has been completed. As discussed previously, social science and criminal justice data often contain interrelated independent variables. When independent variables are interrelated or correlated with each other, there is multicollinearity. Although multicollinearity is present in most criminal justice research, high degrees of it can be a problem. When two independent variables are highly related, they can influence significance tests, standardized coefficients, and unstandardized coefficients. In essence, too much multicollinearity can give distorted results. In this section of the chapter, we explore multicollinearity issues in multiple regression, including how to assess multicollinearity and how to overcome this problem (in some instances). This section is important, as multicollinearity is an issue for most multivariate statistical techniques.

The methods of determining multicollinearity will be used again in Chapter 13 with regard to other forms of regression.

Assessing Multicollinearity

The simplest method of assessing multicollinearity is to use the Pearson's Product Moment Correlation (Pearson's r), discussed in Chapter 9. This will show if there is a strong relationship between two variables. If the Pearson's r is high between two independent variables, say 0.70 or so, you could have problems with multicollinearity in the model. However, simply examining Pearson's r is not enough to fully conclude that there is a severe collinearity issue in the data. The reason for this is that Pearson's r only examines the relationship between two variables. This can cause two problems in a multiple regression when multiple independent variables are included in the model. First, two variables that do not appear to be related in the bivariate analysis can become related by the inclusion of other variables because of interactions. Second, two variables that may appear to be heavily correlated in a bivariate analysis may become disentangled when a number of variables are added into the model. Thus, while a Pearson's Product Moment Correlation is a good start on assessing multicollinearity, it should not be your stopping point.

There are three primary methods of assessing multicollinearity in regression output in SPSS: (1) Tolerance, (2) Variance Inflation Factor (VIF), and (3) the Condition Index Number Test. The procedures for running these tests are shown in the accompanying box.

Tolerance is the first collinearity statistic presented in the SPSS output (**Table 12-11**). Tolerance tells how much of the variance of an independent variable does not depend on other independent variables. Tolerance values greater than 0.25 indicate that there are no problems with multicollinearity. In Table 12-11, all the tolerance values are well above 0.25, indicating that there are no multicollinearity issues in the model.

How do you do that?
Multicollinearity Statistics

1. Open a data set.
   a. Start SPSS.
   b. Select *File*, then *Open*, then *Data*.
   c. Select the file you want to open, then select *Open*.
2. Once the data is visible, select *Analyze*, then *Regression*.
3. Select the variables as described in running a regression analysis.
4. Select the *Statistics* button.
5. Check the box by *Collinearity Diagnostics*, then select *Continue*.
6. Select *OK* to run the regression analysis with the collinearity diagnosis.
7. An output window should appear containing output similar to Table 12-11.

TABLE 12-11   Tolerance and VIF Output for the Number of Months Sentenced in Prison

Coefficients[a]

Model		Unstandardized Coefficients		Standardized Coefficients	t	Sig.	Collinearity Statistics	
		B	Std. Error	Beta			Tolerance	VIF
1	(Constant)	233.701	5.519		42.342	.000		
	Offense Seriousness	–38.228	.696	–.430	–54.902	.000	.477	2.097
	Felony	–33.478	2.625	–.094	–12.754	.000	.542	1.844
	Criminal History Total	3.156	1.022	.017	3.089	.002	.959	1.042
	Total Number of Counts	23.024	.671	.190	34.327	.000	.950	1.053
	Supervision Status	–5.541	3.760	–.008	–1.474	.141	.940	1.064
	Type of Trial	18.661	2.550	.040	7.319	.000	.960	1.042
	Violent Offense	52.109	2.699	.120	19.306	.000	.756	1.323
	Warrant Status	2.009	3.269	.003	.615	.539	.963	1.038
	Drug Offense	–19.241	2.226	–.057	–8.642	.000	.665	1.504
	Delta vs. Other Counties	–7.826	3.505	–.012	–2.233	.026	.959	1.043
	Race	4.166	1.719	.014	2.423	.015	.930	1.076
	Sex	–17.023	2.433	–.038	–6.996	.000	.980	1.020
	Age	.674	.088	.042	7.692	.000	.962	1.039

[a] Dependent Variable: Sentence in Months

The **Variance Inflation Factor** (VIF) is one of the most trusted measures of multicollinearity because it not only indicates if there is a problem, it also indicates which variables are problematic, how severe the problem is, and what happens if the standard error is high. The VIF is also sensitive to linear combinations, which assists in determining the multiple effects of independent variables. A VIF of 4 or less indicates no problem with multicollinearity in the model. A VIF of 5 is acceptable, but anything greater should lead you to explore other collinearity diagnostics to assess problems. In Table 12-11, all VIFs are under 4; thus, there are no problems with multicollinearity.

The **Condition Index Number Test** (CINT) (Belsley, Kuh, and Welsch, 1980) is another way to assess multicollinearity. This output is not provided above as it is very lengthy, but students are encouraged to run the analysis and interpret the CINT in regard to the example. The CINT is a two-pronged analysis. To use the CINT, first examine the *Condition Index Number* column on the output. All of these values should be less than 30. If the value is 30 or greater, follow the row across the different variables. Each of the values in this row should be less than 0.50. If not, the variables that are greater than 0.50 are multicollinear and have to be adjusted for in some way. Methods of correcting multicollinearity are discussed in the next section. For the example used in this chapter, the CINT reveals that there are no multicollinearity issues in this model.

There are several other techniques to discern multicollinearity in a model. First, if the R^2 is significant for a model but none of the independent variables is significant, this can be a result of multicollinearity. Second, if the beta weights have an absolute value greater than 1, there are probably multicollinearity issues. Finally, a powerful method of assessing multicollinearity in multiple regression models is the Haitovsky test (Rockwell, 1975). This test is not available in SPSS; SAS is one of the few programs that offers this procedure. It is worth mentioning here because it is the one test of multicollinearity that is indisputable. If you pass this test, you have no problems with multicollinearity in the model.

Adjusting for Multicollinearity

There are several ways to correct for multicollinearity. The first way is to do nothing. Doing nothing about multicollinearity sometimes makes sense because it will not always reduce or change the significance level of the t-value or change the beta coefficients enough from what is already expected. If you drop a variable from an equation, you are introducing bias into the statistical model that the theoretical model does not call for.

Another way to deal with multicollinearity is simply to increase the total sample size. This decreases the standard error of the sample and makes multicollinearity less of a problem. Although this appears to be an easy fix, in most instances, further data collection is difficult if not impossible. Increasing the sample size is one way of attempting to deal with multicollinearity, but it is usually one of the most difficult ways as well.

A better method of dealing with multicollinearity is to combine the multicollinear variables into an index utilizing factor analysis. This must be supported theoretically, however, before it is attempted. This is somewhat argued against because there is a better statistical technique available that deals with combinations, factors, and indices: *structural equation modeling*. Structural equation modeling (SEM) is a multivariate technique that combines multiple regression and factor analysis. If you want to reduce multicollinearity by using factor analysis, it is more effective to utilize SEM (see Chapter 14 for a discussion of both factor analysis and structural equation modeling).

A fourth way to deal with multicollinearity is to simply drop one of the multicollinear variables. This must also be supported by the theoretical model, however, and will not work if the variable is theoretically central to the model. Finally, you can use part and partial correlations to ascertain which of your multicollinear variables explain more of the variance. Once you have ascertained either one, you can use residuals to separate the multicollinear effects of the two (or more) variables. This procedure is beyond the scope of this book, but further discussion can be found in Lewis-Beck (1980).

■ Conclusion

In this chapter, we have introduced the basics of multiple regression analysis and some of the most common techniques of interpreting statistical output. Multiple regression's two basic jobs are to indicate the combined effect of all independent variables on the dependent variable and to indicate the separate or individual effects of each of the independent variables on the dependent variable. As has been stated, however, this short review of OLS regression does not represent a full understanding of multiple regression analysis; rather, it acts as an introduction to this multifaceted topic. This procedure is very complicated and requires a great deal of time and effort to master. You should expect to take courses on OLS regression and on assessing and correcting multicollinearity in data and to complete several research projects of your own before you begin to truly understand all the inner workings of multiple regression analysis. These procedures are also very computer intensive, so you will need to learn one of the statistical programs, such as SPSS or SAS. This will also require a lot of effort and learning. Once you master these, however (and even before), it can greatly increase your ability to understand the academic articles in criminology, to conduct real world research, and to test theories about the world around you. Chapter 13 deals with extending multiple regression to dependent variables that are limited in nature, i.e., not interval level.

■ Key Terms

adjusted R^2

asterisk

autocorrelation

b coefficient

beta

Condition Index Number Test

dummy coding

dummy variables

error term

error variance

explosive increasing functions

explosive decreasing functions

F-test

first derivative

heteroskedasticity

interaction terms

misspecification

multicollinearity

multiple R

multiple regression

multivariate analysis

OLS regression

parabolic functions

predicted values

R^2

residuals

specification error

standard error

standardized effects

stepwise regression

sum of squares

systematic error

tolerance

transformation

unstandardized effects

unsystematic error

Variance Inflation Factor

■ Key Formulas

Slope/Intercept

$$y = a + bx + e$$

Multiple Regression Equation

$$y = a + b_1x_1 + b_2x_2 + b_3x_3 + \ldots b_nx_n + e$$

Multiple Regression Equation with Parabolically Transformed Variable

$$y = a + b_1x_1 + b_2x_1^2 + \ldots b_nx_n + e$$

Multiple Regression Equation with Log Transformed Variable

$$y = a + b_1x_1 + b_2\ln x_2 + b_nx_n + e$$

Exercises

1. Select three journal articles in criminal justice and criminology that have as their primary analysis Ordinary Least Squares (OLS) regression. For each of the articles:
   a. Trace the development of the concepts and variables to determine if they are appropriate for the analysis. (Why or why not?)
   b. Before looking at the results of the analysis, attempt to analyze the output and see if you can come to the same conclusion(s) as the author(s) of the article.
   c. Discuss the results and conclusions found by the author(s).
   d. Discuss the limitations of the analysis and how it differed from that outlined in this chapter.
2. Using the following regression output, create a presentation table of the results. Once this is complete, interpret the results using the methods outlined in this chapter. The dependent variable for this example is the number of delinquent acts per census tract. Be sure to discuss the significance of the model, the R^2, the significance of the variables in the model, the unstandardized coefficients, and the standardized coefficients.

Model Summary

Model	R	R Square	Adjusted R Square	Std. Error of the Estimate	Change of Statistics				
					R Square Change	F Change	df1	df2	Sig. F Change
1	.751[a]	.564	.481	28.956	.564	6.826	7	37	.000

[a]Predictors: (Constant), OWNER-OCCUPIED HOUSING UNITS, NUMBER OF PEOPLE RECEIVING SOCIAL ASSISTANCE, BLACK HEADS OF HOUSEHOLD, VACANT HOUSING UNITS RENTED OR SOLD BUT NOT OCCUPIED, POPULATION CHANGE, 1980–1990, BOARDED-UP HOUSING UNITS, MEDIAN RENTAL VALUE.

ANOVA[b]

Model		Sum of Squares	df	Mean Square	F	Sig.
1	Regression	40061.840	7	5723.120	6.826	.000[a]
	Residual	31023.138	37	838.463		
	Total	71084.978	44			

[a]Predictors: (Constant), OWNER-OCCUPIED HOUSING UNITS, NUMBER OF PEOPLE RECEIVING SOCIAL ASSISTANCE, BLACK HEADS OF HOUSEHOLD, VACANT HOUSING UNITS RENTED OR SOLD BUT NOT OCCUPIED, POPULATION CHANGE, 1980–1990, BOARDED-UP HOUSING UNITS, MEDIAN RENTAL VALUE.
[b]Dependent Variable: Delinquent Incidents for the Census Tract

Coefficients[a]

Model		Unstandardized Coefficients		Standardized Coefficient			Collinearity Statistics	
		B	Std. Error	Beta	t	Sig.	Tolerance	VIF
1	(Constant)	19.434	29.510		.659	.514		
	NUMBER OF PEOPLE RECEIVING SOCIAL ASSISTANCE	-5.91E-03	.108	−.006	−.055	.957	.902	1.108
	BLACK HEADS OF HOUSEHOLD	.102	.021	.675	4.832	.000	.604	1.654
	VACANT HOUSING UNITS RENTED OR SOLD BUT NOT OCCUPIED	−.133	.248	−.066	−.538	.594	.785	1.273
	POPULATION CHANGE, 1980–1990	−.105	.098	−.134	−1.075	.289	.759	1.317
	BOARDED-UP HOUSING UNITS	.198	.207	.146	.956	.345	.507	1.973
	MEDIAN RENTAL VALUE	-3.13E-02	.094	−.060	−.334	.740	.363	2.758
	OWNER-OCCUPIED HOUSING UNITS	6.07E-04	.008	.011	.077	.939	.541	1.848

[a]Dependent Variable: Delinquent Incidents for the Census Tract

3. As discussed in this chapter and with the example used in Exercise 2, run the same regression, but now obtain the multicollinearity diagnostics using SPSS and census data. Evaluate the tolerance, Variance Inflation Factor, and the Condition Index Number Test. Does the model suffer from multicollinearity?

4. Design a research project as you did in Chapter 1 and discuss how you would incorporate the use of OLS regression into this design. Discuss what limitations or problems you would expect to encounter.

■ References ■

Berry, W. D. (1993). *Understanding Regression Assumption.* Thousand Oaks, CA: Sage Publications.

Galton, F. (1869). *Hereditary Genius: An Inquiry into Its Laws and Consequences.* London: Macmillan.

Galton, F. (1877). Typical laws of heredity. *Nature,* 15:492.

Galton, F. (1885). Opening address. *Nature,* 32:507–510.

Lewis-Beck, M. S. (1980). *Applied Regression: An Introduction.* Thousand Oaks, CA: Sage Publications.

Rockwell, R. C. (1975). Assessment of multicollinearity. *Sociological Methods and Research,* 3:308-320.

Studenmund, A. H. (1997). *Using Econometrics: A Practical Guide.* Reading, MA: Addison-Wesley.

Stolzenberg, R. M. (1980). The measurement and decomposition of causal effects in nonlinear and nonadditive models. In K. F. Scheussler (ed.), *Sociological Methodology*, pp. 459-488. San Francisco, Jossey Bass.

Yule, G. U. (1897). On the theory of correlation. *Journal of the Royal Statistical Society*, 60:812–854.

■ For Further Reading

Belsley, D., E. Kuh, and R. Welsch (1980). *Regression Diagnostics.* Hoboken, NJ: Wiley.

Berry, W. D., and S. Feldman (1985). *Multiple Regression in Practice.* Thousand Oaks, CA: Sage Publications.

Fisher, R. A. (1935). *The Design of Experiments.* Edinburgh: Oliver & Boyd.

Fisher, R. A., and W. A. Mackenzie (1923). The manurial response of different potato varieties. *Journal of Agriculture Science*, 13:311–320.

Fox, J. (1991). *Regression Diagnostics.* Thousand Oaks, CA: Sage Publications.

Garnett, J. C. M. (1919). On certain independent factors in mental measurement. *Proceedings of the Royal Society of London*, 96:91–111.

Goodstein, D. L. (1985). *States of Matter.* New York: Dover.

Hardy, M. A. (1993). *Regression with Dummy Variables.* Thousand Oaks, CA: Sage Publications.

Jaccard, J., R. Turrisi, and C. K. Wan (1990). *Interaction Effects in Multiple Regression.* Thousand Oaks, CA: Sage Publications.

■ Notes

1. Although this formula does not use Greek letters, it should not be assumed that this is sample data rather than population data (see Chapter 1). As stated below, regression is most often a descriptive rather than an inferential statistical procedure. This would typically necessitate the use of Greek letters for formula notation. It is more common, however, to use English notation for regression.

2. For analyses with the dependent variable represented in the original measure (number of crimes, income, etc.), a regression using unstandardized scores is necessary. In other words, do not use rates as dependent variables.

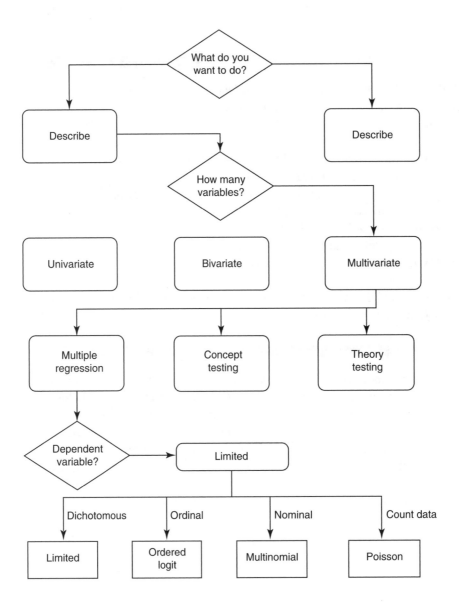

Multiple Regression II: Limited Dependent Variables[1]

CHAPTER 13

In Chapter 12, we explored multiple regression for dependent variables that were continuous and at the interval level. But what happens when the dependent variable you are studying is <u>neither continuous nor at the interval level?</u> Often in research, the dependent variable you are most interested in is dichotomous (e.g., criminal or non-criminal), ordinal level data (strongly agree, agree, disagree, strongly disagree), censored or truncated data, or even nominal level data. In this chapter, we explore alternative regression techniques to address dependent variables you may find in research (both in articles and in studies you might work on) that do not fit the traditional requirement of continuous and interval level data.

■ Dealing with Limited Dependent Variables

As you have learned, statistical techniques and methods have become increasingly complex. As criminal justice data evolved, the belief that a straight line was always an appropriate representation of the relationship between a dependent variable and independent variables began to be questioned and examined. Transformations of independent and dependent variables were perhaps some of the earliest attempts to "linearize" variables (see the discussion in Chapter 12 on transformations). For example, using a logged form of variables (where each data point is multiplied by its natural logarithm value) was introduced to try to linearize those variables that did not fit the traditional regression model (Menard, 1995; Roncek, 1993; Studenmund, 1997).

These techniques did much to linearize data, particularly skewed variables; however, researchers in criminal justice were obtaining more and more data that simply did not fit the requirements of a straight line and could not be forced into a straight line through transformations. As more problems were encountered in attempting to fit data to theoretical explanations, more techniques were designed to accommodate the data that was available. Perhaps some of the most important advances in criminal justice statistical techniques have been techniques involving **limited dependent variables**.

One form of data that does not readily lend itself to linear regression models is that of a dichotomous dependent variable. This type of data has an outcome variable with only two possibilities, typically coded 0 and 1. Recall in the last chapter when the example

---

[1]Special thanks to Rebecca Murray for her dedication in writing this chapter.

focused on the number of months in prison a defendant received. Criminologists are also interested in the determination of guilt, so one of the key outcome measures of a trial might be whether the person was found guilty or not and whether the person received incarceration. In this kind of sentencing data, a variable addressing guilt might be coded 1 for "sentenced to incarceration/guilty" and 0 for "not sentenced to incarceration/not guilty" (Spohn and Holleran, 2002). This type of data does not lend itself to OLS regression, as discussed in Chapter 12, because of important assumptions it violates in terms of the form of the dependent variable and the error term.

OLS Assumptions That Are Violated by Dichotomous Variables

As you saw in Chapter 12, ordinary least squares regression makes several assumptions about the dependent variable. These assumptions are reviewed in **Table 13-1**. Some of these assumptions are violated in the case of dichotomous dependent variables. In this section of the chapter, we show why dichotomized variables cannot be used in OLS regression by showing the assumptions violated by OLS analyses. There is a form of regression, however, that can be used with dichotomized dependent variables. This procedure (Logistic Regression) is then explained in the remainder of the chapter.

The most obvious assumption violated by dichotomized variables is the requirement that a dependent variable be continuous. This is not consistent with a dependent variable being either a "0" or a "1." This becomes problematic when attempting to explain the effect of an independent variable in terms of a "change" in the predicted dependent variable. In short, the only change possible for a dichotomous variable is a change from 0 to 1, or vice versa. This particular assumption may be dealt with (and is dealt with in logistic regression) by altering the examination of the dependent variable to a *probability* of it being 1. In this instance, predicted Y would no longer be either 1 or 0, but would instead be the probability of being 1 (and similarly, the probability of being 0).

Unfortunately, this solution for continuity does not aid another assumption of OLS regression analysis that is violated by dichotomous variables: the assumption that the dependent variable must be unbounded. Altering the dependent variable into a

TABLE 13-1   **Assumptions of OLS Regression**

Assumption
1   The variables are interval level
2   The variables are linearly associated in the coefficients and the error terms
3   There is no specification error in the model
4   Most measurement error (random and systematic) is eliminated
5   Residuals should have a mean of zero (0)
6   No heteroskedasticity in the model
7   No autocorrelation in the model
8   The independent variables are not correlated with the error term
9   There is no perfect multicollinearity between two (2) independent variables

probability by definition bounds this variable between 0 and 1. This boundedness becomes problematic when examining the relationship between an unrestricted independent variable and a restricted dependent variable. In this instance, because the **predicted probabilities** of Y are based on the value of the independent variable (which, again, is unrestricted), these predicted probabilities may take values that are less than 0 or greater than 1, depending on the value of the independent variable. Several techniques have been developed for dealing with this problem, the most promising of which, logistic regression, is explained below.

The use of dichotomous variables and either continuous or dichotomous independent variables also affects assumptions about the error term, specifically by influencing the relationship between independent variables (X) and the dependent variable (Y). Aldrich and Nelson (1984) assert that three requirements of the error term stemming from ordinary least squares regression are problematic when dealing with a dichotomous dependent variable. As discussed in Chapter 12, these requirements include (1) that for each observation, the expected value of the error term is zero; (2) that the error terms are homoskedastic (the variance of the error term is constant for all values of X); and (3) that there is no autocorrelation among the error terms. This is not to say that only these three assumptions exist regarding error terms for OLS regression, but these are the three assumptions most violated when there is a dichotomous dependent variable.

The first requirement about the errors or residuals—that for each observation the expected value of the error term is zero (or that the mean of the errors is zero)—is violated because determining the residual or error depends on the probability of Y being 0 or 1 and on the particular values of X. For instance, when $Y = 0$, then $0 = \Sigma\ bx - M_i$ (the sum of the independent variables times their coefficient plus a residual). Carrying this formula out, M_i would equal $- \Sigma\ bx$. When $Y = 1$, the error term (M_i) would be $1 - \Sigma\ bx$. So the error terms are tied to the independent variable because at one point the errors are equal to $1 - \Sigma\ bx$ and at another they are equal to $- \Sigma\ bx$. To complicate things further, the error terms are also tied to the probability of the dependent variable being either 1 or 0 (they change based on whether Y is 1 or 0). These errors, then, cannot be assumed to have a mean equal to zero, and because of this, they violate one of the assumptions of OLS regression in regard to the expected value of the error term being equal to zero.

The second requirement (regarding homoskedasticity) is also violated with dichotomous dependent variables. Again, the value of the error term changes with the value of the independent variable and with the probability of Y being either 1 or 0. For example, the variance, defined as the expected value of the square of the error term, is equal to PQ (where P is the probability of $X = 1$ and Q is the probability of $X = 0$, or $1 - P$). At its maximum, P and Q would each equal 0.5, and the variance would equal 0.25. As one moves to more extreme values, the variance decreases, because as P goes up, Q goes down, and vice versa. When $P = 0.9$, $Q = 0.1$ and the variance would be $(0.9)(0.1) = 0.09$. Because the variance has a maximum value when $P = 0.5$ and decreases as the value of P goes up or down, the error term is necessarily dependent on the value of Y in a situation where a dichotomous dependent variable is present. This means that, as discussed above, we do not have just one error term with a dichotomous dependent variable, but instead, have an error term that changes based on values of the independent and dependent variables. Because of this, the variance of the error term is not constant, and the errors are in fact inherently heteroskedastic. As you learned in Chapter 12, heteroskedastic errors can cause coefficients that are biased, leading to conclusions that may or may not be accurate—which is very problematic when trying to determine the effect of a host of independent variables on a dependent variable.

This problem was originally dealt with by using a method of weighted b squares or weighted least squares regression. Here, a new variable was created, which was 1 divided by the standard deviation of residuals from the regression on the dependent variable. This new variable was then multiplied by every variable in the regression. This scaled the variable and fixed the standard errors. This is a tempting solution because it only requires correcting the error terms—a fairly straightforward solution that can be accomplished rather simply. Although this did solve the problem of heteroskedasticity, it did not keep predicted values of the dependent variable within the range 0 to 1, and there remained the fact that there was a nonlinear relationship between X and Y (Roncek, 1991). So, although this technique is a quick fix for heteroskedastic errors, it does not account for some of the other problems that go along with this type of limited dependent variable. Because weighted least squares regression is only a temporary fix, we omit a discussion of this analytic technique from this text. To truly account for all problems inherent in having a dichotomous dependent variable, a better model specification is required. This, you will see, is accomplished through logistic regression.

The third requirement related to error terms, autocorrelation among the error terms, is that when the **probability** of a dependent variable being a particular value *is precisely* 1 minus the probability of the dependent variable being another value (which will happen with dichotomized variables because you have only two categories), these probabilities (and the errors associated with them) are correlated absolutely. This can lead to errors in the coefficients but can also be dealt with in a straightforward manner. Because we know the probability that the dependent variable will take on a particular value is 1 minus the probability that it will take on the other value, both probabilities can be interpreted easily. For example, we can use the phrase "There's a 70% chance of rain today" to explain *both* that there is a 70% chance of rain and that there is a 30% chance of no rain.

Due to the particular nature of the dependent variable and the subsequent erratic error terms, the relationship between independent variables and a dichotomous dependent variable is nonlinear. Whereas the relationship in an OLS regression is that Y is exactly equal to the linear combination or sum of the independent variables (Roncek, 1991), this is not the case for a dichotomous variable. As stated earlier, because the effect of X on Y changes with (particularly extreme) values of X, ordinary least squares regression cannot fit a meaningful straight line to this type of data. Instead, graphically, the relationship between continuous independent variables and a dichotomous dependent variable results in a sigmoid, or S-shaped, curve. This violates the regression assumption of a straight-line relationship, and precludes the use of OLS with this type of data.

Though dichotomous dependent variables violate several of the assumptions of OLS regression, many of the assumptions remain the same. **Table 13-2**, provides all of the assumptions for logit regression. This table is divided by the assumptions that are shared between logit and OLS regression as well as those that differ.

Although OLS regression may produce somewhat acceptable results for dichotomous dependent variables for moderate (centered) values of X (because the sigmoid

TABLE 13-2   Assumptions of Logit Regression

Same Assumptions as OLS Regression	
1	Inclusion of all relevant variables in the model
2	Exclusion of all irrelevant variables in the model
3	Error terms are assumed to be independent
4	Does not require linear relationships, but assumes a linear relationship between the logit of the independent variables and the dependent variable
5	Additivity
6	No perfect/excessive multicollinearity between independent variables
7	Large sample size

Assumptions that Differ from OLS Regression	
1	Does not assume a linear relationship between the dependent and independent variables
2	The dependent variable need not be normally distributed
3	The dependent variable need not be homoskedastic for each level of the independent(s)
4	Normally distributed error terms are not assumed
5	No requirement that independent variables be interval level
6	No requirement that the independent variables be unbounded

logit curve is similar to a linear relationship in the middle), particularly for extreme values of the independent variable, OLS regression will yield regression coefficients that are biased or inaccurate, usually underestimating the effect of independent variables. Because OLS regression analysis is inappropriate for dichotomous dependent variables, binary logistic regression, which has relaxed assumptions that are more readily met for these variables, can be employed.

Logistic Regression

Logistic regression (commonly referred to as **logit**, but also known as binary or dichotomous logit) is a technique that works well for dichotomous dependent variables because its assumptions are less restrictive than OLS assumptions. There are other forms of logit regression, as will be discussed below; but if you hear or see the term "logit regression", it is practical to assume that the author or instructor is referring to dichotomous logit regression, or a regression concerning a dichotomous dependent variable.

Logit regression allows a mixture of continuous and dichotomous independent variables and a dichotomous dependent variable. Additionally, the logit technique avoids the "unboundedness problem" by transforming the dependent variable to the log of the odds that a particular score on Y will be 1. So $\ln(P/1 - P) = a + b_1X_1 + b_2X_2$ ..., where ln represents the natural log and P the probability of an event occurring (equaling 1); you will remember the formula to the right of the equal sign as being that for a regression line. In this formula, the dependent variable becomes unbounded, as do the independent variables, ranging from infinity (when $P = 1$) to negative infinity (when $P = 0$). Additionally, whereas the relationship between X and Y is not linear, X is assumed to have a linear relationship with the logit (log of the odds) of P (probability of $Y = 1$).

An Important Note About Odds

We talk about the **odds** of something happening all the time, such as "What are the odds of winning this game?" It is important, however, to remember what odds really are. Odds are the probability that an event will occur divided by the probability that an event will not occur. For example, if I am taking my chances on red on a roulette table, you could say my odds of winning on red are 50/50 (I have a 50% chance of getting red and a 50% chance of getting black). A different way of looking at this is to think of odds as the number of successes over the number of failures. If I were returning to the roulette table but this time was betting on a particular number, my odds of winning would be 1/35 (assuming that there are 36 spaces). It is important to remember that this is different from my *probability* of winning, which would be 1/36, or the total number of possibilities. In the same way, if I roll a die, the odds that I will get a 4 are 1/5, but the probability that I will get a 4 is 1/6.

Although logistic regression finds a best-fitting equation just as linear regression does, the principles on which it does so are different. Instead of using a least squared deviations criterion for the best fit, it uses a **maximum likelihood method**, which maximizes the probability of getting the observed results given the fitted regression coefficients. In other words, the likelihood function expresses the probability of the observed data as a function of the unknown regression parameters, so that the estimated regression coefficients are those that make the observed data most likely. Again, the logit function calculates the natural log of the odds that an event will occur, expressed as $P/P - 1$ (the probability of an event occurring divided by the probability that the event will not occur), where the log of the odds is not bounded by 0 and 1.

In utilizing maximum likelihood, begin by choosing parameters of the model and compute the likelihood of the actual data given those parameters. Then choose parameters that result in the greatest likelihood computed through a series of iterations (Roncek, 1991). In other words, try to pick a set of coefficients and intercepts that maximizes the chances of obtaining the pattern of 0's and 1's observed in the data.

Interpreting Logit Results

Like OLS regression in the previous chapter, we presented here a set manner in which to interpret the statistical tables produced from logit regression analysis. Roncek (1991) developed a straightforward method for interpreting logistic regression results. These steps will be explained according to a data set that was analyzed in SPSS, using logistic regression. This data is from an ICPSR study on gang membership (Esbensen, 2003). Here, we are trying to predict whether a youth will answer yes or no to the question "Have you ever been a gang member?" The answer to this question, coded as "yes" or "no," is used as the dichotomous dependent variable ("no answer" was recoded as a missing value). Independent variables, or predictors of whether someone had ever been a gang member, include sex, race, age, amount of schooling by father, amount of schooling by mother, and whether parents know where the youth is (measured by a Likert scale ranging from "strongly disagree" to "strongly agree").

Step 1: Establishing Statistical Significance

The first step is a determination of whether the model itself is statistically significant. Whereas this is examined through an *F*-test for OLS regression, the significance of a logistic regression model can be determined through maximizing the likelihood of the observed data, given the coefficient estimates. Another way of saying this is that we begin by writing an equation for the likelihood of observing some pattern of 0's and 1's, which will depend on the probability of observing a particular value of the dependent variable. This equation, called the *likelihood equation*, is as follows: $\mathcal{L} = \Pi \pi^{yi} (1 - \pi)^{1 - yi}$, where $\mathcal{L}$ is the likelihood, Π is a symbol meaning to multiply for all cases (much as Σ means to add all cases), π (pi) is a constant (3.16), and y_i means you have selected a particular case from the data. It is not crucial to be able to calculate the likelihood equation by hand, but at least knowing what it is can be helpful. When a program such as SPSS determines the likelihood of the patterns of 0's and 1's in the data, it begins with start values for the coefficient and intercept (start values are often guessed differently by different programs) and then tweaks these values slightly to increase the likelihood of getting the final patterns of 0's and 1's. Called an *iterative procedure*, the tweaking continues until the coefficients and intercept that are most likely to predict the data have been reached.

TABLE 13-3   Steps for Interpreting Logit Regression Results

Step	Key Information Needed	Statistic
1	What is the probability that the results of the model could have happened by chance?	$p < 0.05$ (–2 Log Likelihood)
2	How good are the results?	Nagelkerke R^2; Cox and Snell R^2
3	Are the logit coefficients statistically significant?	$p < 0.05$ (Wald Test)
4	What are the size of the effects of the logit coefficients? (Several options available)	1. Direction (Sign of b) 2. Odds Ratios (Exp (B)) 3. Maximum and Minimal for "Meaningful Values" 4. Predicted Probabilities ($P = e^{a=bx}/1+ e^{a=bx}$)
5	Interpret the standardized effects of the independent variables and rank by importance.	$b * \sigma^2$

How do you do that?
Logistic Regression

1. Open a data set.
   a. Start SPSS.
   b. Select *File*, then *Open*, then *Data*.
   c. Select the file you want to open, then select *Open*.
2. Once the data is visible, select *Analyze*, then *Regression*, then *Binary Logistic* . . . .
3. Select the variable you wish to include as your dependent variable and press the ▶ next to the *Dependent* window.
4. Select the variables you wish to include as your independent variables and press the ▶ next to the *Covariates* window.
5. Leave the method as *Enter*.
6. Click *OK*.
7. An output window should appear containing tables similar in format to Tables 13-4 and 13-5.

In SPSS output, this likelihood ratio is reported as the −2 log likelihood. In effect, this function compares the model with all independent variables specified to a base log-likelihood model that contains an intercept-only model. To test the significance of the model, you would test the change in the log likelihood twice. The difference between the two models is calculated with degrees of freedom $n - k$, where n is the number of cases and k is the number of parameters to be estimated in the model. As with OLS regression, the alpha level is typically set at 0.05. If the difference is statistically significant, the conclusion can be reached that the model as a whole is statistically significant.

In the first part of **Table 13-4**, the Omnibus Tests of Model Coefficients from the SPSS output, you can see the significance of the overall model (given in the stepwise block under the Sig. column). In the second part of the table, the Model Summary, the −2 Log Likelihood for the model is given. If the significance was not given in SPSS output, you would use this number to determine if the overall logistic model was statistically significant. You would simply compare this value to a critical value taken from the table for Chi-Square in the Appendix B Table B-2 with K-1 degrees of freedom and significance level of 0.05. Here, if you look under 6 degrees of freedom (number of variables in the model minus one) for 0.05 (0.95), the obtained value of 4560.254 greatly exceeds 12.592. This value (as well as the Chi-square values in the top part of the table) are so large because we are dealing with over 5000 cases. As stated previously, Chi-square is heavily influenced by large sample sizes, and this is an example of where the sample size may be influencing the statistical significance of the relationship. Using either the −2 Log Likelihood or the Chi-square value in the Sig. column shows that the overall model is statistically significant. The table footnote lets you know how many iterations it took to reach the greatest likelihood of the patterns of 1's and 0's given the parameters of the independent variables. In this model, the estimation was terminated at iteration 5 because the parameter estimates did not get any better at predicting whether someone had been a gang member.

Step 2: Determining the Strength

Now that you know the overall model is significant, the second step in interpreting this model is an assessment of the strength of the model (i.e., how much was explained). As you have learned, for OLS regression this is examined through the R^2. Unfortunately, there is no mathematical counterpart to R^2 in logistic regression, so a relative or pseudo-R^2 value must be used. Luckily, SPSS provides two options for pseudo-R^2 to use in logistic regression. Either the **Cox and Snell R^2** or the **Nagelkerke R^2** can be used in interpreting the strength of the overall model. Each of these has its supporters, and there are advantages and disadvantages to each that are not pursued here. As each of these can be considered a proportional reduction in error, this statistic is analogous to the R^2 value obtained in an OLS regression. In the model in **Table 13-4**, we account for between 7% and just under 12% of the error in predicting if a child was ever in a gang, depending on whether you use the Cox and Snell R^2 or the Nagelkerke R^2. While there are two pseudo-R^2's associated with logit regression, unlike OLS regression, which only generates one true R, the most frequently used R^2 in logit regression at present is the Nagelkerke R^2.

Step 3: Examining Specific Variables

The third step in interpreting logistic regression results is to look at which of the independent variables is statistically significant (Roncek, 1991). Here, a Chi-square test is employed where the coefficients are divided by their standard errors to obtain the Chi-square value called a *Wald Test* (Wald, 1949). These are calculated by taking b/se_b^2 and comparing this number to a critical value of Chi-square obtained from utilizing the appropriate degrees of freedom. If the calculated Chi-square exceeds the table

TABLE 13-4 **Logistic Regression Output**

Omnibus Tests of Model Coefficients

	Step	Chi-square	df	Sig.
1	Step	397.572	6	0.000
	Block	397.572	6	0.000
	Model	397.572	6	0.000

Model Summary

Step	−2 Log Likelihood	Cox and Snell	Nagelkerke
1	4560.254[a]	0.070	0.118

[a]Estimation terminated at iteration 5 because parameter estimates change by less than 0.001.

value, that variable can be considered statistically significant. As with almost all other analyses discussed in this book, hand calculation is not necessary because all the information you need is provided in the printouts. The same is true here. In **Table 13-5** the significance associated with the **odds ratio**, exp(B), is given. Here all of the independent variables except the level of the mother's schooling are significant ($p < 0.05$).

Step 4: Determining the Size of Effects

The fourth step in interpreting logistic regression results is determining the size of the effects for the significant independent variables. As shown above, SPSS displays a B coefficient similar to the one in OLS regression. Here it is referred to as a *logit coefficient.* It is important to keep in mind, however, that this coefficient is not the amount of change in the dependent variable (Y) given a unit change in the independent variable (X), but is, instead, the *log of the odds* of a change in Y for a change in X. Remember that we are working with a dichotomous dependent variable, so we cannot discuss change in terms of this variable because there are only two possible answers. Instead, we explain the results in terms of the change in the log of the odds given by the coefficient listed. In this case, it is the odds ratio, given by Exp(B), that a particular youth would be in a gang given his or her scores on the other variables. In the example above, we can say that with each increased level of education of the father, the odds of having been a gang member decreases by 0.932 (because the B is negative, we have a negative effect on the odds of being a gang member). This is actually not very helpful in the analysis of gang membership. In terms of odds, anything less than 2 is too small to be of much help in the analysis; we would like to see something at least 2 times as likely to occur when we are talking about odds.

Even though this is a fairly straightforward way of interpreting the results, odds are not a common way to interpret the effect of X on Y. Because the effects of independent variables on dichotomous dependent variables can change depending on the value of the particular independent variable, interpreting logit coefficients or odds ratios can be misleading, because they give effects for all values of the independent variable even though the individual values may be different. A better method is to determine the effects *for particular values of the independent variable*. In the example above, then, the effect of father's schooling on being in a gang could be different for different values of schooling. For example, if the father had only 8.5 years of school, there may be a slight effect on whether the child was in a gang; but if the father had 20 years of school, there might be a strong effect on whether the child was a gang member. Because it is not likely that you would want to calculate the odds ratio or the probability for every value of X, choosing meaningful values can be helpful. This being said, choosing "meaningful values" leaves some ambiguity. What is a meaningful value for you may not be meaningful for someone else. Before choosing effects of any value of X, therefore, you should have a good reason for choosing that value. Choosing a meaningful value has some drawbacks, so an alternative technique involves utilizing maximum and minimal effects. Note the term is minimal, not minimum. This gives a range of the effects of continuous independent variables on dependent variables.

Maximum and minimal effects for continuous variables are often of particular interest because they give the range of the effect of the independent variable. The maximum possible effect of a continuous variable is always one-fourth of its logit coefficient (B value), because the maximum occurs for cases with a predicted probability of

0.5 (see assumptions above). Although a fixed minimum does not exist, the logistic curve never reaches a probability of 1 or 0, so we can obtain a **minimal effect** by contrasting its maximum with its effect when the predicted probability is 0.1 or 0.9, 0.09 times the coefficient (Roncek, 1991). In the example in **Table 13-5**, for continuous independent variables such as age, the maximum effect is the logit coefficient of (0.524)(0.25), or 0.131, and the minimal effect is (0.524)(0.09), or 0.047. Again, these maximum and minimal effects are often some of the most important "meaningful values" for logistic models. Obtaining effects at meaningful values, however, is only part of the analysis. Keep in mind that these effects are still in the clumsy *log of the odds*. Taking effects at meaningful values and making the effects themselves meaningful often requires relating them back to probabilities.

Interpretation of logit coefficients can also be related back to probabilities, because odds are often confusing to interpret. Referring back to the equation $\ln[P/(1-P)] = a + bx$, exponentiating the right side of the equation and carrying out some simple algebra will yield the probability:

$$P = \frac{e^{a+bx}}{1 + e^{a+bx}}$$

This gives a probability for an individual case from the logit model, and tends to be more easily interpreted than the log of the odds, or even the odds ratio. This can also be performed at meaningful values of X. Looking at the example in Table 13-5, you can determine the effect of age in terms of probability by plugging the appropriate values into the formula above. Because the mean age is about 14, use this as the meaningful value for X. Plugging this into the formula would result in the calculation

$$P = \frac{e^{-6.850+(0.524)(14)}}{1 + e^{-6.850+(0.524)(14)}}$$

where a is the constant (from Table 13-5), b is from the B column, and x is the mean. The result is 0.619, so we know that the effect of age equaling 14 means a 0.619 higher probability of being in a gang than that of other ages.

Another way to think about the effects of continuous independent variables is through derivatives. As with OLS regression, the effect of continuous independent variables on a dichotomous dependent variable is nothing more than the first-order

TABLE 13-5 Coefficients of the Variables

	B	Std. Error	Wald	df	Sig.	Exp(B)
Sex	−0.495	0.077	40.813	1	0.000	0.610
Race	0.175	0.022	62.521	1	0.000	1.191
Age	0.524	0.059	78.225	1	0.000	1.689
Father's schooling	−0.071	0.023	9.341	1	0.002	0.932
Mother's schooling	−0.030	0.026	1.376	1	0.241	0.970
Parents know whereabouts	−0.360	0.032	125.464	1	0.000	0.697
Constant	−6.850	0.872	61.695	1	0.000	0.001

(partial) derivative. In OLS regression, this is seen as $\partial Y/\partial X$. The difference with logistic regression is that unlike OLS, in which the first-order derivative is a constant, the first-order derivative in logit yields an equation that must be solved for by inputting different values of X: $\partial P/\partial X_i = b_i e^{a+bx}/(1 + e^{a+bx})^2$. Thus, inputting meaningful values as discussed above is nothing more than obtaining the first-order derivative for certain values of X. Just as in OLS, this derivative, will establish the effect of X on Y, but it will be the effect of X on Y assuming X is a certain value.

Interpreting the effects of dichotomous independent variables is a bit easier than continuous variables because, again, there are only two choices for the independent variable (e.g., male or female). This eliminates the need to figure odds or probabilities at meaningful values of the independent variable, because there are only two values. For dichotomous independent variables, it is important to remember that the effects of being in one category of the independent variable also give you the effects of being in the other category, because these effects are 1 minus the first effects. For example, the probability or odds that a male is a gang member is 1 minus the probability or odds that a female is a gang member. Looking to the example again,

$$P = \frac{e^{-6.850 - (0.495)(1)}}{1 + e^{-6.850 + (0.495)(1)}}$$

will yield 0.00065, which means that a male has only a slightly higher probability of being in a gang than a female (assuming that female is the reference category).

Step 5: Determining Standardized Effects

The fifth, and final, step in interpreting logistic regression is to determine the standardized effects of the logit coefficients and, consequently, the rank order of the independent variables in terms of importance. In OLS regression, standardized beta is calculated as

$$\beta = \frac{b(\sigma X)}{(\sigma y)(\sigma N)}$$

Because a logit regression does not have a standard Y but instead is a form of a probability, the standard deviation of a logistic distribution is not 1 as it is with a normal distribution, but instead, is $\pi/\sqrt{3}$. Typically, then, the logit coefficients are standardized by dividing these coefficients by the standard deviation of the logit distribution. This is calculated by the formula:

$$\beta = \frac{b(\sigma X)}{\pi/\sqrt{3}}$$

This method is valid as long as only one dependent variable is used and as long as all of the effects of the independent variables are compared using the same levels of probability (Roncek, 1991).

An alternative and simpler method was developed by Roncek (1997). This may be calculated simply by multiplying the logit coefficient for a particular variable by its corresponding standard deviation. Although this is not a fully standardized procedure, it is acceptable in determining the rank order of standardized effects. This may also be helpful when using SPSS, because standardized coefficients are not reported for logistic regression in this program. At this point, variables can be ranked according to beta size, and ratios of coefficients will be valid. In the example we have been using, running a frequency (see Chapter 3) and including the standard deviation tells us that the standard deviation for "father's schooling" is 1.815 and for "parents know whereabouts" is 1.125. Taking each of these times their logit coefficients gives **semi-standardized coefficients** of −0.129 for father's schooling and −0.405 for parent's knowledge of whereabouts. It is fairly clear that although these variables are both significant, parents' knowledge of the whereabouts of their children seems to have a stronger influence on whether their children have ever been in a gang than the level of the father's schooling.

Interactive Effects and Other Types of Logit

The explanations above deal with basic dichotomous logistic regression, including why it is needed and how it is used. Logistic regression, however, can be utilized for more than working with dichotomous dependent variables. Specifically, it can incorporate interactive terms and can be used for dependent variables that are ordered or are nominal but not dichotomous. These additional functions, discussed below, demonstrate how flexible and useful logistic regression is for many cases of limited dependent variables.

Interactive Terms

Interactive terms can also be used with logit, and are not much more difficult than their use with OLS regression. As with OLS regression, interactive terms among two variables can be added to the model by taking the product of the value of each variable. For example, if you wanted to test an interaction between age and sex, simply take the product of these and add it to the model as another variable. Determining the effects of interactive terms is a bit different than the explanation above, however, because some of the independent variables are functions of other independent variables. Determining effects at particular values of X for interactive terms means holding constant *both* terms that comprise interactive terms.

Ordered/Ordinal Logit

Often in research, you are dealing with survey data, which almost always captures ordinal level data. Until relatively recently, ordinal level dependent variables did not lend themselves to regression analysis. This changed with the development of logit regression analysis. **Ordered logit** regression, also called ordinal logit regression, is similar to dichotomous logit, except that, rather than having a dependent variable with only two outcomes (0 or 1), there is more than one outcome and they are ordinal in nature. This type of dependent variable is often seen in research using Likert scales or other data that is ordered rather than interval. You are probably most familiar with this in terms of instructor evaluations you complete at the end of academic semesters; an example of values here for an ordinal scale range across strongly agree, agree, neutral, disagree, and strongly disagree. The process for analyzing ordered logit is similar to dichotomous logit, with some small exceptions. As with dichotomous logit, you can use Table 13-3 for the steps in interpreting ordinal logit results.

How do you do that?
Ordered Logit Regression

1. Open a data set.
   a. Start SPSS.
   b. Select *File*, then *Open*, then *Data*.
   c. Select the file you want to open, then select *Open*.
2. Once the data is visible, select *Analyze*, then *Regression*, then *Ordinal . . . .*
3. Select the variable you wish to include as your dependent variable and press the ▶ next to the *Dependent* window.
4. Select the variables you wish to include as your independent variables and press the ▶ next to the *Factor(s)* window.
5. Click *OK*.
6. An output window should appear containing tables for your ordered logit regression results.

Ordered logit utilizes an additional step in determining whether its use is appropriate. This is the test for the **proportional odds assumption**, a test that is necessary because there are different categories for the dependent variable (i.e., strongly disagree, disagree, agree, strongly agree). These different categories will have different intercepts, but the slopes of these categories must be parallel to utilize ordered logit. This is a test that determines if the slopes of the various models are equal, or in other words, that the proportional odds are not proportionately different. In this test the null hypothesis states that there is no difference in the slopes or the proportions of the odds. Unlike other significance tests, the null hypothesis in this case must fail to be rejected to justify the use of ordered logit. In other words, to use this technique, you want a nonsignificant Chi-square for this test. Once you have it, you can proceed with calculating ordered logit (see the How Do You Do That box for ordinal logit regression) and interpreting the effects of the independent variables on choosing one category or another just as you did for dichotomous dependent variable logit regression above.

Multinomial Logit

Multinomial logit regression should be employed if the dependent variable is nominal but is not dichotomous. For instance, multinomial logit could be employed for data such as religious affiliation (Catholic, Baptist, Presbyterian, etc.) as the dependent variable. As with dichotomous logit, probabilities are compared to each other, such that the probability of event A is compared with that of event C and the probability of event B is compared with that of event C: $\ln(P_A/P_C)$ and $\ln(P_B/P_C)$. Essentially, this procedure breaks the data down into a number of dichotomous dependent variables and performs a normal binary logistic regression on the various combinations. The number of logit equations is equal to the number of categories minus one. In this example, C is a special category called a *reference category*. This is the category we are most interested in finding out about because it represents a base for all the analyses.

Because multinomial logit employs a series of dichotomous logits, there are some considerations that must be accounted for (Roncek, 1991). First, all categories have to be real alternatives to each other (none can be subcategories of one another). Second, you should not use a reference category that is smaller than the other categories. Once these issues are addressed, the five steps discussed above for logistic regression can be utilized, with the exception that step 4 is omitted for multinomial logistic regression because the unstandardized effects are incalculable.

How do you do that?
Multinomial Logit Regression

1. Open a data set.
   a. Start SPSS.
   b. Select *File*, then *Open*, then *Data*.
   c. Select the file you want to open, then select *Open*.
2. Once the data is visible, select *Analyze*, then *Regression*, then *Multinomial Logistic* . . . .
3. Select the variable you wish to include as your dependent variable and press the ▶ next to the *Dependent* window.
4. Select the variables you wish to include as your independent variables and press the ▶ next to the *Factor(s)* window.
5. Click *OK*.
6. An output window should appear containing tables for your multinomial logit regression results.

Criticisms of Logistic Regression

As with all techniques, logistic regression has some important shortcomings. A common criticism of logistic regression is that the coefficients are based on the values of the independent variables (Liao, 1994). Although this may not be problematic for larger samples, the result may be biased coefficients for smaller sample sizes. A good rule of thumb for logistic regression is that the smaller of the two outcome categories for the dependent variable should have at least 10 times the number of independent variables used in the model. Current advances, such as Exact Logit Techniques (Metah and Patel, 1995), seek to overcome sample size restrictions, although these are not explained here. Other problems include multicollinearity, similar to OLS regression. As discussed in Chapter 12, this can be checked by using the Haitovsky test, looking at VIF's, examining the tolerance, and/or calculating the Condition Index Number. Another potential problem for logistic regression models relates to influential cases (outliers, or cases far outside the rest of the distribution) that may alter results.

Overall, logistic regression procedures deal well with the problem of categorical or ordinal dependent variables. Perhaps the biggest limitation of logit is that effects depend on particular values of independent variables. This limitation can be eased through finding maximum and minimum effects, but remains an important disadvantage of the procedure. Nonetheless, whereas OLS falls short for this particular type of data, logistic techniques have developed methods of compensating for violations of OLS assumptions. Another method of dealing with limited dependent variables is Poisson regression. This procedure is discussed in the next section.

■ Poisson and Negative Binomial Regression

Whereas logit dealt with categorical dependent variables, **Poisson regression** deals primarily with **count data**. Count data is data that is somehow bounded. For example, crime, although a continuous variable, is still bounded by zero. You can have no less than zero crimes in a given place. Although OLS regression is still used to evaluate this type of data, the debate is not yet settled as to whether OLS is the most appropriate analytic technique. Poisson regression appears to be the best method for examining counts of rare events such as crime (Osgood, 2000), so it is beginning to be utilized extensively within criminal justice research (Gardner, Mulvey, and Shaw, 1995; Land, McCall, and Nagin, 1996).

Although count data is not as bounded as a dichotomous variable, it still violates assumptions of both unboundedness and continuity for OLS regression. As such, this type of data is also inherently heteroskedastic, in that the variance of the error term becomes more skewed the closer the dependent variable gets to its lower bound of zero. In the case of primarily criminal justice count data, then, the relationship between the independent variables and dependent variable does not form a straight line, but instead, often forms an L-shaped curve, indicating a high number of zero values.

Like all of the different logit regression analyses discussed above, Poisson regression relaxes the assumption of OLS regression that the independent and dependent variables must be linear in their coefficients. Instead, it assumes that for any given unit, a rate parameter λ (lambda) represents the predicted number of events within a certain period of time. Poisson assumes a linear relationship between the log of lambda and the linear sum of the independent variables: $\ln(\lambda) = \Sigma bx$. In determining the predicted number of events, Poisson uses the following equation to show the probability that P is equal to p (a certain count of events):

$$(P = p) = \frac{e^{-\lambda}\lambda^{y}}{Y!}$$

where $Y = 0, 1, 2, \ldots, \ln(\lambda)$. Again, λ is known as the rate parameter, but when dealing with only one time period, it can also be thought of as the predicted number of events. Here, the log of lambda is a linear function of the independent variables: $\ln(\lambda) = a + \Sigma bx$. By using the log of λ, the independent variables can take on any value, but Σbx will never be less than zero.

Although Poisson regression is more appropriate than OLS regression for count data, it makes some important (and severe) assumptions of its own. First, Poisson regression assumes that events are positive integers. That is, working with count variables such as "number of murders" precludes the use of negative numbers (i.e., there is no such thing as a "negative" count of assault, murder, etc.) Second, Poisson regression assumes that events occur through independent Poisson processes. This is a difficult assumption to meet because it asserts that each event of the dependent variable that occurs (e.g., each assault or each murder) occurs independent of all the other assaults or murders. The third assumption is that the mean is the true rate for each unit. In other words, this model assumes that the mean is the best estimate for each unit of analysis. Finally, it assumes that the mean is exactly equal to the variance, also known as the rate parameter (λ).

This final assumption is particularly difficult to meet for criminal justice data; and in fact, it is often the case with data such as crime counts that the variance exceeds the mean, so the data are considered **overdispersed**. In the case of overdispersion, a different statistical technique is necessary. This procedure is called **negative binomial regression** and accounts for overdispersion in data. Here, the idea is that the number of events experienced by each unit is assumed to follow a Poisson process, but negative binomial regression has another assumption: that the Poisson parameter itself has a **gamma distribution** (Land, McCall, and Nagin, 1996). Thus, it adds an error term and allows for greater variance in the data. Determining whether particular types of data are over- or underdispersed (when the variance is smaller than the mean) cannot be done using SPSS, but can be done easily using PROC GENMOD in SAS and looking at the deviance and Pearsons' statistic, or by calculating a t statistic as outlined in Gardner, Mulvey, and Shaw (1995).

Table 13-6 provides a reference of the assumptions associated with Poisson and Negative Binomial regressions.

A Note About Dispersion in Poisson and Negative Binomial Regression

Criminal justice and criminological data oftentimes suffer from overdispersion. This is especially true for continuous variables where the variance is oftentimes larger than the mean. This dispersion can bias the coefficients and give the illusion of statistical significance when there is none. The most common manner of dealing with overdispersion in data involves the use of negative binomial regression. Again, underdispersion is possible, but usually not a problem with criminological data.

There are two primary methods of determining if there is overdispersion in data. The first is to examine the results of a test of significance of the dispersion parameter, also known as alpha (α). The second is to use the Cameron and Trevidi test. Both tests are not required as the outcome of each is always the same. Based on the conclusion of either of these two tests, you would use either Poisson regression (not overdispersed data) or negative binomial regression (overdispersed data).

Interpreting Poisson and Negative Binomial Regression

The same steps used to interpret logistic regression can be used to interpret Poisson and negative binomial regressions (see Table 13-3). Indeed, steps 1 to 4 are the same as for binary logistic regression, highlighting the similarities in these two techniques for limited dependent variables. In **Table 13-7**, a data set is examined that is attempting to determine the effects of several independent variables on the number of violent crimes across census blocks (Sabol, 2004). Because this data is, in fact, overdispersed, a Negative Binomial regression model has been performed. Because this cannot be analyzed in SPSS, this model has been examined in SAS. If you have access to SAS, it is suggested that you try it to get a better understanding of the various types of statistical software available. Here you can see that the model itself is significant (step 1) because the probability of F (Pr $> F$) is significant to 0.0001 (the same as with OLS regression).

TABLE 13-6   Assumptions of Poisson and Negative Binomial Regression

Assumptions	
1	The dependent variable must have a positive, integer value
2	The values are generated by independent Poisson processes
3	The mean is the true rate for each case
4	Explanatory variables account for all variation among the population
5	The mean is exactly equal to the variance

Step 2, determining how much variance is explained by the model, is also straightforward with a Poisson or negative binomial procedure. SAS gives an R^2 (which is actually a pseudo-R^2) and an adjusted R^2. Both indicate the proportional reduction in error from introducing the independent variables into the model as compared to a model with just an intercept. As discussed previously, the adjusted R^2 accounts for model parsimony. This can be used to determine how much is gained by adding the independent variables into the model.

Determining which independent variables are significant (step 3) is also straightforward (see **Table 13-8**), as SAS gives both the effects and the probability associated with t in its output. As with other procedures, it is simple to determine which independent variables are significant by which have a probability associated with t that is 0.05 or less. In this example, all the independent variables are statistically significant because in the Pr > |t| column, each of the independent variables has a probability equal to 0.05 or less.

Step 4, explaining the effects of particular independent variables, is the same as interpretation for an OLS model in which the dependent variable is logged (Stolzenberg, 1979). As with logistic regression, however, it is important to keep in mind the various interpretations for continuous independent variables and dichotomous independent variables. In addition, for Poisson/negative binomial regression, the interpretation is slightly different for any independent variables that are logged. As you learned earlier, it is sometimes necessary to log independent variables that are highly skewed. Because Poisson/negative binomial is a procedure that uses a logged form of the dependent variable, the interpretation for the effects of logged and unlogged independent variables

TABLE 13-7   Negative Binomial Regression Output

Source	df	Sum of Squares	Mean Square	F Value	Pr > F
Model	7	680,186	97,169	336.46	<0.0001
Error	2390	690,238	288.80238		
Corrected total	2397	1,370,424			

Root MSE	16.99419	R^2	0.4963
Dependent mean	32.95371	Adjusted R^2	0.4949
Coefficient of variation	51.56987		

is slightly different. We start with an interpretation for logged and unlogged continuous independent variables, then look at how to interpret dichotomous independent variables.

For *logged continuous independent variables*, the first-order derivative can be seen as $(\partial Y/Y)/(\partial X/X)$ and interpreted directly as the proportional change in Y for a proportional change in X, or multiplied by 100 and interpreted as the percentage change in Y for a percentage change in X. Population is a variable that is often logged, so it can be used for explaining this interpretation. In **Table 13-8**, interpreting "2000 Population" would mean that you can expect a 0.00650 (or a 0.6%) increase in the number of crimes for each unit increase in the population (often measured in thousands).

For *unlogged continuous independent variables*, the first-order derivative can be seen as $(\partial Y/Y)/\partial X$ and interpreted as the proportional change in Y for a unit change of ∂X in X, or multiplied by 100 and interpreted as the percentage change in Y for a unit change in X. In the example above, you would expect a 0.14525 (or a 14%) increase in the number of crimes for each increase in the rate of vacant homes.

For *dichotomous independent variables*, effects can be seen as $e^b - 1$ and interpreted as the change in the number of events for a change in switching from one category of the independent variable to the other. In the example above, there are no dichotomous independent variables, but for clarification, if the independent variable "percent minority" were simply a black/white dichotomous variable, you would be able to interpret the findings as such: $e^{0.13257} - 1 = -3.02$. This can be interpreted as an expected decrease of 3.02 crimes by switching from white to black, or vice versa (depending on which category is the reference).

Step 5, standardizing effects of independent variables, can be completed in the same way as logistic regression. In other words, utilizing a semi-standardized coefficient by taking the b coefficient times its standard deviation is sufficient to yield an adequate standardized effect. Keep in mind, however, that just like interpreting unstandardized effects in step 4, interpreting standardized effects must be consistent with the type of independent variable you are interpreting. Specifically, standardized effects are still specific to whether the independent variable is logged or unlogged, continuous or dichotomous.

Including several different interpretations for Poisson/negative binomial regression models does add a degree of complexity in explaining the effects of independent variables in the model. These different interpretations, however, also give important

TABLE 13-8 Coefficients of the Variables

Variable	df	Parameter Estimate	Std. Error	t Value	Pr > |t|	Standardized Estimate	Variance Inflation
Intercept	1	−2.54830	1.41434	−1.80	0.0717	0	0
1990 population	1	0.01317	0.00123	10.70	<0.0001	0.70946	20.85343
2000 population	1	0.00650	0.00124	5.24	<0.0001	0.33561	19.47684
Owner-occupied	1	−0.07240	0.00963	−7.52	<0.0001	−1.09699	101.00049
Percent minority	1	0.13257	0.00946	14.02	<0.0001	0.24144	1.40776
Vacancy Rate	1	0.14525	0.05328	2.73	0.0065	0.05324	1.80975
Poverty Rate	1	0.10816	0.02957	3.66	0.0003	0.09440	3.16052

distinctions as to how logged–unlogged or continuous–dichotomous variables influence a count dependent variable. In short, although the explanations of this particular type of model may be cumbersome, they allow for much more precise predictions than possible using a less sophisticated model such as OLS regression.

Criticisms of Poisson and Negative Binomial Models

As with all techniques, Poisson/negative binomial models have received a certain amount of criticism. Perhaps the biggest criticism of this technique is the somewhat severe assumptions of Poisson that were discussed here. Meeting assumptions such as independent processes can be difficult at best and are often simply ignored when calculating this procedure. Other problems have been somewhat addressed, but still leave lingering problems. For instance, though the condition of overdispersion is dealt with through Negative Binomial, it adds the constraint that the error term is assumed to be gamma distributed in the population. This additional assumption has been dealt with only through computationally intensive techniques such as Semi-Parametric Mixed Poisson. Although techniques like this attempt to address the problems of Poisson/ negative binomial regression, they are also very complex and, as such, are not widely used. The technique of Semi-Parametric Mixed Poisson is not explained here, but a good explanation can be found in Land, McCall, and Nagin (1996).

Even with these criticisms and problems that have no easy solution, there is still a great deal of merit to using Poisson or negative binomial models in criminal justice research. Poisson and negative binomial procedures are the most common and straightforward ways of dealing with dependent variables that are count variables. They are fairly simple to calculate using software such as SAS, and are not much more difficult to interpret than OLS procedures that use logged variables. In short, they can be very useful for data that is often utilized in criminal justice research, and very powerful tools for explanation when used correctly.

■ Other Multiple Regression Techniques

Although perhaps the most widely used regression procedures that deal with limited dependent variables have been discussed, you must be aware that the techniques described above are by no means exhaustive for techniques dealing with limited dependent variables. To demonstrate them all comprehensively would take much more space than is available in a single chapter. There are a few techniques, however, that are worth mentioning as important in dealing with limited dependent variables. Although they are not explained in depth here, they are techniques that have been used successfully by many researchers to accommodate limited dependent variables.

Probit Regression

Probit regression is similar to logistic regression (logistic regression and Probit regression produce predicted probabilities that are similar to one another) with some important differences. The difference between logit and Probit is that whereas logit predicts the log of the odds of a dependent variable, Probit uses the cumulative normal probability distribution. The equation for a Probit model is as follows: $\Pr(Y = 1|X) = \Phi(xb)$, where Φ is the standard cumulative normal probability distribution and xb is called the *Probit score* or *index*. Whereas a logit model predicts the log of the odds that a case falls into one category on Y versus another, the Probit model can be thought of as being a Z score. In the Probit model, a unit change in X produces a b-unit change in the cumulative normal probability, or Z score, that Y falls into a particular category. For example, the Probit model would express the effect of a unit change in X on the cumulative normal probability that a crime would be committed on a particular block in any particular year. Both the logit and Probit regression models are estimated by maximum likelihood, so goodness of fit and inferential statistics are based on the log likelihood and Chi-square test statistics.

An example of a dependent variable that would require Probit regression is the number of months sentenced to prison. Generally, only the cases where an offender was sentenced to incarceration will be analyzed using OLS regression. The case outcomes for many individuals are not evaluated because around 60% of all convicted defendants receive probation. If all of the cases where an offender did not receive a sentence behind bars were added in (those cases where the defendant received probation or some community alternative to prison/jail), it would be more appropriate to use a Probit regression model.

Tobit Regression

Tobit regression is often used for data that is censored or truncated. Censored data occurs when the values of the data are clustered around some threshold, either lower (left-censored) or upper (right-censored). For example, if you wanted to measure the amount of serious crime that people engaged in over their lifetimes, most people would answer that they have never engaged in serious crime, so the values of crime participation would be clustered around zero and would be considered left-censored. Truncated data occurs when part of the data is missing. For example, age data for most criminal sentencing stops at 18 years old. That is because people under that age are tried in juvenile court for all but the most serious crimes. This means the data on age and criminal sentencing is missing below 18 years old.

Tobit regression is helpful in that it allows a prediction of censored and truncated data that accounts for a clustered distribution. It does this by using a probabilistic formulation, like other techniques discussed in this chapter, through maximum likelihood, to estimate the regression coefficients for all cases. In the example above it would allow you to estimate the amount of crime that people would have engaged in had they engaged in any crime. Although SPSS does not compute Tobit analyses, Tobit regression can be conducted using other software, such as LIMDEP or SAS.

Multicollinearity and Alternative Regression Techniques

As with OLS regression in the last chapter, multicollinearity is still an issue that must be addressed when utilizing the regression techniques explored in this chapter. Recall, multicollinearity is present whenever two independent variables are closely related. This makes it difficult to determine which variable is having the most influence on the dependent variable.

Although multicollinearity is still an issue with alternative regression techniques for limited dependent variables, there is oftentimes not a procedure to evaluate the extent of the multicollinearity in these models. Because this is the case, you must run an OLS regression in addition to the regression technique utilized for limited dependent variables. The only information you want from the OLS regression analysis is the multicollinearity diagnostics. Remember that to ascertain multicollinearity for OLS regression results in SPSS, there are three statistics: the tolerance, the variance inflation factor (VIF), or the condition index number test (CINT). If there is no multicollinearity in the model, the tolerance values will be greater than 0.25, the VIF values will be equal to 4 or less, and the CINT will be less than 30.

While the use of OLS multicollinearity diagnostics may be counterintuitive when dealing with a dependent variable that is limited in some manner, it is appropriate and accepted. You are not evaluating the relationship of the independent variables to the dependent variable(s). Rather, you are examining the relationship of the independent variables to one another. Thus, multicollinearity is calculated the same across all versions of multiple regression.

■ Conclusion

Dealing with limited dependent variables is certainly a challenge in understanding statistical effects in criminal justice data. This challenge, however, is not insurmountable. Although it may be tempting to force data into the constraints of OLS regression, doing so could have a profoundly negative influence on the results and subsequent explanations of research. With this in mind, then, the most important part of any statistical procedure, including those that use limited dependent variables, is to know your data. As discussed in the procedures outlined in this chapter, one must have knowledge regarding what kind of data comprises both the dependent and independent variables in any research endeavor. This will allow for an analysis that is appropriate to the data you want to use, and results that will be both clear and precise.

This chapter concludes our discussion of statistical regression techniques. The next chapter further expands on the relationship of independent and dependent variables, as well as the relationship of variables and criminological theory, through two other multivariate statistical techniques: factor analysis and Structural Equation Modeling (SEM).

■ Key Terms

count data	odds
Cox and Snell R^2	odds ratio
gamma distribution	ordered logit regression
limited dependent variable	overdispersion
−2 Log Likelihood	Poisson regression
Logistic regression	predicted probabilities
logit	probability
maximum likelihood method	Probit regression
minimal effects	proportional odds assumption
multinomial logit regression	semi-standardized coefficient
Nagelkerke R^2	Tobit regression
negative binomial regression	

■ Exercises

1. Select three journal articles in criminal justice and criminology that have as their primary analysis Logit regression, multinomial regression, and Poisson regression or negative binomial regression (one article for each analysis). For each of the articles:
   a. Trace the development of the concepts and variables to determine if they are appropriate for the analysis. (Why or why not?)
   b. Before looking at the results of the analysis, attempt to analyze the output and see if you can come to the same conclusion(s) as those in the article.

c. Discuss the results and conclusions found by the author.

d. Discuss the limitations of the analysis and how it differed from that outlined in this chapter.

2. Design a research project as you did in Chapter 1 and discuss how you would incorporate the use of logit regression into this design. Discuss what limitations or problems you would expect to encounter.

References

Aldrich, J. H., and F. D. Nelson (1984). *Linear Probability: Logit and Probit Models*. Thousand Oaks, CA: Sage Publications.

Belsley, D., E. Kuh, and R. Welsch (1980). *Regression Diagnostics*. Hoboken, NJ: Wiley.

Esbensen, F. (2003). *Evaluation of the Gang Resistance Education and Training (GREAT) Program in the United States, 1995–1999*. ICPSR Study No. 3337.

Gardner, W., E. P. Mulvey, and E. C. Shaw (1995). Regression analysis of counts and rates: Poisson, overdispersed Poisson, and negative binomial models. *Psychological Bulletin*, 118:392–404.

Land, K. C., P. L. McCall, and D. S. Nagin (1996). A comparison of Poisson, negative binomial, and semiparametric mixed Poisson regression models: With empirical applications to criminal careers data. *Sociological Methods and Research*, 24:387–442.

Liao, T. F. (1994). *Interpreting Probability Models: Logit, Probit, and Other Generalized Linear Models*. Thousand Oaks, CA: Sage Publications.

Menard, S. (1995). *Applied Logistic Regression Analysis*. Thousand Oaks: Sage Publications.

Metah, C. R., and M. R. Patel (1995). Exact logistic regression: Theory and examples. *Statistics in Medicine*, 14:2143–2160.

Osgood, W. D. (2000). Poisson-based regression analysis of aggregate crime rates. *Journal of Quantitative Criminology*, 16:21–43.

Roncek, D. (1991). Using logit coefficients to obtain the effects of independent variables on changes in probabilities. *Social Forces*, 70:509–518.

Roncek, D. (1993). When will they ever learn that first derivatives identify the effects of continuous independent variables, or "Officer, you can't give me a ticket, I wasn't speeding for an entire hour." *Social Forces*, 71:1067–1078.

Roncek, D. (1997). Interpreting the relative importance of negative binomial and Poisson regression coefficients. Paper presented at the 1997 annual meetings of the American Society of Criminology, San Diego, CA.

Sabol, W. J. (2004). *Drug Offending in Cleveland, Ohio Neighborhoods, 1990–1997 and 1999–2001*. ICPSR Study No. 3929.

Spohn, C., and D. Holleran (2002). The effect of imprisonment on recidivism rates of felony offenders: A focus on drug offenders. *Criminology*, 40:329–358.

Stolzenberg, R. (1979). The measurement and decomposition of causal effects in nonlinear and nonadditive models. In K. E. Scheussler (ed.) *Sociological Methodology*, 1980, pp. 459–488. San Francisco, Jossey Bass.

Studenmund, A. H. (1997). *Using Econometrics: A Practical Guide*, Third Edition. New York: Addison-Wesley.

Wald, A. (1949). Note on the consistency of the maximum likelihood estimate. *Annals of Mathematical Statistics*, 20:595–601.

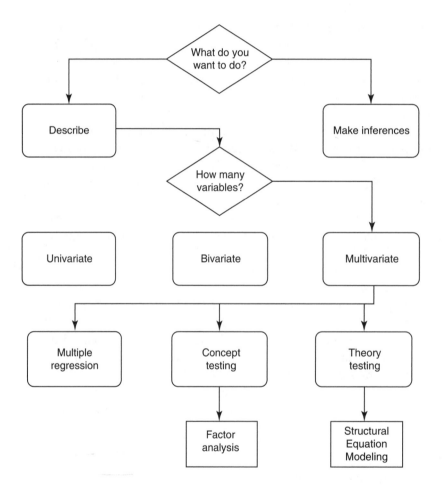

Factor Analysis and Structural Equation Modeling

CHAPTER

14

■ Introduction

Up to this point in the discussion of multivariate statistics, we have focused on the relationship of isolated independent variables on a single dependent variable at some level of measurement. In criminological theory, theorists are often concerned with a collection of variables, not with observable individual variables. In these cases, regression analysis alone is often inadequate or inappropriate. As such, when testing theoretical models and constructs, as well as when examining the interplay between independent variables, other multivariate statistical analyses should be used.

This chapter explores two multivariate statistical techniques that are more effective and more appropriate for analyzing complex theoretical models. The statistical applications examined here are factor analysis and structural equation modeling. SEM

Factor Analysis

Factor analysis is a multivariate analysis procedure that attempts to identify any underlying "factors" that are responsible for the covariation among a group of independent variables. The goals of a factor analysis are typically to reduce the number of variables used to explain a relationship or to determine which variables show a relationship. Like a regression model, a factor analysis is a linear combination of a group of variables (items) combined to represent a scale measure of a concept. To successfully use a factor analysis, though, the variables must represent indicators of some common underlying dimension or concept such that they can be grouped together theoretically as well as mathematically. For example, the variables income, dollars in savings, and home value might be grouped together to represent the concept of the economic status of research subjects.

Factor analysis originated in psychological theory. Based on the work undertaken by Pearson (1901) in which he proposed a "... method of principal axes ...", Spearman (1904) began research on the general and specific factors of intelligence. Spearman's two-factor model was enhanced in 1919 with the development by Garnett of a multiple-factor approach. This multiple-factor model was officially coined "factor analysis" by Thurstone in 1931.

There are two types of factor analyses: **exploratory** and **confirmatory**. The difference between these is much like the difference discussed in regression between testing a model without changing it and attempting to build the best model based on the data utilized.

Exploratory factor analysis is just that: exploring the loadings of variables to try to achieve the best model. This usually entails putting variables in a model where it is expected they will group together and then seeing how the factor analysis groups them. At the lowest level of science, this is also commonly referred to as "hopper analysis," where a large number of variables are dumped into the hopper (computer) to see what might fit together; then a theory is built around what is found.

At the other end of the spectrum is the more rigorous confirmatory factor analysis. This is confirming previously defined hypotheses concerning the relationships between variables. In reality, probably the most common research is conducted using a combination of these two where the researcher has an idea which variables are going to load and how, uses a factor analysis to support these hypotheses, but will accept some minor modifications in terms of the grouping.

It is common for factor analysis in general, and exploratory factor analysis specifically, to be considered a **data reduction** procedure. This entails placing a number of variables in a model and determining which variables can be removed from the model, making it more parsimonious. Factor analysis purists decry this procedure, holding that factor analysis should only be confirmatory; confirming what has previously been hypothesized in theory. Any reduction in the data/variables at this point would signal weakness in the theoretical model.

There are other practical uses of factor analysis beyond what has been discussed thus far. First, when using several variables to represent a single concept (theoretically), factor analysis can confirm that concept by its identification as a factor. Factor analysis can also be used to check for multicollinearity in variables to be used in a regression analysis. Variables that group together and have high factor loadings are typically multicollinear. This is not a common method of determining multicollinearity as it adds another layer of analysis.

There are two key concepts to factor analysis as a multivariate analysis technique: **variance** and **factoral complexity**. Variance is discussed in the four paragraphs below, followed by a discussion of factoral complexity.

Variance comes into play because factor analysis attempts to identify factors that explain as much of the common variance within a set of variables as possible. There are three components to variance: **communality**, **uniqueness**, and **error variance**.

Communality is the part of the variance shared with one or more other variables. Communality is represented by the sum of the squared loadings for a variable (across factors). Factor analysis attempts to determine the factor or factors that explain as much of the communality of a set of variables as possible. All of the variance in a set of variables can be explained if there are as many factors as variables. That is not the goal of factor analysis, however. Factor analysis attempts to explain as much of the variance as possible with the least amount of variables (parsimony). This will become important in the interpretation of factor analysis.

Uniqueness, on the other hand, is the variance specific to a particular variable. Part of the variance in any model can be attributed to variance in each of the component variables (communality). Part of the variance, however, is unique to the specific factor and cannot be explained by the component variables. Uniqueness measures the variance that is reflected in a single variable alone. It is assumed to be uncorrelated with the component factors or with other unique factors.

Error variance is the variance due to random or systematic error in the model. This is in line with the error that was discussed in Chapter 11, and the error associated with regression analysis discussed in the preceding chapters.

Factoral complexity is the number of variables loading on a given factor. Ideally a variable should only load on one factor. This means, theoretically, that you have accurately determined conceptually how the variables will group. The logical extension of this is that you have a relatively accurate measure of the underlying dimension (concept) that is the focus of the research. Variables that load (cross-load) on more than one factor represent a much more complex model, and it is more difficult to determine the true relationships between variables, factors, and the underlying dimensions.

■ Assumptions

As with the other statistical procedures discussed in this book, there are several assumptions that must be met before factor analysis can be utilized in research. Many of these are similar to assumptions discussed in previous chapters; others are unique to factor analysis.

Like regression, the most basic assumption of factor analysis is that the data is interval level and normally distributed (linearity). Dichotomized data can be used with a factor analysis, but it is not widely accepted. It is acceptable to use dichotomized, nominal level data for the principle component analysis portion of the factor analysis. There is debate among statisticians, however, whether dichotomized, nominal level data is appropriate after the factors have been rotated because of the requirement of the variables being a true linear combination of the underlying factors. It has been argued, for example, that using dichotomized data yields two orthogonal (uncorrelated) factors such that the resulting factors are a statistical artifact and the model is misspecified. Other than dichotomized nominal level data, nominal and ordinal level data should not be used with a factor analysis. The essence of a factor analysis is that the factor scores are dependent upon the data varying across cases. If it cannot be reasoned that nominal level data varies (high to low or some other method of ordering) then the factors are uninterpretable. It is also not advised to use ordinal level data because of the non-normal nature of the data. The possible exception to this requirement concerns fully ordered ordinal level data. If the data can be justified as approximating interval level (as some will do with Likert-type data), then it is appropriate to use it in a factor analysis.

The second assumption is that there should be no specification error in the model. As discussed in previous chapters, specification error refers to the exclusion of relevant variables from the analysis or the inclusion of irrelevant variables in the model. This is a severe problem for factor analysis that can only be rectified in the conceptual stages of research planning.

Another requirement of factor analysis is that there be a sufficient sample size so there is enough information upon which to base analyses. Although there are many different thoughts on how large the sample size should be for an adequate factor analysis, the general guidelines follow those of Hatcher (1994), who argued for sample sizes of at least 100, or 5 times the number of variables to be included in the principle component analysis. Any of these kinds of cut-points are simply guides, however, and the actual sample size required is more of a theoretical and methodological issue for individual models. In fact, there are many who suggest that factor analysis is stable with sample sizes as small as 50.

One major difference in the assumptions between regression and factor analysis is multicollinearity. In regression, multicollinearity is problematic; in factor analysis, multicollinearity is necessary because variables must be highly associated with some of the other variables so they will load ("clump") into factors. The only caveat here is that all of the variables should not be highly correlated or only one factor will be present (the only criminological theory that proposed one factor is Gottfredson and Hirschi's work on self-control theory). It is best for factor analysis to have groups of variables highly associated with each other (which will result in those variables loading as a factor) and not correlated at all with other groups of variables.

Analysis and Interpretation

As with most multivariate analyses, factor analysis requires a number of steps to be followed in a fairly specific order. The steps in this process are outlined in **Table 14-1** and discussed below.

Step 1—Univariate Analysis

As with other multivariate analysis procedures, proper univariate analysis is important. It is particularly important to examine the nature of the data as discussed in Chapter 10. The examination of the univariate measures of skewness and kurtosis will be important in this analysis. If a distribution is skewed or kurtose, it may not be normally distributed and/or nonlinear. This is detrimental to factor analysis. Any variables that are skewed or kurtose should be carefully examined before being used in a factor analysis.

Step 2—Preliminary Analysis

Beyond examining skewness and kurtosis, there are a number of preliminary analyses that can be used to ensure the data and variables are appropriate for a factor analysis. These expand upon the univariate analyses and are more specific to factor analysis.

TABLE 14-1   Steps in the Factor Analysis Process

Step 1	Examine univariate analysis of the variables to be included in the factor analysis
Step 2	Preliminary analyses and diagnostic tests
Step 3	Extract factors
Step 4	Factor extraction
Step 5	Factor rotation
Step 6	Use of factors in other analyses

How Do You Do That?
Obtaining Factor Analysis Output in SPSS

1. Open a data set.
   a. Start SPSS
   b. Select *File*, then *Open*, then *Data*
   c. Select the file you want to open, then select *Open*
2. Once the data is visible, select *Analyze, Data Reduction, Factor . . .*
3. Select the independent variables you wish to include in your factor analysis and press the ▶ next to the *Variables* window.
4. Select the *Descriptives* button and click on the boxes next to *Anti-image* and *KMO and Bartlett's Test of Spericity* and select *Continue*.
5. Select the *Extraction* button and click on the box next to *Scree Plot* and select *Continue*.
6. Select *Rotation* and click on the box next to the type of *Rotation* you wish to use and select *Continue*.
7. Click *OK*.
8. An output window should appear containing tables similar in format to the following tables.

First, a **Bartlett's test of sphericity** can be used to determine if the correlation **matrix** in the factor analysis is an **identity matrix**. An identity matrix is a correlation matrix where the diagonals are all 1 and the off-diagonals are all 0. This would mean that none of the variables are correlated with each other. If the Bartlett's Test is not significant, do not use factor analysis to analyze the data because the variables will not load together properly. In the example in **Table 14-2**, the Bartlett's Test is significant, so the data meets this assumption.

It is also necessary to examine the **anti-image correlation matrix** (**Table 14-3**). This shows if there is a low degree of correlation between the variables when the other variables are held constant. Anti-image means that low correlation values will produce large numbers. The values to be examined for this analysis are the off-diagonal values; the diagonal values will be important for the KMO analysis below. In the anti-image matrix in Table 14-3, the majority of the off-diagonal values are closer to zero. This is

TABLE 14-2   **KMO and Bartlett's Test**

Kaiser-Meyer-Olkin Measure of Sampling Adequacy.		.734
Bartlett's Test of Sphericity	Approx. Chi-Square	622.459
	df	91
	Sig.	.000

TABLE 14-3   Anti-image Correlation Matrix*

	ARR ESTR	CHU RCH	CLUBS	CUR FEW	GANG R	GRND DRG	GUN REG	OUT DRUG	OUT GUN	GRUP GUN	UCON VICR	U_GUN _CM	SKIP PEDR	SCH_ DRUG
ARRESTR	.545(a)													
CHURCH	-.010	.648(a)												
CLUBS	-.0412	-.166	.734(a)											
CURFEW	-.055	-.215	-.059	.695(a)										
GANGR	-.141	-.054	.029	.017	.849(a)									
GRND_DRG	.000	.233	-.066	-.160	-.104	.598(a)								
GUN_REG	.076	.041	-.022	.051	-.250	-.125	.768(a)							
OUT_DRUG	.082	.092	-.021	-.024	-.154	.063	-.338	.733(a)						
OUT_GUN	.032	-.004	-.133	.152	-.243	.076	-.036	-.009	.819(a)					
GRUP_GUN	-.015	-.026	.015	.154	-.161	-.121	.063	.052	-.229	.787(a)				
UCONVICR	-.286	.022	.181	.175	-.070	-.252	.128	.087	-.051	.052	.667(a)			
U_GUN_CM	.196	-.116	.052	-.082	-.119	.166	-.316	-.049	-.153	-.022	-.514	.682(a)		
SKIPPEDR	.079	-.209	-.125	-.014	.106	.035	.043	-.023	-.012	.106	.047	-.011	.842(a)	
SCH_DRUG	-.208	-.022	.163	-.042	.038	-.081	-.039	-.323	-.153	-.034	-.062	.085	.03	.727(a)

*See Appendix D for full description of the variables used in Table 14–3.

what we want to see. If there are many large values in the off-diagonal, factor analysis should not be used.

Additionally, this correlation matrix can be used to assess the adequacy of variables for inclusion in the factor analysis. By definition, variables that are not associated with at least some of the other variables will not contribute to the analysis. Those variables identified as having low correlations with the other variables should be considered for elimination from the analysis (dependent, of course, on theoretical and methodological considerations).

In addition to determining if the data is appropriate for a factor analysis, you should determine if the sampling is adequate for analysis. This is accomplished by using the Kaiser-Meyer-Olkin Measure of Sampling Adequacy (Kaiser, 1974a). The KMO compares the observed correlation coefficients to the partial correlation coefficients. Small values for the KMO indicate problems with sampling. A KMO value of 0.90 is best; below 0.50 is unacceptable. A KMO value that is less than 0.50, means you should look at the individual measures that are located on the diagonal in the anti-image matrix. Variables with small values should be considered for elimination. In the example in Table 14-2, the KMO value is 0.734. This is an acceptable KMO value, although it may be useful to examine the anti-image correlation matrix to see what variables might be bringing the KMO value down. For example, the two diagonal values that are between 0.50 and 0.60 are ARRESTR (Have you ever been arrested?) and GRND_DRG (Have you ever been grounded for drugs or alcohol?). ARRESTR cannot be deleted from the analysis because it is one of the dependent variables, but it might be necessary to determine whether GRND_DRG should be retained in the analysis.

Step 3—Extract the Factors

The next step in the factor analysis is to extract the factors. The most popular method of extracting factors is called a **principle components analysis** (developed by Hotelling, 1933). There are other competing, and sometimes preferable, extraction methods (such as **maximum likelihood**, developed by Lawley in 1940). These analyses determine how well the factors explain the variation. The goal here is to identify the linear combination of variables that account for the greatest amount of common variance.

As shown in the principal components analysis in **Table 14-4**, the first factor accounts for the greatest amount of common variance (26.151%), representing an Eigenvalue of 3.661. Each subsequent factor explains a portion of the remaining variance until a point is reached (an Eigenvalue of 1) where it can be said that the factors no longer contribute to the model. At this point, those factors with an Eigenvalue above 1 represent the number of factors needed to describe the underlying dimensions of the data. For Table 14-4, this is factor 5, with an explained variance of 7.658 and an Eigenvalue of 1.072. All of the factors below this do not contribute an adequate amount to the model to be included. Each of the factors at this point are not correlated with each other (they are orthogonal as described below).

Note here that a principal components analysis is **not** the same thing as a factor analysis (the rotation part of this procedure). They are similar, but the principal components analysis is much closer to a regression analysis, where the variables themselves are examined and their variance measured. Also note that this table lists numbers and not variable names. These numbers represent the factors in the model. All of the variables represent a potential factor, so there are as many factors as variables in the left

TABLE 14-4   Total Variance Explained

Total Variance Explained

	Initial Eigenvalues			Extraction Sums of Squared Loadings		
Component	Total	% of Variance	Cumulative %	Total	% of Variance	Cumulative %
1	3.661	26.151	26.151	3.661	26.151	26.151
2	1.655	11.823	37.973	1.655	11.823	37.973
3	1.246	8.903	46.877	1.246	8.903	46.877
4	1.158	8.271	55.148	1.158	8.271	55.148
5	1.072	7.658	62.806	1072	7.658	62.806
6	.961	6.868	69.674			
7	.786	5.616	75.289			
8	.744	5.316	80.606			
9	.588	4.198	84.804			
10	.552	3.944	88.748			
11	.504	3.603	92.351			
12	.414	2.960	95.311			
13	.384	2.742	98.053			
14	.273	1.947	100.000			

Extraction Method: Principal Components Analysis

columns of the table. In the right 3 columns of the table, only the factors that contribute to the model are included (factors 1-5).

The factors in the principal components analysis show individual relationships, much like the beta values in regression. In fact, the factor loadings here are the correlations between the factors and their related variables. The **Eigenvalue** used to establish a cutoff of factors is a value like R^2 in regression. As with regression, the Eigenvalue represents the "strength" of a factor. The Eigenvalue of the first factor is such that the sum of the squared factor loadings is the most for the model. The reason the Eigenvalue is used as a cutoff is because it is the sum of the squared factor loadings of all variables (the sum divided by the number of variables in a factor equals the average percentage of variance explained by that factor). Because the squared factor loadings are divided by the number of variables, an Eigenvalue of 1 simply means that the variables explain at least an average amount of the variance. A factor with an Eigenvalue of less than 1 means the variable is not even contributing an average amount to explaining the variance.

It is also common to evaluate the **scree plot** to determine how many factors to include in a model. The scree plot is a graphical representation of the incremental variance accounted for by each factor in the model. An example of a scree plot is shown in **Figure 14-1**. To use the scree plot to determine the number of factors in the model, look at where the scree plot begins to level off. Any factors that are in the level part of

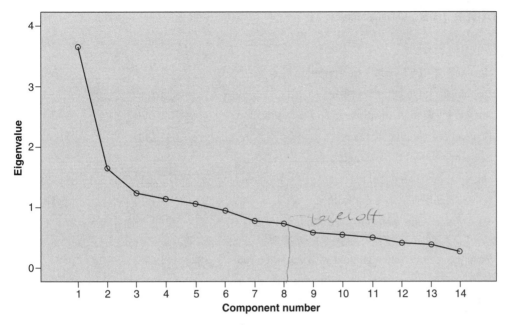

Figure 14-1 Scree Plot

the scree plot may need to be excluded from the model. Excluding variables this way should be thoroughly supported both theoretically and methodologically.

Figure 14-1 is a good example of where a scree plot may be in conflict with the Eigenvalue cutoff, and where it may be appropriate to include factors that have an Eigenvalue of less than 1. Here, the plot seems to level off about the eighth factor, even though the Eigenvalue cutoff is reached after the fifth factor. This is somewhat supported by **Table 14-5**, which shows the initial Eigenvalues. This shows that the Eigenvalues keep dropping until the eighth factor and then essentially level off to Eigenvalues in the 0.2 to 0.5 range. This shows where the leveling occurs in the model. In this case, it may be beneficial to compare a factor model based on an Eigenvalue of 1 cutoff to a scree plot cutoff to see which model is more theoretically supportable.

It is also important in the extraction phase to examine the communality. The communality is represented by the sum of the squared loadings for a variable across factors. The communalities can range from 0 to 1. A communality of 1 means that all of the variance in the model is explained by the factors (variables). This is shown in the "Initial" column of **Table 14-6**. The initial values are where all variables are included in the model. They have a communality of 1 because there are as many variables as there are factors. In the "Extraction" column, the communalities are different and less than 1. This is because only the five factors used above (with Eigenvalues greater than 1) are taken into account. Here the communality for each variable as it relates to one of the five factors is taken into account. Although there are no 0 values, if there were, it would mean that variable (factor) contributed nothing to explaining the common variance of the model.

At this point and after rotation (see below), you should examine the **factor matrix** (**component matrix**) to determine what variables could be combined (those that load together) and if any variables should be dropped. This is accomplished through the

TABLE 14-5   Communalities

	Initial	Extraction
Recoded ARREST varible to represent yes or no	1.000	686
Do you ever go to church?	1.000	.693
Are you involved in any clubs or sports in school?	1 000	600
Do you have a curfew at home?	1.000	.583
Scale measure of gang activities	1.000	.627
Have you ever been grounded because of drugs or alcohol?	1.000	.405
Do you regularly carry a gun with you?	1.000	694
Have your parents ever threatened to throw you out of the house because of drugs or alcohol?	1.000	.728
Have you ever carried a gun out with you when you went out at night?	1.000	.636
Have you ever been in a group where someone was carrying a gun?	1.000	.646
Recoded UCONVIC variable to represent yes or no	1.000	.774
Have any of your arrests involved a firearm?	1.000	.785
Recoded SKIPPED variable to yes and no	1.000	.410
Have you ever been in trouble at school because of drugs or alcohol?	1.000	.528

Extraction Method: Principal Components Analysis

TABLE 14-6   Component Matrix

	Initial	Extraction
ARRESTR	1.000	.686
CHURCH	1.000	.693
CLUBS	1.000	.600
CURFEW	1.000	.583
GANGR	1.000	.627
GRND_DRG	1.000	.405
GUN_REG	1.000	.694
OUT_DRUG	1.000	.728
OUT_GUN	1.000	.636
GRUP_GUN	1.000	.646
UCONVICR	1.000	.774
U_GUN_CM	1.000	.785
SKIPPEDR	1.000	.410
SCH_DRUG	1.000	.528

Extraction Method: Principal Components Analysis

TABLE 14-7   Component Matrix with Blocked Out Values

	Component				
	1	2	3	4	5
GANGR	.751				
GUN_REG	.643	.424			
U_GUN_CM	.637				−.429
UCONVICR	.634				−.405
OUT_GUN	.627				
SCH_DRUG	.494		.442		
SKIPPEDR	−.452	.413			
CHURCH		.525		.514	
GRND_DRG		−.403			
OUT_DRUG	.520		.534		
ARRESTR		−.473		.631	
CURFEW			.433	.463	
CLUBS					.527
GRUP_GUN	.467				.523

Extraction Method: Principal Components Analysis

Note: 5 components extracted.

Factor Loading Value. This is the correlation between a variable and a factor where only a single factor is involved or multiple factors are orthogonal (in regression terms, it is the standardized regression coefficient between the observed values and common factors). Higher factor loadings indicate that a variable is closely associated with the factor. Look for scores greater than 0.40 in the factor matrix. In fact, most statistical programs allow you to block out factor loadings that are less than a particular value (less than 0.40). This is not required, but it makes the factor matrix more readable, and was done in **Table 14-7.**

Step 4—Factor Rotation

It is possible, and acceptable, to stop at this point and base analyses on the extracted factors. Typically, however, the extraction in the previous step is subjected to an additional procedure to facilitate a clearer understanding of the data. As shown in Figure 14-2, the variables are related to factors seemingly at random. It is difficult to determine clearly which variables load together. Making this interpretation easier can be accomplished by rotating the factors. That is the next step in the factor analysis process. **Factor rotation** simply rotates the ordinate plane so the geometric location of the factors makes more sense (see **Figure 14-2**). As shown in this figure, some of the factors (the dots) are in the positive, positive portion of the ordinate plane and some are in the positive, negative portion. This makes the analysis somewhat difficult. By rotating the plane, however, all of the factors can be placed in the same quadrant. This

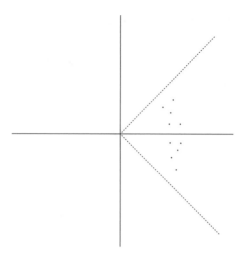

Figure 14-2 Factor Plot in Coordinate Planes

makes the interpretation much simpler, but the factors themselves have not been altered at all.

This step involves, once again, examining the factor matrix. Values should be at least 0.40 to be included in a particular factor. For those variables that reach this cutoff, it is now possible to determine which variables load (group) with other variables. If this is an exploratory factor analysis, this is where a determination can be made concerning which variables to combine into scales or factors. If this is a confirmatory factor analysis, this step will determine how the theoretical model fared under testing.

There are two main categories of rotations: **orthogonal** and **oblique.** There are also several types of rotations available within these categories. It should be noted that, although rotation does not change the communalities or percentage of variation explained, each of the different types of rotation strategies may produce different mixes of variables within each factor.

The first type of rotation is an orthogonal rotation. There are several orthogonal rotations that are available. Each performs a slightly different function, and each has its advantages and disadvantages. Probably the most popular orthogonal procedure is **varimax** (developed by Kaiser in his Ph.D. dissertation and published in Kaiser 1958). This rotation procedure attempts to minimize the number of variables that have high loadings on a factor (thus achieving the goal of parsimony discussed above). Another rotation procedure, **quartermax**, attempts to minimize the number of factors in the analysis. This often results in an easily interpretable set of variables, but where there are a large number of variables with moderate factor loadings on a single factor. A combination or compromise of these two is found in the **equamax** rotation, which attempts to simplify both the number of factors and the number of variables.

An example of a varimax rotation is shown in **Table 14-8**. As with the principal components analysis, the values less than 0.40 have been blanked out. Here, the five factors identified in the extraction phase have been retained, along with the variables from the component matrix in Table 14-7. The difference here is that the structure of the model is clearer and more interpretable.

This table shows that the variables OUT_DRUG (respondent reported being out with a group where someone was using drugs), GUN_REG (respondent carries a gun

TABLE 14-8   Rotated Component Matrix

	Component				
	1	2	3	4	5
OUT_DRUG	.848				
GUN_REG	.778				
GANGR	.503				
GRUP_GUN		.776			
OUT_GUN		.687			
UCONVICR			.789		
U_GUN_CM			.741		
CURFEW				.447	
CHURCH				.819	
SKIPPEDR				.539	
CLUBS				.539	
ARRESTR					.783
SCH_DRUG					.524
GRND_DRG					.524

Extraction Method: Principal Component Analysis. Rotation Method: Varimax with Kaiser Normalization.

regularly), and GANGR (respondent reported being in a gang) are the variables that load together to represent Factor 1; GRUP_GUN (respondent reported being out with a group where someone was carrying a gun) and OUT_GUN (respondent has previously carried a gun) load together to represent Factor 2; UCONVICR (respondent had been convicted of a crime) and U_GUN_CM (respondent had been convicted of a crime involving a gun) load together to represent Factor 3; CURFEW (respondent has a curfew at home), CHURCH (respondent regularly attends church), SKIPPEDR (respondent had skipped school because of drugs), and CLUBS (respondent belongs to clubs or sports at school) load together to represent Factor 4; and ARRESTR (respondent has been arrested before), SCH_DRUG (respondent has been in trouble at school because of drugs), and GRND_DRG (respondent has been grounded because of drugs) load together to represent Factor 5.

A second type of rotation is an oblique rotation. The oblique rotation used in SPSS is **oblimin.** This rotation procedure was developed by Carroll in his 1953 work and finalized in 1960 in a computer program written for IBM mainframe computers. Oblique rotation does not require the axes of the plane to remain at right angles. For an oblique rotation, the axes may be at almost any angle that sufficiently describes the model (see **Figure 14-3**).

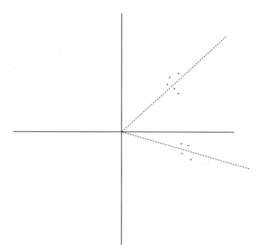

Figure 14-3 Graph of Oblique Rotation

Early in the development of factor analysis, oblique rotation was considered unsound as it was a common perception that the factors should be uncorrelated with each other. Thurstone began to change this perception in his 1947 work, in which he argued that it is unlikely that factors as complicated as human behavior and in a world of interrelationships such as our society could truly be unrelated such that orthogonal rotations alone are required. It has since become more accepted to use oblique rotations under some circumstances.

The way an oblique rotation treats the data is somewhat similar to creating a least-squares line for regression. In this case, however, two lines will be drawn, representing the ordinate axes discussed in the orthogonal rotation. The axes will be drawn so that they create a least-squares line through groups of factors (as shown in Figure 14-3).

As shown in this graph, the axis lines are not at right angles as they are with an orthogonal rotation. Additionally, the lines are oriented such that they run through the groups of factors. This illustrates a situation where an oblique rotation is at maximum effectiveness. The closer the factors are clustered, especially if there are only two clusters of factors, the better an oblique rotation will identify the model. If the factors are relatively spread out, or if there are three or more clusters of factors, an oblique rotation may not be as strong a method of interpretation as an orthogonal rotation. That is why it is often useful to examine the factor plots, especially if planning to use an oblique rotation.

Although there are similarities between oblique and orthogonal rotations (i.e., they both maintain the communalities and variance explained for the factors), there are some distinct differences. One of the greatest differences is that, with an oblique rotation, the factor loadings are no longer the same as a correlation between the variable and the factor because the factors are not independent of each other. This means that the variable/factor loadings may span more than one factor. This makes interpretation of what variables load on which factors more difficult because the factoral complexity is materially increased. As a result, most statistical programs, including SPSS, incorporate both a factor loading matrix (**pattern matrix**) and a factor **structure matrix** for an oblique rotation.

The interpretation of an oblique rotation is also different than with an orthogonal rotation. It should be stressed, however, that all of the procedures up to the rotation phase are the same for both oblique and orthogonal rotations. It should also be noted that an orthogonal rotation can be obtained using an oblique approach. If the best angle of the axes happened to be at 90 degrees, the solution would be orthogonal even though an oblique rotation was used.

The difference in the output between an orthogonal and an oblique rotation is that the pattern matrix for an oblique rotation contains negative numbers. This is not the same as a negative correlation (implying direction). The angle in which the axes are oriented is measured by a value of delta (f) in SPSS. The value of f is at 0 when the factors are most oblique, and negative value of f means the factors are less oblique. Generally, negative values are preferred, and the more negative the better. This means that, for positive values, the factors are highly correlated; for negative values, the factors are less correlated (less oblique); and when the values are negative and large, the factors are essentially uncorrelated (orthogonal).

Table 14-9 does not contain a lot of negative values. There are also a fairly large number of values much greater than zero. These variables should be carefully examined to determine if they should be deleted from the model; and the value of using an oblique rotation in this case may need to be reconsidered.

TABLE 14-9   Pattern Matrix

	Component				
	1	2	3	4	5
GANGR	.418	.048	.433	.144	−.253
GUN_REG	.092	−.026	.760	−.141	−.164
U_GUN_CM	.127	.210	.313	−.164	−.730
UCONVICR	.101	−.031	−.078	.276	−.792
OUT_GUN	.666	.143	.244	−.038	−.146
SCH_DRUG	−.023	−.087	.480	.483	.008
SKIPPEDR	−.152	.510	−.050	−.174	.101
CHURCH	−.003	.836	−.090	.018	−.111
GRND_DRG	.050	−.304	.123	.487	.075
OUT_DRUG	−.046	−.070	.869	−.040	.077
ARRESTR	.069	.084	−.229	.801	−.168
CURFEW	−.516	.421	.152	.311	.149
CLUBS	.281	.492	.016	−.005	.550
GRUP_GUN	.800	−.080	−.059	.129	.112

Extraction Method: Principal Components Analysis. Rotation Method: Oblimin with Kaiser Normalization.

Note: Rotation converged in 20 iterations.

TABLE 14-10   **Structure Matrix**

	Component				
	1	**2**	**3**	**4**	**5**
GANGR	.568	−.088	.576	.229	−.441
GUN_REG	.274	−.089	.798	−.041	−.329
U_GUN_CM	.326	.088	.451	−.099	−.777
UCONVICR	.296	−.199	.128	.340	−.826
OUT_GUN	.732	.030	.395	.061	−.330
SCH_DRUG	.114	−.185	.528	.542	−.131
SKIPPEDR	−.256	.572	−.151	−.273	.238
CHURCH	−.089	.822	−.121	−.107	.034
GRND_DRG	.126	−.378	.187	.542	−.045
OUT_DRUG	.112	−.103	.845	.050	−.091
ARRESTR	.110	−.053	−.107	.783	−.187
CURFEW	−.547	.447	.026	.216	.282
CLUBS	.098	.543	−.069	−.098	.556
GRUP_GUN	.780	−.170	.095	.183	−.087

Extraction Method: Principal Components Analysis. Rotation Method: Oblimin with Kaiser Normalization.

The structure matrix for an oblique rotation displays the correlations between factors. An orthogonal rotation does not have a structure matrix because the factors are assumed to have a correlation of 0 between them (actually the pattern matrix and the structure matrix are the same for an orthogonal rotation). With an oblique rotation, it is possible for factors to be correlated with each other and variables to be correlated with more than one factor. This increases the factoral complexity, as discussed above, but it does allow for more complex relationships that are certainly a part of the intricate world of human behavior.

As **Table 14-10** shows, GANGR loads fairly high on Factors 1 and 3. This is somewhat supported by the direct relation between GANGR and these factors shown in the pattern matrix; but the values in the structure matrix are higher. This is a result of GANGR being related to both factors; thus part of the high value between GANGR and Factor 3 is being channeled through Factor 1 (which essentially serves as an intervening variable).

Step 5—Use of Factors in Other Analyses

After completing a factor analysis, the factors can be used in other analyses, such as including the factors in a multiple regression. The way to do this is to save the factors as variables; therefore the factor values become the values for that variable. This creates a kind of scale measure of the underlying dimension that can be used as a measure of the concept in further analyses. Be careful of using factors in regression, however. Not that it is wrong, but factored variables often have a higher R^2 than might be expected in the model. This is because each of the factors is a scale measure of the underlying dimension. A factor with three variables known to be highly correlated will naturally have a higher R^2 than separate variables that may or may not be correlated.

Another problem with factors being utilized in multiple regression analysis is the interpretation of a factor that contains two variables, if not more. In this case, the b-coefficient is rendered useless. Because of the mathematical symbiosis of the variables, the only coefficient that can adequately interpret the relationship is the standardized coefficient (**beta**).

Factor analysis is often a great deal of work and analysis. Because of this, and the advancement of other statistical techniques, factor analysis has fallen largely into disuse. While originally factor analysis' main competition was **path analysis**, now structural equation modeling (SEM) has taken the advantage over both statistical approaches.

■ Structural Equation Modeling

Structural equation modeling (SEM) is a multi-equation technique in which there can be multiple dependent variables. Recall that in all forms of multiple regression, we used a single formula ($y = a + bx + e$) with one dependent variable. Multiple equation systems allow for multiple indicators for concepts. This requires the use of **matrix algebra**, however, which greatly increases the complexity of the calculations and analysis. Fortunately, there are statistical programs such as AMOS or LISREL that will perform the calculations, so we will not address those here. Byrne (2001) explores SEM using AMOS and Hayduk (1987) examines SEM using LISREL. Though the full SEM analysis is beyond the scope of this book, we will address how to set up a SEM model and understand some of the key elements.

Structural equation models consist of two primary models: Measurement (null) and structural models. The measurement model pertains to how observed variables relate to unobserved variables. This is important in the social sciences as every study is going to be missing information and variables. As discussed in Chapter 2, there are often confounding variables in research that are not included in the model or accounted for in the analysis, but which have an influence on the outcome (generally through variables that are included in the model). Structural Equation Models deals with how concepts relate to one another and attempts to account for these confounding variables. An example of this is the relationship between socioeconomic status (SES) and happiness. In theory, the more SES you have, the happier you should be; however, it is not the SES that makes you happy but the confounding variables like luxury items, satisfactory job, etc. that may be making this relationship.

SEM addresses this kind of complex relationship by creating a theoretical model of the relationship that is then tested to see if the theory matches the data. As such, SEM is a confirmatory procedure; a model is proposed, a theoretical diagram is generated, and an examination of how close the data is to the model is completed. The first step in creating a SEM is putting the theoretical model into a path diagram. It may be beneficial to review the discussion of path models presented in Chapter 3 as these will be integral to the discussion of SEM that follows. In SEM, we create a path diagram based on theory and then place the data into an SEM analysis to see how close the analysis is to what was expected in the theoretical model. We want the two models to not be statistically significantly different.

Variables in Structural Equation Modeling

Like many of the analysis procedures discussed throughout this book, SEM has a specific set of terms that must be learned. Most importantly in SEM, we no longer use the terms dependent and independent variables because there can be more than one dependent variable in an analysis. For the purposes of SEM, there are two forms of variables: exogenous and endogenous. Exogenous variables are always analogous to independent variables. Endogenous variables, on the other hand, are variables that are at some point in the model a dependent variable, though at other points they may be independent variables.

SEM Assumptions

Five assumptions must be met for structural equation modeling to be appropriate. Most of these are similar to the assumptions of other multivariate analyses. First, the relationship between the coefficients and the error term must be linear. Second, the residuals must have a mean of zero, be independent, be normally distributed, and have variances that are uniform across the variable. Third, variables in SEM should be continuous, interval level data. This means SEM is often not appropriate for censored data. The fourth assumption of SEM is no specification error. As noted above, if necessary variables are omitted or unnecessary variables are included in the model, there will be measurement error and the measurement model will not be accurate. Finally, variables included in the model must have acceptable levels of kurtosis. This final assumption bears remembering. An examination of the univariate statistics of variables will be key when completing a SEM. Any variables with kurtosis values outside the acceptable range will produce inaccurate calculations. This often limits the kinds of data most used in criminal justice and criminology research. Though scaled ordinal data is sometimes still used in SEM (continuity can be faked by adding values), dichotomous variables often have to be excluded from SEM as binary variables have a tendency to suffer from unacceptable kurtosis and non-normality.

Advantages of SEM

There are three primary advantages to SEM. These relate to the ability of SEM to address both direct and indirect effects, the ability to include multi-variable concepts in the analysis, and inclusion of measurement error in the analysis. These are addressed below.

First, SEM allows researcher to identify direct and indirect effects. Direct effects are principally what we look for in research—the relationship between a dependent variable we are interested in explaining (often crime/delinquency) and an independent variable we think is causing or related to the dependent variable. This links directly from a cause to an effect. For example:

The hallmark of a direct effect is that an arrow goes from one variable only into another variable. In this example, delinquent peers are hypothesized to have a direct influence on an individual engaging in crime.

An indirect effect occurs when one variable goes through another variable on the way to some dependent or independent variable. For example:

A path diagram of indirect effects indicates that age has an indirect influence on crime by contributing the type of peers one would associate with on a regular basis.

Complicating SEM models, it is also possible for variables in structural equation modeling to have both direct and indirect effects at the same time. For example:

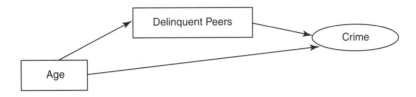

In this case, age is hypothesized to have a direct effect on crime (age-crime curve), but it also has an indirect effect on crime through the number of delinquent peers with which a juvenile may keep company. These are the types of effects that are key in SEM analysis.

The second advantage of SEM is that it is possible to have multiple indicators of a concept. The path model for a multivariate analysis based on regression might look like **Figure 14-4.**

With multiple regression analysis we only evaluate the effect of individual independent variables on the dependent variable. Although this creates a fairly simple path

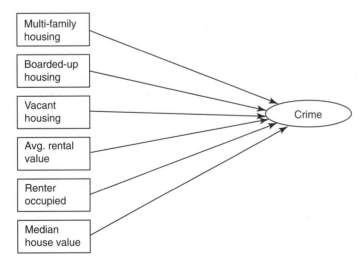

Figure 14-4 Path Model of Regression Analysis on the Effect of Social Disorganization on Crime at the Block Level

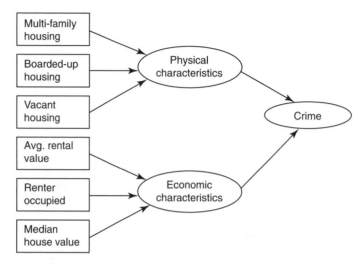

Figure 14-5 Path Model of SEM Analysis on the Effect of Social Disorganization on Crime at the Block Level

model, it is not necessarily the way human behavior works. SEM allows for the estimation of the combined effects of independent variables into concepts/constructs. **Figure 14-5** illustrates the same model as Figure 14-4 but with the addition of theoretical constructs linking the independent variable to the dependent variable.

As shown in Figure 14-5, the variables are no longer acting alone, but in concert with like variables that are expected conceptually to add to the prediction of the dependent variable. In the case of social disorganization, physical and economic characteristics of a neighborhood are predicted to contribute to high levels of crime.

The final advantage of SEM is that it includes measurement error into the model. As previously discussed, path analyses often use regression equations to test a theoretical causal model. A problem for path analyses using regression as its underlying analysis method is that, while these path analyses include error terms for prediction, they do not adequately control for measurement error. SEM analyses do account for measurement error, therefore providing a better understanding of how well the theoretical model predicts actual behavior. For example, unlike the path analysis in Figure 14-5, **Figure 14-6** indicates the measurement error in the model.

The e represents the error associated with the measurement error of each measured variable in the structural equation model. Notice there are no error terms associated with the constructs of physical and economic characteristics. This is because these concepts are a combination of measured variables and not measures themselves.

SEM Analysis

There are six steps in conducting a structural equation model analysis. These are listed in **Table 14-11.**

The first part of any SEM is specifying the theoretical/statistical model. This part of the process should begin even prior to collecting data. Proper conceptualization, operationalization, and sampling are all key parts of Step 1.

Once conceptualization and model building are complete, the second step is to develop measures expected to be representative of the theoretical model and to collect

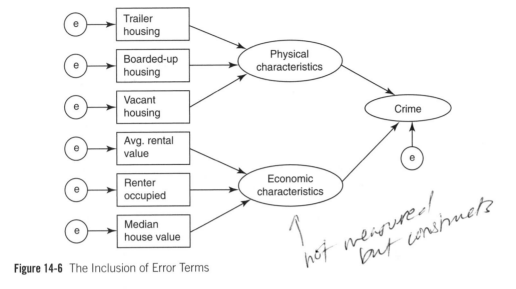

Figure 14-6 The Inclusion of Error Terms

(handwritten annotation: "not measured but constructs")

TABLE 14-11 Steps in SEM Analysis Process

Step 1	Specify the model
Step 2	Select measures of the theoretical model and collect the data
Step 3	Determine whether the model is identified
Step 4	Analyze the model (χ^2)
Step 5	Evaluate the model fit (How well does the model account for your data?)
Step 6	If the original model does not work, respecify the model and start again

data. This is more of an issue for research methods than statistics, so you should consult a methodology book for guidance here.

The third step in structural equation modeling is determining if the model is identified. Identification is a key idea in SEM. Identification deals with issues of whether there is enough information (variables) and if that information is distributed across the equations in a way that allows estimation of coefficients and matrices that are not known (Bollen, 1989). If a model is not identified in the beginning of the analysis, the theoretical approach must be changed or the analysis abandoned. There are three types of identified models: overidentified, just identified, and underidentified. Overidentified models permit tests of theory. This is the desired type of model identification. Just identified models are considered not interesting. These are the types of models researchers deal with when using multiple regression analyses. Underidentified models suggest that we can not do anything with the model until it has been re-specified or additional information has been gathered.

There are several ways to test identification. First, is the t-Rule. This test provides necessary, but not sufficient, conditions for identification. If this test is not passed, the model is not identified. If the test is passed, the model still may not be identified. The Null B Rule applies to models in which there is no relationship between the endogenous variables, which is rare. The Null B Rule provides sufficient, but not necessary conditions for identification. This test is not used often, with the exception of no relationships existing between endogenous variables. The Recursive Rule is sufficient but not necessary for identification. To be recursive, there must not be any feedback loops among endogenous variables. An OLS regression is an example of a non-recursive model as long as there are no interaction terms involved. The Recursive rule does not help establish identification for models with correlated errors. The Order Condition Test suggests that the number of variables excluded from each equation must be at least P–1, where P is the number of equations. If equations are related, then the model is underidentified. This is a necessary, but not a sufficient, condition to claim identification. The final test of identification is the Rank Condition test. This test is both necessary and sufficient, making it one of the better tests of identification. Passing this test means the model is appropriate for further analyses.

The fourth step in SEM analysis is analyzing the model. Unlike multiple regression analyses, SEM utilizes more than one coefficient for each variable. SEM is therefore based on matrix algebra with multiple equations. The primary matrices used in SEM are the **covariance** matrix and variance-covariance matrix. These are divided into four matrices of coefficients and four matrices of covariance. The four matrices of coefficients are: 1) a matrix that relates the endogenous concepts to each other (β); 2) a matrix that relates exogenous concepts to endogenous concepts (γ); 3) a matrix that relates endogenous concepts to endogenous indicators (λy); and, 4) a matrix that relates exogenous concepts to exogenous indicators (λx). Once these relationships have been examined for the endogenous and exogenous variables, the covariance among the variables must be examined. This is accomplished through the four matrices of covariance, which are: 1) covariance among exogenous concepts (Φ); 2) covariance among errors for endogenous concepts (Ψ); 3) covariance among errors for exogenous indicators ($\Theta\delta$); and 4) covariance among errors for endogenous indicators ($\Theta\epsilon$). Within these models, we are interested in three statistics: the mean, variance, and covariance.

Although much of SEM analysis output is similar to regression output (R^2, b-coefficients, standard errors, and standardized coefficients), it is much more complicated. In SEM, researchers have to contend with, potentially, four different types of coefficients and four different types of covariances. This typically means having to examine up to twenty pages of output. Due to space limitations and the complexity of the interpretation, a full discussion of SEM output is not included in this book. Anyone seeking to understand how to conduct an SEM should take a class strictly on this method. As was indicated above, however, the most important element of SEM is its ability to evaluate an overall theoretical model. This portion of the anlaysis is briefly discussed here.

The key equation in SEM is the basis for the structural model. This equation is used in each of the eight matrices. The formula to examine the structural model is:

$$\eta = \beta\eta + \gamma\xi + \zeta$$

Where η is the endogenous (latent) variable(s), ξ is the exogenous (latent) variable(s), β are coefficients of endogenous variables, γ are coefficients of exogenous variables, and ζ is any error among endogenous variables. This is the model that underlies structural equation model analysis.

The two other equations that need to be calculated are the exogenous measurement model and the endogenous measurement model. The formula for the exogenous measurement model is:

$$X = \lambda_X \xi + \delta$$

Where X is an exogenous indicator, λ_x is the coefficient from exogenous variables to exogenous indicators, ξ is the exogenous latent variable(s), and δ is error associated with the exogenous indicators. The endogenous measurement model is:

$$Y = \lambda_y \eta + \epsilon$$

Where Y is an endogenous indicator, λ_y is the coefficient from endogenous variables to endogenous indicators, η is the endogenous latent variable(s), and ϵ is error associated with the endogenous indicators.

These matrices and equations will generate the coefficients associated with the structural equation model analysis. After all of these have been calculated, it is necessary to examine which model, the structural model or the measurement model, can better predict the endogenous variables. This is accomplished by using either Chi-Square or an Index of Fit. Remember that when using Chi-Square as the method to determine the validity of the model, you want a value associated less than 0.05. If the Chi-Square is significant, it means that the hypothesized model is better than if it had been "just identified."

Besides Chi-Square, there are ten different **indices of fit** to choose from to determine how well the theoretical model is at predicting endogenous variables. **Table 14-12** provides the list of Indices of Fit used in SEM.

A full discussion of these is beyond this book. Suffice it to say that GFI and the AGFI produce a score between 0 and 1. A score of 0 implies poor model fit; a score of 1 implies perfect model fit. These measures are comparable to R^2 values in regression. The NFI,

TABLE 14-12   **SEM Indices of Fit**

Shorthand	Index of Fit
GFI	Goodness of Fit
AGFI	Adjusted Goodness of Fit
RMR	Root Mean Square Residual
RMSEA	Root Mean Square Error of Approximation
NFI	Normal Fit Index
NNFI	Non-Normal Fit
IFI	Incremental Fit
CFI	Comparative Fit
ECVI	Expected Cross Validation
RFI	Relative Fit

NNFI, IFI, and CFI all indicate the proportion in improvement of the proposed model relative to the null model. A high value here indicates better fit of the model.

The sixth step in the SEM analysis is a decision of the contribution of the model. If the model has been shown to be adequate, the theoretical model is supported and you have "success." If the model was not adequate (underspecified), you must either abandon the research or go back and re-conceptualize the model and begin again.

There is not enough room in this chapter to provide a full example of SEM output. Even in a simplistic example, the amount of output is almost twenty pages in length. This discussion is meant to give you a working knowledge of structural equation modeling such that you can understand its foundation when reading journal articles or other publications.

■ Conclusion

This chapter introduced you to multivariate techniques beyond multiple regression analysis, principally factor analysis and structural equation modeling. This short review does not represent a full understanding of these multivariate statistical analyses. These procedures are complicated and require a great deal of time and effort to master. You should expect to take a course on each of these and complete several research projects of your own before you begin to truly understand them.

This chapter brings to an end the discussion of multivariate statistical techniques. Although other analytic strategies are available, such as hierarchal linear modeling (HLM) and various time-series analyses, this book is not the proper platform for a description of these techniques. In the next chapters of the book, we address how to analyze data when you do not have a population and must instead work with a sample. This is inferential statistical analysis.

■ Key Terms

anti-image correlation matrix
Bartlett's test of sphericity
beta
communality
component matrix
confirmatory factor analysis
covariance
data reduction
Eigenvalue
endogenous variables
equamax
error variance
exogenous variables
exploratory factor analysis
factor analysis
factor loading value
factor matrix
factor rotation
factorial complexity
identification

identity matrix
indices of fit
matrix
matrix algebra
maximum likelihood
oblimin
oblique
orthogonal
path analysis
pattern matrix
principal components analysis
quartermax
scree plot
structural equation modeling
structure matrix
uniqueness
variance
varimax

■ Key Equations

Slope-Intercept/Multiple Regression Equation

$$y = a + bx + e$$

The Structural Model

$$\eta = \beta\eta + \gamma\xi + \zeta$$

The Exogenous Measurement Model

$$X = \lambda_X \xi + \delta$$

The Endogenous Measurement Model

$$Y = \lambda_y \eta + \epsilon$$

■ Questions and Exercises

1. Select journal articles in criminal justice and criminology that have as their primary analysis factor analysis, path analysis, or structural equation modeling (it will probably require three separate articles). For each of the articles:
   a. Discuss the type of procedure employed (for example, an oblique factor analysis).
   b. Trace the development of the concepts and variables to determine if they are appropriate for the analysis (why or why not?).
   c. Before looking at the results of the analysis, attempt to analyze the output and see if you can come to the same conclusion(s) as the article.
   d. Discuss the limitations of the analysis and how it differed from that outlined in this chapter.
2. Design a research project as you did in Chapter 1 and discuss which of the statistical procedures in this chapter would be best to use to analyze the data. Discuss what limitations or problems you would expect to encounter.

■ References

Bollen, K. A. 1989. *Structural Equations with Latent Variables*. New York: John Wiley and Sons.

Byrne, B. M. 2001. *Structural Equation Modeling With AMOS: Basic Concepts, Applications and Programming*. New Jersey: Lawrence Erlbaum Associates, Publishers.

Carroll, J. B. 1953. Approximating Simple Structure in Factor Analysis. *Psychometrika*, 18:23–38.

Duncan, O. D. 1966. Path analysis: Sociological Examples. *American Journal of Sociology* 73:1–16.

Garnett, J. C. M. 1919. On Certain Independent Factors in Mental Measurement. *Proceedings of the Royal Society of London*, 96:91–111.

Hatcher, L. 1994. *A Step-by-Step Approach to Using the SAS System for Factor Analysis and Structural Equation Modeling.* Cary, NC: SAS Institute, Inc.

Hayduk, L. A. 1987. *Structural Equation Modeling with LISREL: Essentials and Advances.* Baltimore: The Johns Hopkins University Press.

Hotelling, H. 1933. Analysis of a Complex of Statistical Variables into Principal Components. *Journal of Educational Psychology,* 24:417.

Kaiser, H. F. 1958. The Varimax Criterion for Analytic Rotation in Factor Analysis. *Psychometrika,* 23:187–200.

Kaiser, H. F. 1974a. An Index of Factorial Simplicity. *Psychometrika,* 39:31–36.

Kaiser, H. F. 1974b. A Note on the Equamax Criterion. *Multivariate Behavioral Research,* 9:501–503.

Kim, J. O., and C. W. Mueller. (1978a). *Introduction to Factor Analysis.* London: Sage Publications.

Kim, J. O., and C. W. Mueller. (1978b). *Factor Analysis: Statistical Methods and Practical Issues.* London: Sage Publications.

Kline, R. B. 1998. *Principles and Practice of Structural Equation Modeling*. New York: The Guilford Press.

Lawley, D. N. 1940. The Estimation of Factor Loadings by the Method of Maximum Likelihood. *Proceedings of the Royal Society of Edinburgh*, 60:64–82.

Long, L. S. 1983. *Confirmatory Factor Analysis*. London: Sage Publications.

Pearson, K. 1901. On Lines and Planes of Closest Fit to System of Points in Space. *Philosophical Magazine*, 6(2):559–572.

Spearman, C. 1904. General Intelligence, Objectively Determined and Measured. *American Journal of Psychology*, 15:201–293.

Wright, S. 1934. The Method of Path Coefficients. *Annals of Mathematical Statistics*, 5: 161–215.

Thurstone, L. L. 1931. Multiple Factor Analysis. *Psychological Review*, 38:406–427.

Thurstone, L. L. 1947. *Multiple Factor Analysis*. Chicago: University of Chicago Press.

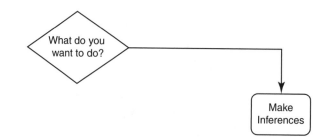

Describe

What do you
want to do?

Make
Inferences

Introduction to Inferential Analysis

In the first stop on the flowchart of statistics back in Chapter 1, you decided which of the three goals you wanted to accomplish: describe a data set (telling something about the general characteristics of criminals); make inferences from a sample to a population (using characteristics about criminals in one city to draw conclusions about criminals across the United States); or doing both. Earlier in the book, we focused on describing the characteristics of a known population. Now we deal specifically with those techniques that allow researchers to examine the characteristics of a *sample* (a small group drawn from the population) and make inferences to the *population*. These procedures are called *inferential analyses*.

Inferential analyses are useful because, in many cases, collecting data on an entire population is either too costly, too time consuming, or perhaps even impossible. For example, even with the smallest portion of the criminal population (those in prison), there are over 1 million people in the United States for which data would have to be collected to study with descriptive statistics. This would take a great deal of money. Additionally, researchers could spend the rest of their professional lives attempting to study all of the people in the United States who are on probation and parole, but because this number changes continually, the researchers still would not have an accurate portrait of this population. Finally, we have no idea who all of the people are in the United States who have committed a crime (even a felony), so there would be no way to examine this population.

An even more concrete example of how descriptive statistics would be too costly, time consuming, or impossible can be drawn from the automotive field. It is easy to see that crash-testing every car produced would be quite counterproductive to the automobile industry. It is possible, however, and more economically feasible, to crash-test a small number of each type of car and infer the results of those tests to all cars of that type. This is the process of **inferential analysis**.

Inferential analysis takes advantage of specific sampling techniques that permit researchers to make inferences from a sample to a population. The information gleaned from the sample is generalized to the larger population from which it was selected. Inferential analyses permit a decision to be made with a known probability of error about whether a sample characteristic is different from a population characteristic and whether these differences are large enough to allow the conclusion that the population represented by the sample differs on a certain characteristic.

A Bit of History

Inferential analyses are among the oldest of all statistical analyses. The roots of inferential analyses were first developed in the Orient around 200 BC (Dudycha and Dudycha, 1972). This was a form of probability analysis used in assessing whether an expected child was likely to be male or female. This use of probability grew and became popular initially as a method of playing games of chance. Eventually, more scientific uses of probability arose as the normal curve and other developments were made. By the 1700s, Bernoulli had developed his **law of large numbers**, which was the forerunner to concepts such as the central limit theorem and sampling distributions. The law of large numbers states that as the sample size increases, the variability of data, or standard error, decreases. Since then, inferential analyses have become one of the most important areas of statistical analysis.

■ Terminology and Assumptions

The concepts in inferential analysis are the same as those of descriptive analysis. The terminology is changed, however, to reflect the need to determine whether one is speaking of the sample (statistic) or the population (parameter). For example, in inferential analyses, a frequency distribution can be compared to a sampling distribution, the measure of central tendency (mean) is the expected value, and the measure of dispersion (standard deviation) is the standard error. Other terms and definitions important to inferential analysis are listed below to familiarize you with the discussion that is to follow. In most cases, these definitions are expanded later in the chapter.

- **Population**: an entire group of study. A population may be small, such as "all prisoners in the North Dakota Department of Corrections"; or the population can be very large, such as "all criminals in the United States."
- **Sample**: a subset of the population drawn to allow statistical analysis. The basis of inferential analyses is addressing the issue of whether a selected sample can reasonably be concluded to belong to the population or is different from it.
- **Parameter**: a characteristic of the population: for example, the average income of all criminals in the United States.
- **Statistic**: a characteristic of the sample for which researchers may make inferences to the population parameter: for example, the average income of all criminals in Indiana.
- **Inference**: performing an analysis on the sample, then concluding that the findings apply to the population.

- **Estimate**: a result of sample analysis of the population parameters. For example, we know it is unlikely that the average income of a sample of criminals will be the exact average income of all criminals in the United States. We may argue, however, that it is an accurate *estimate* of that parameter.
- **Expected value**: the estimated population parameter. Most of the time, the estimated value will be the mean of the population. Similar to the mean, the expected value does not relate how closely a particular statistic is to the parameter. It simply assures us that *on average*, the statistic will be an accurate estimate of the parameter.
- **Standard error**: a measure of the variation of a statistic around the parameter it is estimating. Typically, the standard error measures the variability of sample means around the population mean. The standard error may be interpreted in the same manner as the standard deviation. It is influenced by sample size such that as sample size increases, the standard error decreases.

As with any statistical procedure, inferential analyses have a number of assumptions that must be met before they can be considered reliable. The assumptions for inferential analyses are perhaps more important because we do not know the population parameter. Because we cannot confirm conclusions made concerning the data, as is possible with descriptive analyses, it is imperative that the assumptions of the inferences be met so we can rely on the mathematical proofs of the analyses to support the conclusions.

The first assumption of inferential analyses is that the sample must be taken from the same population for which inferences are to be made. This is actually the watermark of inferential analyses. The bottom line of any inferential analysis is rejecting or failing to reject the null hypothesis that there is no difference between the sample and a population.

The second assumption is that the population and sample must be normally distributed. There are methods of dealing with data that may not be normally distributed, but the foundation of inferential analyses is that a normal distribution is assumed. This assumption is necessary because the components of inferential analyses—the normal curve, central limit theorem, and sampling distribution—all require normality. The robustness of these measures, however, often allows researchers to assume that the data is normally distributed.

The final assumption is that inferential analyses require that a *random sample* of the data has been taken. As discussed below, researchers must assume that there is no systematic error in the sample drawn. We know that error will be present in any sample—that it will not be a true measure of the population parameter. We must assume, however, that the error is random (some error may be above the mean and some below the mean), such that the errors will essentially cancel each other out. If there is error in the sample that is not random, we cannot assume sample statistics will be proper estimates of the population.

These assumptions are put into practice through components of inferential analyses that are essential to their use as a statistical tool. These are the normal curve, sampling distributions, and the central limit theorem, all discussed below.

■ Normal Curve

The concept of the normal curve was introduced in Chapter 6. Here, the normal curve is discussed in terms of its utility for inferential analysis. A review of Z scores will accompany a discussion of probability to introduce the importance of a normal curve in estimating the expected value. This information is also useful as it relates to the central limit theorem, sampling distributions, and confidence intervals.

DeMoivre, a French astronomer, is credited with developing the concept of a normal curve. Although discussed for many years, with parts published in different languages, DeMoivre's "curve of error" was not printed in English until 1718 in his work *The Doctrine of Chances* (DeMoivre, 1967). Although Pearson credits DeMoivre with developing what is now known as a normal curve, many still believe it was Gauss (1809) who was the originator of this principle. In fact, the *normal distribution* is often referred to as a *Gaussian distribution*.

Now that you know how to calculate the mean and standard deviation, you can begin to look at how data distributions are arranged, and to make inferences (conclusions) about the data. One way to make these conclusions is through the use of the normal curve. It is important to remember that this is the **standard normal curve**, which assumes the data is arranged such that there is a mean of 0 and a standard deviation of 1. With this knowledge, *Z* scores can be used as standard deviations from the mean to examine where a particular sample value may fall in the normal distribution of a population.

As discussed in Chapter 6, if a score falls exactly one, two, or three standard deviations from the mean, it is easy to calculate the percentage under the normal curve. If, however, a score does not fall on a standard deviation, it must be converted to standard deviation units to find the area of the curve it occupies. The process of converting a score to a standard deviation is that of calculating a *Z* score. The only difference between calculating a *Z* score here and the calculations performed in Chapter 6 is that we are now attempting to work with data in a sample and use it to estimate population data. As such, notice in the formula below that *X* is now the mean of the sample, $\overline{X}$ has changed to μ (the population mean), and *s* has changed to σ (the population standard deviation):

$$Z = \frac{\overline{X} - \mu}{\sigma}$$

The obvious question here is: If we know the population mean and standard deviation, why don't we just use them for the study? That brings up an interesting point for criminal justice and criminology research. There are times when researchers know the population parameters, or they believe they have valid estimates of them. In these instances, it is useful to compare samples to the population. For example, it has been established that IQs are normally distributed, with a mean of 100 and a standard deviation of 15. Research can be conducted on prisoners, then, that compares this sample to the population as a whole. It is much more difficult, if not impossible, however, to determine the population parameters in other areas. In these cases, the use of *Z* scores is limited at best.

If research on prisoners found that they have an average IQ of 80, would this mean that they are substantially (significantly) different from the "normal" population with average IQs of 100? The *Z* score formula above can be used to examine this hypothesis.

Using the information on IQs above and the knowledge of the average IQ of prisoners would produce the following calculations:

$$Z = \frac{\overline{X} - \mu}{\sigma}$$

$$= \frac{80 - 100}{15}$$

$$= \frac{-20}{15}$$

$$= -1.333$$

Here, the mean IQ score of the prisoners (80) is subtracted from the population IQ score (100). This value is divided by the standard deviation for the population IQ (15). The result of the calculation is a Z score of -1.333.

Using the knowledge of Z scores and the area under the normal curve, this value falls between the mean and 1.96 standard deviations below the mean. If 0.05 is used as the cutoff of a significant difference, we would fail to reject the null hypothesis: in this case that prisoners have significantly different IQs than those of the normal population. In this case, we would conclude that prisoners and nonprisoners come from the same population.

The normal curve is important because there is typically a disconnect between the sample and the population. The sample may or may not be normal, but it is probably not an exact duplicate of the population. The principles of sampling distributions (which are normally distributed) can be used, however, to state that the sample drawn should be representative of the sampling distribution, which is normal. Because the sampling distribution is normal, and a theoretically large number of samples would create a sampling distribution that is very close to the population, which is also normal, it can then be assumed that the sample drawn is an approximation of the population we are attempting to examine. This allows use of the normal curve to establish the probability of a sample expected value estimating the parameter value we are attempting to examine. This allows inferences to be made about the population based on the sample data.

■ Probability

Probability is a concept that is central to research in criminal justice and criminology. In fact, most criminological theories can be reduced to a measure of probability. When researchers attempt to explain criminal behavior, they are essentially trying to establish the probability that a person will commit a crime because of the conditions, abnormalities, or other characteristics that are the subjects of the research. Probability is also used in other places in the criminal justice system. The concept of probable cause is central to police work, judges decide on the probability of a person committing more crimes or fleeing from justice when they set bail, and juries base their decisions on the probability that a person has committed a crime. All of these rely on probability in making decisions. Probability does the same thing for inferential analysis and interpretation.

The concept of probability is central to statistical inference. Even with all the advantages of statistical theory, the end result is that researchers do not know the true answer. In this way, research is much like the initial use of probability theory in gambling. We do not know the true outcome, so we work to make the best estimate as to what the actual outcome may be. Using the principle of the normal curve, researchers can estimate the probability of an event occurring and the probability that a certain sample statistic will match a population parameter.

Probability refers to the relative likelihood of the occurrence of an event or the number of times an event can occur relative to the number of times *any* event can occur. Probability values range from 0 to 1 and are expressed as either a decimal or a percent. A probability of 0 means that something is impossible; a probability of 1 means that it is imminent.

Probability is important for inferential analysis because it represents the probability of being wrong on a decision to reject or fail to reject a null hypothesis. In Chapter 8, Chi-square was used to examine the existence/significance of a relationship. There, 0.05 was used as a general cutoff. This means that you would be correct in rejecting the null hypothesis 95% of the time. It also means, however, that you would be wrong 5% of the time. The probability of being wrong, then, is 5%. It is this probability that is the foundation of decisions made in inferential analysis. The proportion of the area under the normal curve can also be used to establish probability. The probability that a Z score will be less than or equal to -1.96 is 0.025 (0.05/2), and likewise for greater than or equal to 1.96. The probability that a Z score will be less than or equal to -1.96 *or* greater than or equal to 1.96 is 0.05.

The probability is high that you have studied probabilities in another math or, perhaps, psychology class. The typical examples used to examine probability are coin tosses, rolls of dice, or drawing from a deck of cards. Inferential analyses make use of these principles to examine the likelihood of being right or wrong in rejecting or failing to reject the null hypothesis.

Look at a coin toss to examine the issue of probability. A null hypothesis of a coin toss could be that "there is no statistically significant difference in a coin landing on heads and tails." This means that if a coin is tossed 100 times, we would expect there to be 50 heads and 50 tails. We also know, however, that getting exactly 50/50 is pretty remote. What can be established, however, is an acceptable ratio of heads to tails—perhaps a 1% or a 5% difference. This can also be translated to Z scores and the properties of the normal curve evoked to examine how far from normal a given set of 100 tosses would be. Finally, multiple tests of 100 tosses could be used to establish a sampling distribution as described below.

There are two rules that are central to the use of probabilities: the addition rule and the multiplication rule. These are important when examining multiple situations that may occur, such as the various life events that may influence a person's probability of committing a crime.

The **addition rule** is used when establishing the probability of one event *or* another event occurring. For example, what is the probability of getting an ace of spades *or* an ace of diamonds? This rule assumes that the possible outcomes are mutually exclusive; they cannot occur simultaneously. One card drawn cannot be *both* an ace of spades *and* an ace of diamonds. To use this rule, simply add each of the probabilities together: 1/52 + 1/52 = 2/52 or 0.038.

The **multiplication rule** is used when establishing the probability of one event *and* another event occurring. This rule assumes that the events are independent, that is, that one event is not dependent on previous events (if a card is drawn from a deck, it is replaced). The multiplication rule examines the probability of getting an ace of spades *and* an ace of diamonds in two draws (or the probability of getting two heads in a row in a toss of a coin). This rule states that the probability of *both* events occurring is the product of their separate probabilities. To use this rule, simply multiply the probabilities of each event occurring together. In looking at the probability of drawing an ace of spades *and* an ace of diamonds in two draws, the probability would be $(1/52)(1/52) = 1/2704$ or 0.00037. You can multiply as many events together as you like. For example, to determine the probability of a coin tossed four times being all heads, simply multiply the values together. With a 50/50 probability of getting a head, the probability of getting four heads in a row would be $(0.50)(0.50)(0.50)(0.50) = 0.0625$. Whether a 0.05 or 0.01 was used as a cutoff, this would not be enough to reject the null hypothesis that the coin toss was evenly distributed.

■ Sampling

The aim of social research is to draw conclusions and make generalizations about a population, as discussed in the chapters concerning inferential analysis. This is often not practical and sometimes impossible. Researchers must sometimes take smaller groups (samples) to study and then make inferences back to the population. To do this, they must ensure that the sample is representative of the population. The sample should contain the same variations that exist in the population for those characteristics that are relevant to the study.

It cannot be overstressed how big a role sampling plays in the ability to make inferences from sample data. The sampling plan and operation must ensure that the sample is taken from the population for which inferences are to be made; you would not want to make inferences to all criminals based only on the observations of murderers. Also, the sample must be taken with an absolute minimum of bias or error (**sampling error,** coding error, or other error).

Most of the issues of sampling are more appropriate for a methodology course than for a statistics course; however, it is easy to see where there is applicability for the discussion of inferential analysis. Before addressing the issues of sampling that influence the ability to make inferences, a short discussion of the terminology of sampling is needed.

Although the population has been discussed as a single entity, there are actually three populations that are important for inferential analysis. A **parent population** is a population from which one or more samples are drawn. In reality, this is only the theoretically accessible population. This is tantamount to stating that we will be drawing from the population of "all criminals." We know this is not possible, but it is the theoretical population from which to draw. The **target population** is the group to which findings are to be generalized. This is the population we are attempting to make inferences about and from which we have hopefully drawn our sample. Finally, the **study population** is the group from which the sample is actually drawn. Regardless of whether it matches the parent population or target population, this is the sample and data on which the analysis will be based.

Once the study population has been established, there are units of sampling that represent the data and cases that will be examined. The actual list of sampling units in the study population from which the sample is selected is called the **sampling frame**. For the probability samples below, this is the study population/ target population from which a list will be created and samples taken. In essence, the sampling frame represents the population that has been identified to study. From the sampling frame, a **sampling unit** is selected. A sampling unit is all of the elements of the sample that are chosen for selection in some stage of sampling. In a single-stage sample, the sampling unit will equal each element. In a stratified, quota, or multistage sample, different levels of sampling units may be employed. For example, a research project may have *primary sampling units*, which consist of all counties in a four-state area. From these primary sampling units, *secondary sampling units* may be selected, which are cities within those counties. *Final sampling units*, which are people within each of the cities, may then be selected. The same can be done by stratifying a sample based on race and then selecting certain elements within each strata. The final portion of a sampling design is an **element**. This is a unit about which information is collected, and that provides the basis of analysis. This is a single member of the population that is being studied. If a researcher is studying burglars in a city, an element is one of those burglars.

The type of sampling units and the ability to make inferences based on the samples drawn rests in the type of sample procedures used. There are two types of sample collection strategies: probability samples and nonprobability samples. Each of these is discussed below.

Probability Sampling

The preferred method of sampling for inferential analysis is a probability sample. This is because it is assumed that selection bias and other researcher introduced bias has been eliminated from the process. The only variability assumed to remain in the sample is random variation or variation as a result of the variables to be examined. So researchers will always use probability samples when conducting research, right? Not necessarily.

Probability samples are generally quite difficult to draw except from the smallest of populations. The problem is that for a sample to be a true probability sample, researchers must know the probability that each element in the sample has of being selected. This means that all elements in the population must be identified or there must be at least a supportable estimate of the number in the population. This is not always possible. Probability samples also require that the probability of each element being drawn is nonzero. This seems easy enough in theory, but it is much more difficult in practice. For example, if you were studying burglars in the city, there are at least some burglars who have not yet been caught and identified. These people are elements in the population, but they have a zero probability of being selected for the sample because they have not yet been identified. The final issue that must be addressed in probability sampling is that the selection of each element should be independent of the selection of other elements. This is typically not a great challenge to drawing the sample, but be aware of the requirement. Some of the other-than-random probability samples have the potential to violate this assumption. For example, stratified samples are typically considered probability samples, whereas quota samples are not. The only practical difference between these two techniques is that the stratified sample allows each item within the stratified sampling frame to be chosen relatively at random. A quota sample, however, draws elements based at least in part on their need in the sample. In this case, some of the elements are selected for the sample based on previous selections, not independent of them.

Although not everyone agrees on the type of sampling that represents a probability sample, it is generally accepted that simple random samples and stratified samples are probability samples. Two sampling strategies—stratified and multistage—that may also be used as probability samples are discussed in this section.

Simple Random Sample

The ideal method of drawing a sample is **simple random sampling** (SRS). Here, each element in the population has an equal and known probability of being chosen for the sample. Researchers usually assume that a sample is random and that they can make generalizations about the population; but few actually take the time to conduct a pure random sample. To utilize an SRS requires a complete listing of all the elements in the population, numbered consecutively from 1 to N. This is the difficult part because researchers often do not know all of the elements, or there are too many of them to list. Once the list is completed, a table of random numbers is generated and the sample is taken using the random numbers to choose the sample. Simple random sampling also requires that elements drawn at random be replaced in the population after being identified for the sample. Known as *sampling with replacement,* this ensures that each element being drawn is independent of all other elements being drawn.

The advantages of an SRS for inferential analyses are obvious; it employs the use of probabilities necessary in many statistical procedures. The disadvantages are also obvious, however; simple random samples require a complete list of the population to be sampled. This is generally not feasible except with the simplest sampling frame. Also, sampling in this way does not guarantee a representative sample; it only allows the assumption of a representative sample.

Systematic Sampling

An alternative to random sampling is **systematic sampling**. In this method, after the sampling frame is listed, as with an SRS, elements are chosen in a systematic fashion, generally choosing elements at certain intervals in the sampling frame. For example, a telephone book could be used as the sampling frame for a survey. The first name from the phone book could be selected at random by rolling a die and using that number; then every tenth name could be used. The interval of selection depends on the desired size of the sample. If a researcher wanted 1,000 names and there were 10,000 people in the phone book, every tenth name could be drawn so all of the phone book would be covered with the sample. Purists argue that this is a nonprobability sample because various patterns, such as ethnic surnames, exist that destroy representativeness. This is a concern, and samples must be checked for element cycles that may coincide with the sampling interval in any stratified sampling strategy employed. For example, researchers must be aware when conducting neighborhood research if corner houses occupy a fixed position in the list because they are generally more expensive.

There are two other sampling strategies that may or may not use probability sampling: stratified and multistage strategies. Because most researchers work to use probability samples, and these are often employed as probability sampling strategies, they are discussed here.

Stratified Sampling

Stratified sampling uses known information about the sample or population to attempt to ensure consistency in the samples drawn. It is based on the assumption that a stratified group requires a smaller sample because characteristics may be identified and controlled for. This strategy typically ensures that the population characteristics to be studied are represented in the sample. Obviously, it requires knowledge of the population characteristics to ensure representativeness. The procedure involves dividing the population into groups based on the variables to be stratified, then taking a random or nonrandom sample of those characteristics. If random or systematic sampling is used to extract the elements from the strata, it is generally considered a probability sample.

An example of stratified sampling can be drawn from research on correctional officers. There are generally not enough female correctional officers to be able to conduct a random sample and ensure that enough females are chosen at random. It is often necessary, then, to stratify the population between males and females and conduct a random sample of each. If the random sample is based on the proportion of each of the two groups in the population, it is called a *proportionate stratified sample.* If more females are selected than males, it is a *disproportionate stratified sample* (often called a *quota sample*, as described below).

Cluster or Multistage Sampling

Cluster or multistage sampling methods are often used to reduce the costs of large surveys. Cluster and multistage are actually two methods of sampling, but they are often used together. **Cluster sampling** involves dividing the sampling frame into clusters, generally based on a geographic representation. The sample is then drawn based on this cluster stratification. Often, these clusters are subdivided into smaller clusters: primary, then secondary, and then final sampling units. In **multistage sampling**, a sampling frame is divided into clusters or is stratified and then secondary sampling units are chosen. *Multistage* simply means that the sampling frame is divided and then subdivided, sometimes a number of times, before the final sampling unit is actually selected.

Both of these strategies involve creating levels of sampling units from which probability or nonprobability samples are chosen. The best example of multistage cluster samples is the strategy used by the National Crime Victimization Survey. This strategy involves dividing the United States into regions from which counties are chosen at random; within the counties, cities are chosen; within these sampling units, households are then chosen. It is the households that are surveyed and the results inferred to the United States as a whole. This is also the method used by most polling entities, such as Gallup and Harris. This strategy allows these polling companies to take a sample of as few as 1,000 people and make an accurate estimate of, for example, the presidential election results.

Aside from the reduction in cost and time when conducting research, cluster or multistage samples are useful when complete lists are not available or are impractical to obtain. For example, it may not be possible to get a complete list of everyone in the United States, and if we did, the list would be unwieldy. It is certainly easy, though, to obtain a list of the states in the United States and the counties in those states. Once these have been sampled, it is also easy to get a list of the cities within each county.

Then, within those cities, a list of the people can be obtained for the research. You should be aware, however, that although complete lists are not necessary, reliable estimates of population sizes are necessary to ensure the probability of elements being drawn at each stage. Also be aware that this method is less accurate than SRS or stratified sampling strategies because sampling error occurs at each stage.

Nonprobability Sampling

Sometimes it is simply not possible to use probability sampling, or it is so time consuming or costly that the research would probably not get done if probability sampling was required. In these instances, many researchers turn to nonprobability samples. These sampling strategies are more convenient, less expensive, and easier to collect. The problem with nonprobability samples is that the probability of an element being drawn is not known, so there is no way to tell if the sample chosen is representative of the population. Furthermore, inaccuracies cannot be controlled for through randomization, which makes generalization difficult.

There are at least four types of nonprobability sampling strategies. Some of these approximate probability samples are drawn in such a way that they may be used for inferential analysis. Others are so far removed from a probability sample that it is ill advised to use them for inferential analysis. These sampling strategies are discussed below.

Purposive Sampling

In **purposive sampling**, sampling units or elements are chosen based on the researcher's knowledge that the sample is representative of the population to be studied. For example, when Pate, Wycoff, Skogan, and Sherman (1986) began their study of the fear of crime, they looked to two cities where fear of crime was likely to be high. They chose Newark, New Jersey, and Houston, Texas because of their high crime rates. They also chose these cities because they felt that they anchored the ends of metropolitan areas in the United States. Newark was chosen because of its high population density and declining economy; Houston was chosen because it was a relatively new city that had a substantial population but had low population density. Knowing the characteristics of these cities allowed the researchers to examine the target population in different cities that were hypothesized to be representative of the population of large, metropolitan areas without the need to take a sample from every metropolitan area in the United States.

Quota Sampling

A **quota sample** is a nonprobability, stratified sample. Quota sampling involves dividing the population (or study population) into groups based on desired characteristics. Selection of elements is then made based on that stratification. Quota sampling attempts to ensure that sample proportions of variables resemble population parameters. Sometimes, the stratifications are proportionate to their representation in the population (i.e., dividing sample elements based on the proportion of each race in the population being studied). Sometimes, the proportions of elements chosen are not in accordance with their proportion in the population. This is often referred to as *oversampling*, a method by which more elements are chosen than their proportion in the population. This ensures there is enough of this element chosen to allow analysis (e.g., keeping the expected value above 5 for a Chi-square analysis). An example would be to oversample Hispanics in an area where there are not enough to be sure that they will be adequately represented in the sample. The problem is that those chosen from each group may not be representative of that group.

Snowball Sampling

Snowball sampling is employed when it is difficult to identify or find subjects. It is not a probability sampling strategy, but sometimes it is the only way to find elements for research. This strategy is often employed in qualitative research and other research where the subjects are not readily identifiable. For example, in Sutherland's *The Professional Thief* (1937), the only way he could get enough interviews with thieves was to ask those interviewed to recommend or introduce others that they knew to be like them. This sampling strategy does not allow inferences to be made to the population, but it does provide an in-depth understanding of the sample chosen. As such, its value to understanding a small group of those difficult to identify and for which information was previously unknown often overshadows the lack of ability to generalize the results.

Accidental or Convenience Sampling

The final type of nonprobability sampling is the most removed from probability sampling. This is an **accidental** or *convenience* **sample**. These are actually two different types, but both represent the same strategy. An *accidental sampling* strategy simply means accepting whoever happens by during the research: for example, standing at the doors to a mall administering a survey and accepting those who enter the mall and agree to the survey on a particular day. A *convenience sample* involves using people who are nearby to be the subjects of research. You have probably all read articles in which the sample was a "class of college freshmen." Both of these sampling strategies are based exclusively on what is convenient for the researcher, such as asking students in a class to fill out a questionnaire. This strategy is often used to pretest versions of surveys. This strategy is the easiest and least time consuming, but it is very difficult to make inferences to any population other than the very narrow one involved in the sample.

This brief overview serves as an introduction to a particular type of sampling, the data it represents, and its usefulness in inferential analysis. This type of sampling is known as a *sampling distribution*.

■ Sampling Distributions

In our examination of probabilities above, we could have developed a *probability distribution* of the coin tosses or card draws. When probability distributions are used specifically for inferential analysis, they are often called **sampling distributions**. A sampling distribution would be created in this case if a coin was flipped 100 times and the number of heads and tails recorded; then this procedure was repeated 99 times. Each of these 100 samples might have some aberrations from the expected 50/50 ratio, but we would expect the overall result of these samples to be very close to 50% heads and 50% tails.

There are actually three broad types of distributions. The first type, discussed in the first part of the book, is a population distribution that describes the characteristics of the population. The second type of distribution is a sample distribution, which describes the characteristics of a sample drawn from a population. The final type of distribution is a sampling distribution. This distribution differs from the others in that it is a theoretical ideal; that is, you do not have to have all of the data to utilize it. A sampling distribution is used to describe the *expected* characteristics of a large number of samples drawn from the same population.

The real value of a sampling distribution is that a researcher does not have to draw a large number of samples to examine their characteristics. One sample can be drawn and the principles of the central limit theorem (see below) used to make inferences about the sample to the population, assuming, of course, that a probability sample was drawn.

A useful characteristic of a sampling distribution is that if a sufficiently large number of samples is drawn (between 30 and 120, depending on what source is consulted), and a characteristic such as the mean is calculated for each sample, the distribution of these means will approximate a normal curve. If these samples are taken from a normal population, the sampling distribution will also be normal. Even if the samples are taken from a nonnormal population, the sampling distribution will still approximate normality. The number and size of the samples needed to approximate a normal curve depend on how far the population is from normal. An extremely non-normal population may require many fairly large samples, whereas a nearly normal population will require fewer samples. If enough samples were gathered such that all the values in a population were collected and then put in a sampling distribution, the parameters of the population (mean, standard deviation, etc.) could be determined. Because we know that a smaller number of samples approximate the normal curve, however, we do not have to use the entire population to get an estimate of the population parameters.

All parameters of a population and their associated sample statistics have sampling distributions. For example, there are sampling distributions of the variance of a distribution (actually, this is the χ^2 distribution used in Chapter 8), there are sampling distributions of skewness, and others. More important for inferential analysis, there are sampling distributions of the measure of central tendency. A *sampling distribution of the means* is important because as discussed with ANOVA in Chapter 18, the means are often used to estimate population characteristics from a sample. A sampling distribution of the means is often displayed as a *t* table, as discussed in Chapter 17. It can be shown that the expected value of the sample mean approximates the population mean regardless of the form of the population distribution or the sample size. In statistical terms, the sample mean is an unbiased estimate of the population mean. In reality, it is generally assumed that the mean of the sampling distribution of the means is equal to the population mean. The sample mean is used to estimate the population mean when the values of the parameter are unknown.

This measure of the central tendency of a sampling distribution is often referred to as the *expected value*. Typically, this is the overall mean of the sampling distribution of the means. This is the value that researchers would expect or would rely on if they did not know the true value of the population (which they do not). The variation of sample statistics around their parameter is called the *standard error*. The standard error is essentially the same as the standard deviation in descriptive statistics. If a distribution was a sampling distribution of the means, the standard error would be the variability of the sample means around the population mean. Because the standard error is similar to the standard deviation in descriptive statistics, it can be treated the same for use with a normal curve. For example, 95% of the sampling distribution will fall between ± 1.96 standard errors from the population mean.

It should be noted that the distribution of the sampling distribution of the means is not the same as the population distribution. For example, because the sampling distribution of the means is based on a large number of random samples, the variance of this distribution will be smaller than the population distribution. Mathematically, the variance of the sampling distribution of the means is the population variance divided by the sample size (and the standard deviation would be the population standard deviation divided by the square root of N). This is the standard error.

■ Central Limit Theorem

Populations that are not normal in distribution could present a potential problem to the researcher, particularly regarding any inferences made from a sample of that nonnormal population. Most of the populations studied in criminology and criminal justice are not normal. This could cause great problems for the ability to use the normal curve and inferential analyses. Fortunately, there is another statistical ideal that allows researchers to use the principles of the normal curve with nonnormal populations: the **central limit theorem**. The central limit theorem provides researchers with an empirically proven concept that allows estimates and generalizations based on a sample to be inferred to the population from which the sample was drawn.

The central limit theorem is credited as being developed by Laplace in 1810 (Laplace, 1878–1912), but his development was little more than a generalization of DeMoivre's (1967) limit theorem. Since then, however, this theoretical concept of statistical inference has become one of the central parts of inferential analysis.

The central limit theorem states that given any population (regardless of whether or not it is a normal distribution), as the sample size increases, the sampling distribution of the means approaches a normal distribution. This is because, as discussed above, the standard error of the sampling distribution of the means decreases as the sample size increases (the clustering of the sample means around the population mean grows smaller as the sample size grows larger). This allows the use of a normal probability distribution for testing hypotheses about populations with any form of distribution.

The central limit theorem is based on the foundations of the sampling distribution of the mean and the normal curve: that repeated random samples taken from normal or nonnormal populations have sample means that approximate a normal distribution. Extending the theory of sampling distributions, the central limit theorem allows researchers to conclude that even when they take only one sample, the sample mean will approximate the population mean, and will become a closer estimate of the population mean as the sample size increases. The main significance of the central limit theorem is that it allows researchers to operate with the knowledge that the sampling distribution will be normally distributed, provided the sample is sufficiently large. For almost all populations, virtually without regard to the shape of the original population, the sampling distribution of the means derived from the population will be approximately normally distributed provided that the associated sample size is sufficiently large.

The key to this concept is sample size. Kachigan (1986) points out that when considering a population that is normal in form, any random sample of elements taken from that population will also be normal in form. Additionally, a sampling distribution of the measures of central tendency of a random sample, if taken from a normal population, will also be normally distributed. Furthermore, for large samples, the sampling distribution of the means will be normal even though the population is not normal. Kachigan (1986) proposes that sampling distributions of less than 30 are insufficient to approximate a normal curve. He also proposed that the sampling distribution will approximate a normal distribution only with a sample size of 30. He argues that as the sample size increases to

120 and beyond, the sampling distribution becomes more normal in form, and thus more accurate. The conclusion that can be drawn from this is that the sample size needed to utilize the central limit theorem depends, in part, on how far a distribution is from normal. Distributions that are relatively normal will not require a large sample size. Those that are substantially nonnormal may require a sample size even larger than 120. In the end, you must decide what minimum sample size you will accept.

■ Confidence Intervals

The analyses discussed to this point in this chapter are commonly referred to as **point estimations**. This is a process of attempting to identify the expected value (typically, the mean) of a population based on sample data. Hypothesis testing can also take the form of **interval estimation**. This is the use of confidence intervals to address the sampling error of an analysis and to establish a range of values that, if enough samples were drawn in the sampling distribution, would have a high probability of estimating the true population parameter. Even for smaller sample sizes, researchers can establish, with a certain degree of confidence, where the population parameter might fall.

Researchers cannot always, or sometimes do not want to, know exactly where a population mean is. Sometimes it is enough simply to know the range within which it falls. Confidence intervals are used to estimate the range of mean values drawn from a sampling distribution within which the true population parameter is likely to fall. They are also used to estimate the probability that the population mean actually falls within that range of mean values.

You have seen this form of statistical analysis before in election polls. The polls will say that a certain candidate has a particular percentage of the vote, plus or minus a certain percentage. This predicted percentage of the vote is the point estimate in this case: the expected value of the mean of the population. The plus and minus are the range of values within which they think the true population mean will fall. These are the **confidence intervals**.

This can also be done in reverse. If a researcher knows what margin of error he or she wants at a particular confidence, the required sample size can be determined using power analysis, which is discussed in Chapter 16. The fact that researchers can get within 3% of the sampling error with a sample of less than 1,000 and a 1% sampling error with a sample of 9,600 shows how nationwide samples can be so effective with only a relatively few number of people polled.

Calculating Confidence Intervals

Confidence intervals are calculated using a Z score for the particular range we are attempting to predict. Researchers typically use 95% or 99% confidence when working with confidence intervals in the social sciences. Remember that the Z score corresponding to 95% of the area under the normal curve is 1.96, and the Z score corresponding to 99% of the area under the normal curve is 2.58. These numbers correspond to the .05 and .01 critical values discussed in relation to Chi-square.

The formula for calculating a confidence interval is

$$\overline{X} \pm Z_{ci}\left(\frac{\sigma}{\sqrt{N}}\right)$$

where $\overline{X}$ is the mean of the sample that has been drawn, Z_{ci} is the Z score of the desired confidence interval, and σ divided by the square root of N is the population standard deviation. But how do we calculate the population standard deviation when we do not know its parameters? Again, using the law of large numbers associated with the central limit theorem, the population standard deviation can be estimated by using the sample standard deviation and subtracting 1 from it to account for any bias in the estimation.

The reason that the sample standard deviation can be used in place of the population standard deviation is, again, because of the knowledge of the normal curve, central limit theorem, and sampling distributions. It is known from these that as the sample size increases in a sampling distribution, the standard error decreases. Less sample-to-sample variation within the samples (called the *efficiency* of the estimate or sampling distribution) means that there is less variation in the means of those samples. From this it can be argued that the means of the samples are better estimates of the population mean, and a researcher can have more confidence in these estimations. In this way we can calculate for any sample the information that will be needed to estimate the range within which the population mean is likely to fall, and as the sample size increases, the use of the sample standard deviation as an estimation of the population standard deviation will become more accurate.

For a data set with a $\overline{X}$ of 5.2, a standard deviation of 0.75, a $N = 157$, and for a test at the 95% confidence interval, the calculation would be

$$95\% \text{ CI} = 5.2 \pm 1.96\left(\frac{0.75}{\sqrt{157 - 1}}\right)$$

$$= 5.2 \pm 1.96\left(\frac{0.75}{12.53}\right)$$

$$= 5.2 \pm 1.96(0.06)$$

$$= 5.2 \pm 0.12$$

Here a 95% confidence interval is used. This means that the Z in the formula would be 1.96. Because the mean is 5.2, it will be the base from which the other values are subtracted. The values in parentheses are the standard deviation (0.75) and the N (157). The 1 is a constant in the formula to compensate for bias in the sample. The product of these calculations would have the confidence interval running from 5.08 to 5.32. The two extreme values of this range are called the *confidence limits*. What this calculation means is that we are 95% confident that the mean in the population falls between 5.08 and 5.32.

Use of part of this formula can also show how the central limit theorem works and how larger sample sizes decrease the standard error. If the standard error portion of this formula is taken and the sample size is increased to 400, the new calculations would be

$$95\% \text{ CI} = \frac{\sigma}{\sqrt{N - 1}}$$

$$= \frac{0.75}{\sqrt{400 - 1}}$$

$$= \frac{0.75}{\sqrt{19.97}}$$

$$= 0.038$$

By increasing the sample size to 400, the standard error was reduced from 0.12 to 0.03. This would allow a researcher to make a more refined estimate of the location of the parameter and would thus better estimate that value.

This procedure is essentially the reverse of the types of inferential analyses discussed to this point. Remember from Chapter 8 that 0.05 was used as a cutoff for Chi-square. In that case, if the observed value fell outside the critical value, the conclusion would be that there was a statistically significant difference (that one value was in the tails of a normal distribution such that it could not be said that the two values came from the same population). In this case the middle of the normal curve is being used. Instead of looking for the 5% of values that fall in the tail of the curve, we are looking at the 95% of values that fall under the curve, concluding that the population mean falls within that range. In essence, the rejection region begins where the confidence interval ends.

Interpreting Confidence Intervals

It is common to interpret confidence intervals as "there is a 95% chance that the parameter (population mean) is between these two values." **Figure 15-1** shows a normal curve of the theoretical population associated with the calculations above. It demonstrates this interpretation of confidence intervals. You can see that this normal curve has a population mean μ and that the Z scores of ± 1.96 (corresponding to the confidence intervals above) have been superimposed on the curve. This distribution also contains the results of 10 samples from a sampling distribution that might be associated with this data. Each of the samples has an interval that falls within the Z scores used above. Notice that all but one of these samples (S7) includes the population mean as a element of the sample. This is the essence of confidence intervals. We can state with 95% confidence that the population mean should fall within the interval of the sample drawn. Be cautioned, however, that this does not mean that the particular

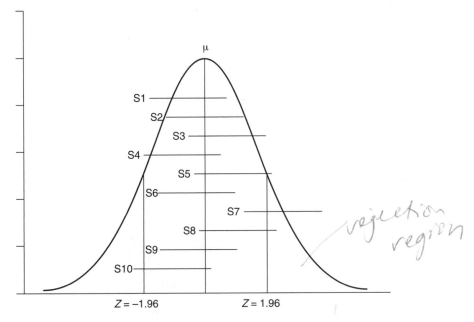

Figure 15-1 Normal Curve Showing Sampling Distribution Results

sample drawn contains the parameter. You could draw the one sample that in this case does not contain the parameter, although the probability is 95% that the single sample drawn contains the parameter.

In all cases of confidence intervals, as in this case, the range of the interval is a function of the level of confidence desired and the sample size. For example, higher confidence requires a wider interval; if a researcher wants to be 99% accurate, he or she must allow for a greater range of scores to estimate the parameter value. Small samples beget wider intervals. If the sample size is small, we must allow for a wider confidence interval because the standard error is large and we are less sure of our results. If the sample size is large, however, we can be more sure of our results because we know the standard error is small, perhaps smaller than the population standard deviation.

■ Conclusion

This chapter has set the stage for inferential analyses, discussed in the remainder of the book. The concepts of the normal curve, probability, and sampling distributions will be applied to hypothesis testing, which is the subject of Chapters 16 through 18. Without knowledge of sampling distributions, hypothesis testing would not be possible because researchers could not have the confidence that the expected value of a sample drawn from a population would closely approximate the population parameter. The normal curve facilitates the use of sampling distributions through probabilities and understanding of the area under the normal curve and Z scores. Confidence intervals expand the ability to estimate the population parameter by moving from a point estimate to an interval estimate, allowing researchers to estimate a population parameter with some allowance for error. All of these concepts will become important in the discussion of hypothesis testing as it applies to inferential analysis.

■ Key Words

accidental sample
addition rule
central limit theorem
cluster/multistage sample
confidence interval
element
estimate
expected value
inference
inferential analysis
interval estimation
law of large numbers
multiplication rule
normal curve
parameter
parent population

point estimation
population
probability
purposive sample
quota sample
sample
sampling distribution
sampling error
sampling frame
sampling unit
simple random sample
snowball sample
standard error
statistic
stratified sample
study population
systematic sample
target population

■ Summary of Equations

Z-Score (Population)

$$Z = \frac{\overline{X} - \mu}{\sigma}$$

Confidence Intervals

$$\overline{X} \pm Z_{ci}\left(\frac{\sigma}{\sqrt{N}}\right)$$

■ Exercises

1. Test the concepts discussed in this chapter on your own.
   a. Flip a coin 10 times and see what percentage of heads you get.
   b. Flip a coin for nine more sets of 10 and calculate the mean number of heads that you get (create a sampling distribution of your coin flips).
   c. What are the expected value and standard error of this distribution?
   d. For the largest run of heads or tails in a row, calculate the probability of that circumstance occurring.
   e. For this experiment, what are your parent population, target population, and study population? How are they different or the same, and why?
   f. Using the mean number of heads from your first set of 10 flips of the coin, construct confidence intervals of what you expect the population mean to be.
      i) How do your results compare to the known population parameter of 50% heads?
      ii) How do your results compare to your sampling distribution?
2. Find a research article in a journal that represents two of the probability sampling strategies and two of the nonprobability sampling strategies.
   a. Discuss why the authors chose this sampling strategy.
   b. What were the stated advantages and disadvantages of this strategy?

■ References

DeMoivre, A. (1967). *The Doctrine of Chances*. New York: Chelsea (reprint of the original 1718 work).

Dudycha, A. L., and L. W. Dudycha (1972). Behavioral statistics: An historical perspective. In R. E. Kirk (ed.), *Statistical Issues: A Reader for the Behavioral Sciences*. Pacific Grove, CA: Brooks/Cole.

Gauss, C. F. (1809). *Theoria motus corporum celestium*. Translated, 1963. Boston: Little, Brown.

Kachigan, S. K. (1986). *Statistical Analysis: An Interdisciplinary Introduction to Univariate and Multivariate Methods*. New York: Radius Press.

Laplace, P. S. (1878–1912). *Oeuvres complètes de Laplace*. Paris: Gauthier-Villars.

Pate, A. M., M. A. Wycoff, W. G. Skogan, and L. Sherman (1986). *Reducing Fear of Crime in Houston and Newark: A Summary Report*. Washington, DC: The Police Foundation.

Sutherland, E. H. (1937). *The Professional Thief*. Chicago: University of Chicago Press.

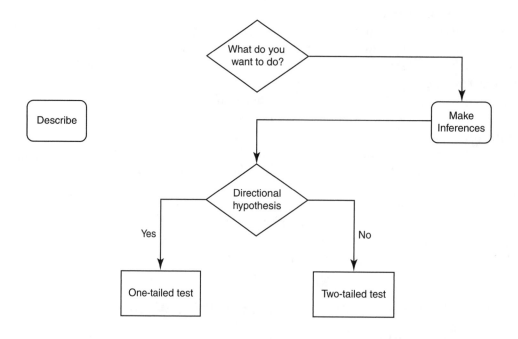

Hypothesis Testing

CHAPTER
16

Although some research is conducted to determine a specific characteristic such as average IQ (univariate analysis), more often researchers are comparing groups of people (bivariate analysis) such as the following:

- Thieves versus normal people
- Police officers who use deadly force versus those who do not
- Judicial sentences in criminal courts versus those in civil courts

Often, researchers do not have the population necessary to use descriptive analyses to conduct the bivariate analysis, but they still want to make the comparisons. As discussed in Chapter 15, inferential analyses may be used to analyze the characteristics of a sample and infer the results to a population.

When comparing these groups, you are trying to answer a question such as: *Are officers who use deadly force different from those who do not?* More specifically, when comparing two groups, you usually have an idea about the characteristics of the groups and you are trying to find out if that idea is true. You formulate a statement, or hypothesis, based on the idea—officers who use deadly force are more aggressive than those who do not—and try to find out if the idea is true. This is the process of **hypothesis testing**. Through inferential analyses and the characteristics discussed in Chapter 15, decisions can be made with a *known probability of error* about whether a sample is different from the population (whether officers who use deadly force are different from regular officers) or if two samples are different from each other (whether judges in one jurisdiction are different in their sentencing patterns from those in another jurisdiction). This process of hypothesis testing begins with carefully thought-out research and null hypotheses and is put into practice through the steps in the hypothesis testing process.

In this chapter, we continue the discussion of inferential analyses and hypothesis testing. We introduce concepts in hypothesis testing such as one- and two-tailed tests, types of error, and **statistical power**. At the conclusion of the chapter, you should be ready to begin using the hypothesis tests discussed in Chapter 17.

■ Null and Research Hypotheses

The null and research hypotheses were discussed in earlier chapters in terms of the research process and certain types of hypothesis tests. This discussion will be a little more specific to inferential analysis and will provide more information about these concepts.

By way of review, remember that there are two types of hypotheses: research hypotheses and null hypotheses. **Research hypotheses** typically state the goals or purpose of the research, whereas **null hypotheses** put research hypotheses in a form that can be validated or refuted statistically by stating that there is no statistically significant difference between the variables to be examined.

Research hypotheses are derived from theory or primary question/research questions and place conceptual language in a format that can be examined empirically. For example, a research hypothesis might read as follows:

Probation officers who have prior criminal justice experience (police officer, corrections officer) are less supportive of rehabilitation than those who have no prior criminal justice experience.

Because it is difficult to prove that there is a difference between these two groups in the population, the null hypothesis is used to facilitate hypothesis testing.

The null hypothesis is important to inferential analyses because we do not know the true parameters of the population, so conclusions or hypotheses cannot be confirmed. Also, samples vary from one sample to the next, and error is present in each sample drawn. The chances of proving a research hypothesis, then, are uncertain. It is even more important, therefore, to examine the data in terms of rejecting a null hypothesis. The null hypothesis associated with the research hypothesis above might read as follows:

There is no statistically significant difference in the attitude toward rehabilitation between probation officers with prior criminal justice experience and those without this experience.

A number of assumptions or requirements are necessary for successful hypothesis testing. Many of these are assumed in the use of inferential analyses, but their importance to hypothesis testing is such that they are restated here.

The first requirement of hypothesis testing is that the hypotheses must be conceptually clear. It is true that primary questions or hypotheses are often vague, to cover a broad range of research; but that does not mean that they should be so vague as to defy an understanding of the topic of the research. Hypotheses should clearly spell out the object of the research or the research goals. Furthermore, the conceptual framework (research questions, concepts, etc.) should be clearly defined in operational terms.

The hypotheses should also be empirically based with no moral judgments. Although it can be found in research, a hypothesis such as "police oppress the poor" is not conducive to scientific inquiry. Hypotheses should be something that can be tested and evaluated empirically and should not be emotionally biased.

The hypotheses should be related to a body of theory and should help refute, qualify, or support existing theories. Although there is always a possibility for completely groundbreaking work, most hypothesis testing is more related to examining the current state of knowledge in a particular area, and it should attempt to either support or refute that knowledge.

Finally, hypotheses should be as specific as possible. The more precise the hypothesis, the greater the validity and the smaller the possibility that the outcome will occur by chance.

Criteria have also been proposed for research hypotheses. Zetterberg (1963) established three criteria for acceptance of a working, or research, hypothesis.

1. The first criterion is that the result of hypothesis testing should disprove the null hypothesis with a certain probability. As outlined in this part of the book, the null hypothesis should be rejected at the critical level previously set before any attempt at accepting the research hypothesis is made. This is simply the beginning of accepting the research hypotheses, however.

2. The second criterion is that the data should be arranged in a manner predicted by the hypothesis. Beyond summary analyses of the hypothesis tests, the data should conform to what was expected in the conceptual model. It is often possible to reject the null hypothesis but have measures of the strength of the association that are so weak that there is little confidence in the ability of the data to provide meaningful information about the relationship. Or perhaps the data is arranged in such a manner that interpretation beyond rejection of the null hypothesis is questionable or not possible. In these cases, interpretation of the research hypothesis may not be advised.

3. The final criterion is that alternative hypotheses should also be rejected. Rejecting the null hypothesis does not automatically mean acceptance of a research hypothesis. There are other (sometimes many other) possible situations that may exist such that rejection of the null hypothesis does not support the research hypothesis. For example, rejecting the null hypothesis that there is no statistically significant difference between college-educated and non-college-educated officers in the number of civil suits filed against a police agency does not mean that employing only college-educated officers will keep a police agency from being sued. A number of other factors must be considered after rejecting the null hypothesis. Differences between college-educated and non-college-educated officers may be because the college-educated officers are predominately in staff or supervisor positions, or there may be a subculture within the department that is producing civil lawsuits independent of the officers. There are a number of explanations that must be explored and rejected before making an argument for accepting the research hypothesis.

■ Steps in Hypothesis Testing

Hypothesis testing can be thought of as a specialized decision-making process. First, assumptions are established through the research and null hypotheses. A sample or samples are then drawn from groups with characteristics the researchers are interested in studying. Once the type of test to use is selected, sample characteristics are then compared to the assumptions made about the unknown population parameters to see if the data drawn supports the assumptions. The results of that comparison allow a decision to be made concerning the rejection or failure to reject the null hypothesis. Although the particular strategy and steps in hypothesis testing, by and large, depend on the type of analyses being conducted, there are some standard procedures that should be undertaken. These steps are identified and discussed in this section of the chapter.

Step 1: Developing the Research Hypothesis

The first step involves putting ideas about the phenomena to be studied in the form of a research hypothesis. Research hypotheses typically are similar to the research/ secondary questions of a project or the propositions of a theory. Although research hypotheses had been used for many years, they were formalized by Neyman and Pearson (1928) as an attempt to clarify the issues of hypothesis testing. They established a structure for developing research questions that were derived from the primary question of the research and that served as a link with the null hypothesis. Using the example research hypothesis above, the proposition can be made that, statistically, the samples drawn from probation officers with prior criminal justice experience and those without prior experience have different sample means. Additionally, there is the assumption that the difference is *so large* that it cannot be attributed to chance. In essence, the researcher has made a set of assumptions about probation officers that he or she will attempt to test empirically.

Step 2: Developing the Null Hypothesis

The next step in the process is to develop the null hypothesis. As stated above, the null hypothesis places the research hypothesis in a form that can be tested empirically. The null hypothesis is essentially the reverse of the research hypotheses. It states that the means of the two samples are equal, or that the difference is so small that it could have occurred by chance or because of sampling error. Hypothesis testing is set up so that, if possible, we can reject the null hypothesis. By rejecting the null hypothesis, we have reason to support the research hypothesis, keeping in mind the guidelines established above for accepting the research hypothesis.

Step 3: Drawing Samples

The third step in the hypothesis testing process is to take samples from the populations to be studied. This is concerned with the probability and nonprobability samples discussed in Chapter 15. It is imperative that the samples drawn be representative of the target population, preferably the theoretical population; otherwise, the ability to generalize from the sample to the population is limited.

Step 4: Selecting the Test

The fourth step involves selecting the appropriate statistical test to properly analyze the data drawn. Two decisions must be made at this point. The first decision is which hypothesis test is most appropriate for the data; this is discussed at length in Chapter 16. The second decision is whether to conduct a one- or a two-tailed test.

One-Tailed and Two-Tailed Tests   In hypothesis testing, the results of tests are discussed in terms of where they fall under the normal curve. At times, researchers want the obtained values to fall toward the middle of the normal curve (confidence intervals) and within the boundaries of ± 1.96 (or 2.58, or whatever level is set). At other times, the goal is to reject the null hypothesis by stating that one variable is so different from another that it is outside the boundaries of ± 1.96. In these cases, scientists refer to the variable of interest as being "in the tails of the distribution." Being "in the tails" is another factor in the hypothesis testing process. For example, if a researcher was conducting research on teaching children about drug use, there are three possible outcomes:

1. The program could result in a reduction in drug use among the children.
2. The program could produce no significant change in behavior.
3. The program could provide the children with information that may make them more likely to use drugs.

Although all of these are possible outcomes, the researcher may not want to test them all. If the goal of the project is to examine the success of the program, the researcher may only want to test the first outcome; or he or she may want to know if there is change but does not care whether the change is positive or negative. These examples represent the two types of hypotheses or tests: one-tailed tests and two-tailed tests.

A **two-tailed test** states that there is a difference between the groups being tested without specifying the direction of the difference. This is the type of hypothesis testing that has been used as examples to this point. For this type of analysis, the normal curve would be divided, essentially, into three sections. The focus of the analysis, then, would be to examine whether the program produced children who used drugs significantly less than normal (more than -1.96 standard deviations from the mean), whether there was no significant difference between those going through the program and those who did not (between ± 1.96 standard deviations from the mean), or whether the program produced children who used drugs significantly more than normal (more than $+1.96$ standard deviations from the mean). Specifically, the researcher would be looking to see if the sample of those who were in the program was significantly different from others in their drug use. If there was a difference, that would be sufficient. It would not be important in this analysis if those in the program used drugs more or less often, only that there was a difference.

That a researcher would not care if those in the program used drugs more than those not in the program is a little absurd. More likely, the success of the program would be based only on a finding that those in the program used drugs less than those not in the program. This determination could be made using a one-tailed test. A **one-tailed test** looks for a directional difference between the two groups, in this case, those in the program used less drugs than others. The hypothesis for a one-tailed test can predict one variable to be greater than or less than the other, but not both.

The advantage of a one-tailed test is that there is more area under the normal curve to work with. The value ± 1.96 has been used throughout this book in connection with hypothesis testing at the 0.05 significance level. This is because the analyses have all been two-tailed tests. Turning to Table B-1 in the appendix, though, you can see that the area beyond this Z score is actually only 0.025. Basing a decision on just one part of the tail, however, allows use of a Z score of 1.65 (actually, it is between 1.64 and 1.65, but convention is to use 1.65 for a 0.05 cutoff). As can be seen from the table, this value corresponds to a two-tailed test of a hypothesis at the 0.90 level. Using a one-tailed test allows researchers to examine more of the area under one end of the normal curve and still determine whether two distributions are different. A one-tailed test, then, is somewhat more precise if it is warranted in the research.

So which type of test is better? This issue has been hotly debated for many years. Probably the most informative debate was that of Jones (1949, 1954), Hick (1952), Burke (1953), and Marks (1953). These articles outlined the issues in much greater detail than is possible here. You are encouraged to read these articles if these tests are of interest to you.

Kimmel (1957) attempted to resolve this debate by outlining three criteria for using a one-tailed test rather than a two-tailed test. The first criterion for use of a one-tailed test is "when a difference in the unpredicted direction, while possible, would be . . . meaningless." An example would be an anger training program for violent offenders in prison. The goal of the research would be to show that the anger training program makes the prisoners less violent. A finding in an unpredicted direction is certainly possible, but it would be meaningless to the research. The second criterion is "when results in the unpredicted direction will, under no conditions, be used to determine a course of behavior different in any way from that determined by no difference at all." You might think of this as analogous to predicting that the scores of a basketball game will have no difference at the end of the game. The research hypothesis in this case might be that the home team wins. Because the prediction is in a specific direction (home team wins) and it is not possible to tie, a finding in the unpredicted direction would be the same as no difference (either way, the home team did not win). The final criterion for using a one-tailed test over a two-tailed test is "when a directional hypothesis is deducible from . . . theory but results in the opposite direction are not deducible from coexisting . . . theory." An example would be that certain antisocial behavior would result in children committing violent crimes. If the assumption is true, this theory would explain that behavior. If the children in the research were no different than normal children, the theory would have no response for that behavior or finding. These are often difficult criteria to meet in social science research. It is good advice, then, to use a two-tailed test unless there is a reason to use a one-tailed test.

Step 5: Calculating the Obtained Value

The fifth step involves using the analysis procedure chosen to calculate the obtained value. This is the process of using, for example, Chi-square (Chapter 8), Z tests, t-tests (Chapter 17), or ANOVA (Chapter 18) to determine the obtained value.

Practically, this step in the process is often skipped because statistical packages calculate the obtained value for you. Obtained values are printed in the output of most statistical programs, but as with Chi-square in Chapter 8, the significance value of the analysis is also printed, so this measure can be used to make a decision (step 7) rather than having to calculate the obtained value and compare it to the critical value.

Step 6: Determining the Significance and Critical Regions

The sixth step involves determining the critical value of the test statistic (the significance level). The critical value is determined by the type of test used in the research, the level of certainty desired, the degrees of freedom, and whether the research is a one-tailed or a two-tailed test. The critical value will also represent the probability of making a Type I error (alpha), as discussed below.

The critical value is determined by the type of test used because each test utilizes a different table of critical values. The table of critical values for Chi-square was used in Chapter 8. There are also tables of critical values for Z test, t-tests, and F-tests, as discussed in Chapter 17.

The level of certainty returns to the decision to use a 0.05, 0.01, or some other significance level for the analysis. It has become a convention within criminal justice and criminology to use a level of 0.05 or 0.01 for hypothesis testing, although other values are available and may be used. The 0.05 level is used throughout this book so you may compare a variety of procedures using similar values; however, you are certainly not limited to the 0.05 level when conducting your own research.

As discussed with Chi-square, the degrees of freedom also play a part in determining the critical value. Degrees of freedom are important because, as discussed in Chapter 15, the size of the sample influences the standard error. Larger samples have smaller standard errors, allowing more accurate estimations of the population parameter. This, in turn, influences the critical value by making it easier to reject the null hypothesis. Alternatively, smaller samples have larger standard errors, allowing less accurate estimations of the population parameter; so the critical value is larger, making it more difficult to reject the null hypothesis.

Finally, the critical value is influenced by the choice of a one-tailed or a two-tailed test. As discussed above, a significance level of 0.05 corresponds to a Z score or confidence interval of 1.96. Also discussed, however, was the fact that a two-tailed test examines the normal curve outside -1.96 and outside $+1.96$; so the 0.05 value is cut in half, with 0.025 outside the negative region and 0.025 outside the positive region. This is important to one-tailed and two-tailed tests. A one-tailed test has a 0.05 rejection region, but it is in only one direction. In a two-tailed test, the rejection region is cut in half, but you have both ends of the distribution to work with.

Again, the reality of this step is that you should determine what level of significance you will accept for the hypothesis test. This should be determined before obtaining the output of the test. Obtaining the critical value from a table, however, is no longer necessary because the statistical output will show exactly the significance value of the analysis.

Step 7: Making a Decision

The final step in the hypothesis testing process is making a decision. This is a very straightforward process in which the obtained value from the statistical test is compared to the critical value as determined above. Using the output from a statistical analysis program, this step involves comparing the significance value obtained in the output to the acceptance level set prior to testing. As discussed in earlier chapters, if the obtained value in the statistical test is greater than the critical value (or if the significance value is less than the level set in step 6), the null hypothesis may be rejected at that significance level. This means that the differences between the two groups are so large that we are 95 or 99% confident that they could not have occurred by chance or sampling error. If the obtained value is less than the critical value, that is, if the significance value is greater than the acceptance level set in step 6, we would fail to reject the null hypothesis at that significance level.

The process leading to this decision comprises a test of a null hypothesis. This is classical statistical inference, in which inferences are made about the characteristics of an unobserved population based on the characteristics of an observed sample from that population. This is not a part of descriptive statistics, only inferential, although descriptive statistics are usually calculated together with inferential statistics.

This process examines the degree of difference between samples to determine whether or not the difference can reasonably be explained by chance. As discussed in Chapter 15, this decision tells us only if a relationship exists. Some argue that significance tests should not be used because they do not provide enough information about the phenomena studied and do not show the strength of a relationship. Detractors of hypothesis testing argue that other statistics, such as measures of association or confidence intervals should be used. These statistics suggest where the relationships lie even if they are not significant. They propose that the results of hypothesis testing are more a reflection of the sample size than of the strength of the relationship. Others argue that significance testing is important because it examines the relationship between sample characteristics and population parameters and allows early determination of justification for future research efforts.

■ Type I and Type II Errors

In inferential analysis, there are several possible outcomes to hypothesis testing, some correct and some incorrect. This can be demonstrated easily by the process of a criminal trial, where we essentially test the null hypothesis that a person is not guilty of the crime: there is no difference between the person on trial and other "innocent" people. The first outcome is to reject a null hypothesis where there truly are differences between the variables: A jury finds a person guilty who truly is guilty. A second outcome is to fail to reject a null hypothesis where there truly is no statistically significant difference between the variables: A jury finds a person not guilty who truly is not guilty. The null hypothesis can also be rejected by mistake—when there actually is no difference between the variables: A jury finds a person guilty who is not guilty. Finally, we can fail to reject a null hypothesis by mistake—when there actually are differences between the variables: A jury finds a person not guilty who is guilty. All of these situations are shown graphically in **Table 16-1**. It is in these last two situations that errors are committed in hypothesis testing. When a null hypothesis is rejected that is true, a Type I error is committed; when we fail to reject a null hypothesis that should have been rejected, a Type II error is committed. Recognition and classification of Type I and Type II errors was developed by Neyman and Pearson (1933).

TABLE 16-1   Outcomes of Hypothesis Testing (Type I and Type II Errors)

Decision	True Population	
	H_0 is true	H_0 is false
Reject H_0	Wrong decision, Type I error	Correct decision
Fail to reject H_0	Correct decision	Wrong decision, Type II error

A **Type I error** refers to the rejection of a null hypothesis that is actually true: for example, when a null hypothesis is rejected when the reality is that there is no difference between the two variables. There are situations (e.g., 1 or 5% of the time for a 0.01 or 0.05 critical level) where it appears that the difference between the variables is so great that it could not be caused by chance, but in fact it was chance or random variation between the two variables and there actually was no association.

The probability associated with committing a Type I error is the significance value of the test. For example, a hypothesis test at the 0.05 level would be correct 95% of the time, but 5% of the time a Type I error would be committed. This is referred to as the **critical probability** and is designated by the Greek letter **alpha** (α). Whenever a null hypothesis is rejected at a given significance value, there is the risk of committing a Type I error, the probability of which is α. The probability of committing a Type I error can be reduced by making the α region small (e.g., using a significance level of 0.00001); however, this increases the risk of committing a Type II error.

A **Type II error** refers to failing to reject a false null hypothesis: for example, failing to reject a null hypothesis when there actually is a difference between the variables. The probability of committing a Type II error is designated by the Greek letter **beta** (β). Beta can be calculated using the formula

$$\beta = \frac{\mu H_0 - \mu_{true}}{\sigma}$$

where μH_0 is the mean of the sample chosen to represent the null hypothesis, μ_{true} is the actual mean if the null hypothesis is true, and σ is the standard deviation of the population.

To illustrate this calculation, take the example used in Chapter 15 concerning the IQ of prisoners. The null hypothesis was that there was no difference between prisoners and "normal" people in the population. When the prisoners were tested, they had an average IQ of 80. The null hypothesis, then, would have been that both prisoners and normal people have an IQ of 80. We know from the population, however, that the average IQ is 100 and that the standard deviation is 15. In this case, the probability of making a Type II error is

$$\beta = \frac{80 - 100}{15}$$

$$= \frac{-20}{15}$$

$$= -1.333$$

This value is actually a Z score. If the research design was a two-tailed test of significance at the 0.05 level, ± 1.96 would be used as the values. These would then be adjusted based on the findings of the formula above. Adding the value above to the lower limit would produce a Z score of -3.29 ($-1.96 + -1.33$). Adding the value to the upper limit would produce a Z score of 0.63 ($1.96 + -1.33$). If the area under the normal curve were calculated for each of these values, they would equal 0.4995 for the lower limit and 0.2357 for the upper limit (see Table B-1 in the appendix). Adding these two together would result in a probability of 0.7352 of making a Type II error in this case. As demonstrated here, the only way to calculate a Type II error is to know the true value of the null hypothesis. Often, this is not possible, so the probability of committing a Type II error, although taken into account, is rarely calculated.

There are three factors generally considered to influence the probability of committing a Type II error. Some of these are real factors; others are more theoretical constructs of hypothesis testing.

The first factor influencing the probability of committing a type II error is the alpha region selected by the researcher (i.e., 0.05 or 0.01). The higher the alpha level, the higher the probability of making a Type II error. This is because the definition of a Type II error is the probability of rejecting a false null hypothesis. If the alpha level is set low, there is less risk of committing a Type II error simply because there is less risk of rejecting the null hypothesis. The risk of committing a Type I error is increased substantially by this strategy, however, and it is suggested that the alpha level not be set artificially low unless there is good reason to avoid making a Type II error.

The second factor is the degree of falseness of the hypothesis. If the null hypothesis is false to a large degree (there is a great difference between the observed and critical values), it is more likely that the correct assumption will be made concerning the outcome of hypothesis testing. If the difference is small, however, hypothesis testing may not reveal the true nature of the difference (see the discussion of power below).

The final factor is the sample size, which is also related to power. By increasing the sample size, the researcher can reduce the probability of making a Type II error. Two factors influence this relationship. First, an increase in the sample size causes the critical value to be located closer to the range of the hypothesized values of the population mean. A second effect of increased sample size is to produce closer clustering of the sample means associated with the population: Increasing the sample size reduces the standard error. Taken together, it is more likely that a Type II error will not occur when using larger samples because we are more likely to obtain an accurate estimate of the population parameter.

Which Is Better, Type I or Type II Error?

A question often asked at this point in the discussion is which type of error is better. It is obvious that for any research project, lessening the probability of committing one type of error leads to an increase in the probability of committing the other type of error. Given this balancing act, which should a researcher attempt to lessen? The answer lies in what the researcher is attempting to do.

Researchers can guard against making a Type I error by setting the significance level (the alpha level) high. If conducting research that had the potential to change the entire criminal justice system in the state, a researcher would want to be very sure of his or her results. In this case, the researcher might set the significance level for rejecting the null hypothesis at 0.001 or even 0.0001. This would set strict standards, and it would be difficult to reject the null hypothesis. The chances of committing a Type I error, however, and changing the entire system based on a false assumption are substantially less. This same logic is often used in medical research, where researchers must guard against putting a drug on the market that might not work or might even be harmful. In these cases, the significance level might even be set at 0.0000001. This would set the risk of making a Type I error at the proverbial "one in a million."

Other times, it may be more important to guard against making Type II errors. Researchers working on the leading edge of research may want to guard more against Type II errors because Type I errors will be brought out in future research, whereas a Type II error might kill the research forever. For example, if conducting research on a new program that had the potential to keep children from joining a gang, the researcher might want to set the rejection level at 0.90, 0.80 or even 0.60. It is true that there would be up to a 40% chance of committing a Type I error, but that may be acceptable. In this case, if the program is making *some* difference in the lives of *some* of the children, the program may be considered a success and should be continued. Future research could use more strict alpha levels and a more rigorous evaluation of the program. It may be more important at this stage, though, to get the program going and give it time to see if it will work. Setting a high significance level may not give the program that opportunity.

Over time, scientists and statisticians have learned to balance Type I and Type II errors by selecting the significance levels that are now taken as standard. Scientists have agreed (as much as they agree on anything) that a 5% or a 1% chance of making a Type I error is a reasonable risk, and at this level, there is an adequate degree of protection from committing Type II errors.

■ Power of Tests

The **power of a statistical test** is the probability that it will reject the null hypothesis if it is false. It is equal to $1 - \beta$. As is obvious from this formula, the power of a test is intimately linked with the probability of making a Type II error.

The power of a test can be traced to Laplace, who in 1823 argued that the data he had on the relationship between barometric changes and the phases of the moon was insufficient to analyze the data accurately. He argued that he would need approximately 72 years' worth of data to be able to analyze the relationship at the 0.01 level.

As stated above, the power of a test is related to the probability of making a Type II error: that is, failing to reject a null hypothesis that is false. Similar to Type II errors, the power of a test is influenced by four factors: sample size, the choice of statistical test, the alpha level, and the falseness of the hypothesis.

Sample size is extremely important to power. In fact, it generally defines the power of a test. As discussed in Chapter 15, the larger the sample size, the smaller the standard error, and the more likely a test is to approximate a true measure of the population parameter (the measure is also more methodologically reliable). How large a sample must be is open to debate. Some hypothesis tests are sensitive to very large sample sizes. Recall from the discussion of Chi-square that a very large sample size could produce significant results when they are not warranted (a Type I error). On the other hand, Marsh, Hau, Balla, and Grayson (1997) have proposed that structural equation modeling is very robust to large sample sizes. Z and t-tests fall somewhere between these two. When conducting your own research, it will be up to you to balance the risk of committing a Type II error and the possibility of having a sample that is too large.

The choice of the statistical procedure relates to whether a one-tailed or a two-tailed test is used. Because of the difference in the rejection regions between one-tailed and two-tailed tests, there is a difference in statistical power between these two tests. Table B-1 in the appendix shows the difference between the obtained value required to reject the null hypothesis for a one-tailed test (1.65 standard deviations from the mean) and that for a two-tailed test (1.96 standard deviations from the mean). Because there is a greater proportion of the normal curve in the rejection region for a one-tailed test, it has more statistical power.

The alpha level, critical probability, or significance level also influences the power of a test; although the true contribution is subject to debate. The definition of statistical power is the probability of making a Type II error. If the significance level is set high in a hypothesis test, there is a greater probability of making a Type II error. Setting the significance level low will lessen the probability of making a Type II error (but increase the probability of making a Type I error). There are those who argue that setting the alpha level lower increases the power of the tests. Although this is mathematically correct, it is methodologically wrong. The goal of the power of the test should be to reduce the probability of making a Type II error *while maintaining an acceptable probability of not making a Type I error*. Simply reducing the alpha level and calling it an increase in power is little more than statistical manipulation. To truly increase statistical power, appropriately high alpha levels should be maintained and the sample size drawn (or increased) such that sufficient power can be achieved.

When a Sample Size of One Is Enough

Three professors (a physicist, a chemist, and a statistician) are called in to see their dean. Just as they arrive, the dean is called out of his office. Soon, the professors notice that there is a fire in the wastebasket. The physicist says, "I know what to do; we must cool the materials until their temperature is lower than the ignition temperature; then the fire will go out." The chemist says, "No! No! We must cut off the oxygen supply so the fire will go out." While the physicist and chemist are arguing, they see with alarm that the statistician is going around the room starting other fires. When they ask what he is doing, he replies, "Trying to get an adequate sample size."

The final factor related to power is the falseness of the null hypothesis. Here, think of statistical tests as a microscope and the power of the test as the power of the microscope examining the difference between the distributions of two variables. For this exercise, assume that both distributions are graphed in a normal curve, with the center of each distribution at the variable's mean. If the means of the distributions are very different from one another (perhaps a difference of 50 on a scale of 100), it would not take a very powerful microscope to see the difference. This would be a situation where the variables are very different, and the hypothesis test would easily show the difference. If the difference was very small, however (a difference of 1 on that same scale), a much more powerful microscope would be necessary to see the difference. Higher power would be required to determine if there was a true difference between the distributions.

All of these come together as a measure of the power of an analysis. This power is typically represented as a probability—the probability of making a Type II error. A research project that has a power level of 0.90 would have only a 10% chance of making a Type II error. A study with a power level of 0.40, however, would have a 60%

chance of making a Type II error. It is generally recommended that research have a power of at least 0.50, or the chances of making a Type II error are considered too great. In the analysis above, β was 0.7352. This means that the power of that analysis was $1 - 0.7352$ or 0.2648. This is a terribly low power, and the findings should be examined carefully to see if the risk of making a Type II error is too high to meet the goals of the research. Unfortunately, most researchers in criminal justice and criminology do not check the statistical power (see Brown, 1989); if they did, it would be much closer to this value than to desired ranges above 0.70.

The best way to minimize both a Type I and Type II error is to set a high critical probability, thus reducing the chances of making a Type I error, and then ensure that the sample size is large enough to maintain the power of the test, thereby reducing the chances of making a Type II error. This can be illustrated with a small and known population. If we took a small sample of this population, we would have to decide on the importance of a Type I or Type II error in our testing. If, however, we took a sample that actually represented the entire population, we could set the critical probability at extremely high levels, thus reducing the chances of making a Type I error and be confident of not making a Type II error.

■ Conclusion

Using the concepts of sampling distributions, the central limit theorem, and the normal curve from Chapter 15, in this chapter we introduced the steps in hypothesis testing. Inherent in proper tests of hypotheses are decisions concerning what type of test to use and whether to use a one-tailed or a two-tailed test. It is also important to keep in mind in this process the issue of statistical power and the probability of making either a Type I or Type II error. With this knowledge, you are now ready to begin testing hypotheses. That is the subject of Chapter 17.

■ Key Terms

alpha
beta
critical probability
critical value
hypothesis testing
null hypotheses
one-tailed test

power of a statistical test
research hypotheses
statistical power
two-tailed test
Type I error
Type II error

■ Summary of Equations

Type II Errors

$$\beta = \frac{\mu H_0 - \mu_{\text{true}}}{\sigma}$$

■ Exercises

1. Find one or more journal articles that deal with hypothesis testing.
   a. Determine the research and null hypotheses of the research.
   b. Determine if the researchers are using a one-tailed or two-tailed analysis. What led you to this conclusion?

   c. Attempt to determine what method of sampling the researchers used.

   d. Attempt to determine what statistical test the researchers used. If you need help, refer to Chapter 17 for a discussion of the various types of tests.

   e. If the values are provided, discuss the obtained and critical values of the analysis. A review of Chi-square or the tests in Chapter 17 may be beneficial.

   f. Discuss the decision of the researchers. Was this decision reasonable given the analyses? Why or why not?

2. Type I and Type II errors are not often discussed in research articles, so it would be counterproductive to require a number of articles explaining these concepts. One good article that addresses this issue is that by Sherman and Weisburd (1995), General deterrent effects of police patrol in crime "Hot Spots": a randomized study, *Justice Quarterly*, 12(4):625–648. Get this article and prepare to discuss in class the advantages and disadvantages of the authors' arguments.

3. Find one or more journal articles that discuss either power analysis or statistical power. What general conclusions about the research can you draw from this analysis or discussion?

■ References

Brown, S. E. (1989). Statistical power and criminal justice research. *Journal of Criminal Justice*, 17:115–122.

Burke, C. J. (1953). A brief note on one-tailed tests. *Psychological Bulletin*, 50:384–387.

Hick, W. E. (1952). A note on one-tailed and two-tailed tests. *Psychological Review*, 59:316–318.

Jones, L. V. (1949). Tests of hypotheses: One-sided vs. two-sided alternatives. *Psychological Bulletin*, 46:43–46.

Jones, L. V. (1954). A rejoinder on one-tailed tests. *Psychological Bulletin*, 51:585–586.

Kimmel, H. D. (1957). Three criteria for the use of one-tailed tests. *Psychological Bulletin*, 54:351–353.

Marks, M. R. (1953). One- and two-tailed tests. *Psychological Review*, 60:207–208.

Marsh, H. W., K. T. Hau, J. R. Balla, and D. Grayson (1997). Is more ever too much? The number of indicators per factor in confirmatory factor analysis. Unpublished manuscript.

Neyman, J., and E. S. Pearson (1928). On the use and interpretation of certain test criteria for purposes of statistical inference. *Biometrika*, 20A:175–240.

Neyman, J., and E. S. Pearson (1933). On the problem of the most efficient test of statistical hypotheses. *Philosophical Transactions of the Royal Society of London*, 231A:289–337.

Zetterberg, H. L. (1963). *On Theory and Verification in Sociology*. Totowa, NJ: Bedminster Press.

■ For Further Reading

Cohen, J. (1962). The statistical power of abnormal-social psychological research: A review. *Journal of Abnormal and Social Psychology*, 65(3):145–153.

Cohen, J. (1992). A power primer. *Psychological Bulletin*, 112(1):155–159.

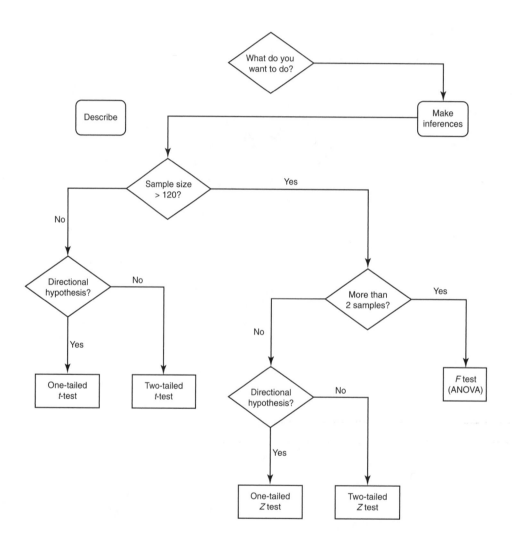

Hypothesis Tests

Statistical Analysis: Mysterious, sometimes bizarre, manipulations performed upon the collected data in order to obscure the fact that the results have no generalizable meaning for humanity. Commonly, computers are used, lending an additional aura of unreality to the proceedings.

—Unknown

In Chapters 15 and 16, we provided the foundations of hypothesis testing and inferential analysis. Now we get down to the business of hypothesis testing. The goal here is the same as in earlier chapters: to make inferences about a population based on data from a sample. There are several tests that can be used to examine the null hypotheses associated with inferential analysis. The test used depends on the data available. It should be stressed that most of the tests used in this chapter are **parametric tests**, tests that make the assumption that a population is normal. Only Chi-square does not make this assumption and thus is a **nonparametric test**. Three tests are discussed in this chapter:

- The Z test for comparing a large sample to a population or two large samples
- The t-test for comparing a small sample to a population or two small samples
- The Chi-square Test of Independence for nonparametric data

Laplace was one of the first people to explore the need to find structure in testing hypotheses. In his 1823 work (1966), Laplace argued that to analyze data properly, one must have "a method for determining the probability that the error in the obtained results is contained in narrow limits, a method without which one risks presenting the effects of irregular causes as law of nature." From this foundation rose one of the oldest and most used of all statistical procedures.

■ *Z* Test

The sum and substance of inferential hypothesis testing is the Z **test**, which examines whether a sample could have come from a known population or whether two samples come from the same population. The Z test is used only for large samples. This is typically taken to mean it should only be used with samples larger than 120.

The Z test draws directly from the normal curve, central limit theorem, and sampling distributions to test whether it is likely that two distributions are similar enough that they could be considered as being from the same population. Essentially, if the means of two distributions are close together and toward the middle of a normal curve, probability theory can be used to state that they are probably from the same population. If the mean of one distribution is in the extreme tail of another distribution, though, it is unlikely that those means share the same population. The Z test is used to determine the relative location of these means.

As discussed above, the Z test is a parametric analysis procedure. As such, there are certain requirements that must be met to use a Z test. First, the dependent variable must be interval level (see Chapter 2). Second, the population must be normal in distribution (see Chapters 6 and 15). Also, as with Analysis of Variance (ANOVA), which is discussed in Chapter 18, the variances of the two samples should be equal (although a Z test is fairly robust for violation of this assumption). The final assumption of a Z test is that the means of the two samples drawn are equal. This is not an assumption that is the goal of the research, however. Researchers actually want to violate this assumption in most hypothesis testing because if the null hypothesis is rejected, it is assumed that this requirement is violated.

The formula for a Z test is virtually the same as the formula for a Z score:

$$Z = \frac{\overline{X} - \mu}{\sigma/\sqrt{N}}$$

This formula is a combination of the formula for a Z score and confidence intervals. In essence, the formula for a Z test takes the formula for the calculation of a Z score and applies it to hypothesis testing. This requires the inclusion of more information about the population and its relationship to the sample—information supplied by the formula for confidence intervals. In this formula, $\overline{X}$ is the mean of the sample, μ the population mean or the mean of the sampling distribution, and σ divided by the square root of N is the population standard deviation. All of these serve the same functions as they did in their original formulas.

As discussed in Chapter 16, researchers typically do not know the population standard deviation, but the sample standard deviation can be used as an approximation if the sample size is large. In these cases, the formula would be modified as follows[1]:

$$Z = \frac{\overline{X} - \mu}{s/\sqrt{N}}$$

This is the operational formula for working with real data. Here s (the sample standard deviation) is inserted in the formula in place of σ (the population standard deviation) because it is a known characteristic of the data.

Calculation and Example

As discussed in Chapter 16, there are some steps in hypothesis testing that should be followed. These steps are outlined below as they apply when conducting hypothesis tests using a Z test.

Step 1: Determining the Research Hypothesis (H_r)

We employ as an example a research project that seeks to determine if juvenile delinquents have a shorter attention span than that of other juveniles their age.

Step 2: Developing the Null Hypothesis (H_o)

In this case, the null hypothesis would be that there is no statistically significant difference in the mean attention span between juvenile delinquents and other juveniles.

Step 3: Drawing the Sample

For this research we might administer an attention span questionnaire to 150 students in a particular high school, and then to 125 juvenile delinquents from the same school chosen at random from those passing through juvenile court in one year. We are thus examining the difference between the population of juveniles in this school and a sample selected from them (the juvenile delinquents).

Step 4: Selecting the Test

The test in this case would probably be a one-tailed test because although it is possible that "normal" juveniles have a lower attention span than juvenile delinquents, it would be meaningless and it would not add to the interpretation of the findings. Also, the critical value for this test will be set at 0.05.

Step 5: Calculating the Obtained Value

When the instrument was administered to measure attention span in the high school, a mean attention span level of 84 was obtained. The same test was then given to a sample of juvenile delinquents, and a mean score of 78 was obtained with a standard deviation of 16. This would be comparing a large sample to the population of the high school. This information can be used to calculate the obtained value of the Z test. In this case, a sample mean score of 78 was obtained from the juvenile delinquents, so the null hypothesis would be $H_o: \mu = 78$, and the research hypothesis would be $H_r: \mu > 78$. Using the formula above, the obtained value would be calculated as

$$Z = \frac{\overline{X} - \mu}{s/\sqrt{N}}$$

$$= \frac{78 - 84}{16/\sqrt{125}}$$

$$= \frac{78 - 84}{16/11.18}$$

$$= \frac{78 - 84}{1.43}$$

$$= \frac{-6}{1.43}$$

$$= -4.19$$

For this formula, the sample mean from the juvenile delinquents ($\overline{X}$) is 78, the population mean from the high school students is 84, the standard deviation of the sample of juvenile delinquents (s) is 16, and the sample size of juvenile delinquents (N) is 125. These calculations resulted in an obtained value of -4.19.

Step 6: Obtaining the Critical Value

Because this is a one-tailed test at the 0.05 level, Table B-3 in the appendix would be used to determine the critical level. Using this table, the critical value will be 1.65. Notice that there are no negative values in this distribution. Negative values do exist but in a different form. A negative number simply means that the sample obtained value is less than the population mean, which means that the sample is on the left side of the normal curve rather than on the right. This does not affect the areas under the normal curve, only the placement of the values. The examination of the difference between the observed and critical values is exactly the same. It still obtains a measure of the distance between the sample mean and the population mean in standard deviation or standard error units. The results indicate how far above or below the population mean the sample mean is. In this case, the sample mean is 4.19 standard deviations below the population mean.

Step 7: Making a Decision

Here, the obtained value is -4.19 and the critical value is 1.65. Using the decision criteria, the null hypothesis would be rejected at the 0.05 level and a conclusion made that there are statistically significant differences in the attention span of juvenile delinquents. In this case, we could conclude that the juvenile delinquents would have a lower attention span than the population as defined in this study. This does not necessarily mean, however, that all juvenile delinquents have a lower attention span than all "normal" juveniles. That conclusion would go beyond the population to which we can generalize. We can, however, generalize to the population within the school and area examined.

Interpretation and Application: Known Probability of Error

The interpretation of a Z test (or any of the inferential analysis procedures discussed in this chapter) relies on the *known probability of error* to determine whether or not two distributions came from the same population. The known probability of error is tied to the normal curve. The area under the normal curve was discussed in Chapters 6 and 15. Recall that the area under the normal curve is consistent and determinable. When comparing two groups (a population with a sample, or two samples), the population or the base sample can be used as a baseline for comparison. This means that the other sample will be compared to this baseline to determine if the two groups could have come from the same population.

Figure 17-1 shows a normal curve that represents the population of the research project above concerning the attention span of juveniles. It is presumed that this curve and distribution are representative of the attention span of "normal" juveniles. This normal curve, however, does not take up the entire axis on which attention span scores of juveniles lie. In fact, it takes up only a small portion of the total possible attention span. This group has a mean of 84. Recall that 95% of the values under a normal curve will fall between ± 1.96 standard deviations of the mean, and 99% of the values will fall between ± 2.58 standard deviations of the mean. It is a relatively simple matter, then, to

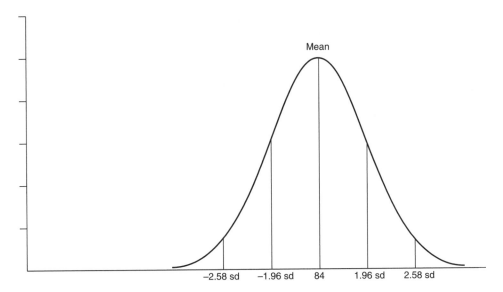

Figure 17-1 Normal Curve for Normal Juveniles' Attention Spans

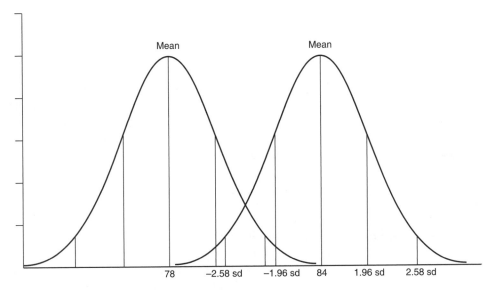

Figure 17-2 Normal Curves of Normal and Delinquent Children's Attention Spans

compare the attention spans of juvenile delinquents with this group to see if they appear to come from the same population.

In **Figure 17-2** the distribution for the juvenile delinquents is placed on the same line as that for the "normal" juveniles. Here, the normal curve containing the distribution of juvenile delinquents is at an entirely different place than the distribution of "normal" juveniles. In fact, the *Z* test calculations above show that the mean of the delinquents' attention span is 4.19 standard deviations below the mean of the "normal" juveniles. Using the area under the normal curve would show that just over 99.99997% of all "normal" juveniles are closer to the mean than the mean of juvenile delinquents. Does that mean that juvenile delinquents and "normal" juveniles are absolutely different? No—there are juvenile delinquents who have scores much closer than this. But based on the placement of the means of the two distributions, it is extremely unlikely that the two distributions are in the same population. In fact, there is only 1 chance in 100,000 that this is the case—pretty good odds.

In **Figure 17-3** the juvenile delinquents had a mean attention span of 82. This is very close to the mean of the "normal" juveniles (84). In fact, if this value were to be subjected to a Z test, the obtained value would be

$$Z = \frac{\overline{X} - \mu}{s/\sqrt{N}}$$

$$= \frac{82 - 84}{16/\sqrt{125}}$$

$$= \frac{82 - 84}{16/11.18}$$

$$= \frac{82 - 84}{1.43}$$

$$= \frac{-2}{1.43}$$

$$= -1.39$$

Using the critical value used in the previous example (1.65 for a test at the 0.05 level), it would probably be determined that these two distributions are so close together that they are likely to come from the same population. After all, the mean of the juvenile delinquents is closer than many of the actual values of the "normal" juveniles.

One more note here: All curves in Figures 17-2 and 17-3 are mesokurtic. The standard deviation (variance) is important here. What if the standard deviation of the juvenile

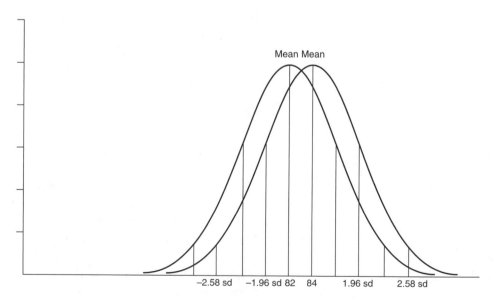

Figure 17-3 Normal Curves of Normal and Delinquent Children's Attention Spans

delinquents had been large, perhaps 50? This would have produced a platykurtic curve. In that case, there would have been a much wider spread of the delinquents, and more of the delinquents would have been closer to the mean of the "normal" juveniles. Would that have changed the findings? Perhaps, but such distributions are accounted for in hypothesis tests (see Figure 17-5 and the discussion of *t*-tests below).

In hypothesis testing, the known probability of error is controlled for in two ways. First, the smaller the sample size, the more extreme the scores will have to be before they can be considered statistically significant. For example, in **Figure 17-4** there are only a few values in the distribution. As a result, most of the values are represented more accurately as outliers from the mean rather than as values close to each other in a normal distribution. In cases such as this, it is difficult to draw definite conclusions about whether another distribution that is close to this one is a part of the same population or different from it. To lessen the probability of drawing errant conclusions, hypothesis testing compensates for smaller sample sizes by requiring larger results from the tests, thereby making it more difficult to reject the null hypothesis. Many of the formulas for hypothesis tests also compensate for small sample sizes. For example, the variance discussed in Chapter 5 used the formula

$$\sigma^2 = \frac{\Sigma(X - \overline{X})^2}{N}$$

to represent a variance descriptive of the population. The formula for an estimate of the variance for samples (often improperly called a sample variance) is

$$s^2 = \frac{\Sigma(X - \overline{X})^2}{N - 1}$$

Here $N - 1$ is used in the denominator of the formula to compensate for unknown error that is always a result of drawing inferences from samples to populations (or between samples) that do not exist when simply describing a population. Having $N - 1$ in the denominator reduces the value of dispersion in the numerator, thus making the quotient of the formula smaller. When compared to the critical value, then, it makes it more difficult to reject the null hypothesis.

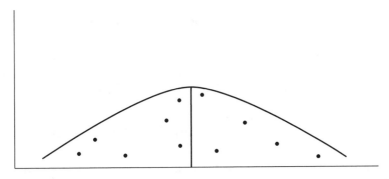

Figure 17-4 Platykurtic Curve and Standard Deviation

One- versus Two-Sample *Z* tests

The example above was treated as if a sample of juvenile delinquents was being drawn and compared to a population. You can see, though, that two samples were actually drawn and compared: one sample representing "normal juveniles"—the high school; and one sample representing the research (experimental) group—the juvenile delinquents. *Z* tests (and *t*-tests) can also be used when the goal is specifically to compare two samples.

A two-sample *Z* test has the same assumptions as a one-sample test, but there are two other assumptions for the two-sample test. For the central limit theorem and other concepts essential to inferential analysis to be fully applicable, both samples must meet the sample size requirements of a *Z* test. If only one of the samples is sufficiently large, a *t*-test is more appropriate (see below). Furthermore, the two samples should be independent and random. This ensures that sampling units in each sample have an equal probability of being selected for the sample. This assumption can also be met if one sample is drawn and then divided into two groups, such as selecting one sample from the population and then dividing between male and female to examine the differences between these two samples.

Although not a difference in assumptions, there is one other difference between one- and two-sample tests. In a two-sample test, the goal is not to estimate the population parameter. Generally, the goal is only to determine if there is a difference between the two groups (two-tailed test) or if there is a specific difference between the two groups (one-tailed test). Ultimately, however, we want to infer that any differences between the means of the two groups will be reflected in the population(s) of the groups.

■ *t*-test

Whereas a *Z* test is appropriate for large samples, it can give inaccurate results when sample sizes are small. When the population standard deviation is known, or when the sample size is large enough to give confidence that the sample standard deviation is an accurate estimate of the population standard deviation, a *Z* test can be used with some confidence. When the sample size is small, however, the sample standard deviation is a more biased estimate of the population standard deviation than is allowable for use with the central limit theorem.

In these cases, a *Z* test should not be used. A statistical consultant to a beer brewery in Ireland faced a similar problem. Working at the Guinness brewery, he found that he could not take 120 or more samples of beer because of the variability of the mix of ingredients and the susceptibility of the beer to small temperature changes (not to mention the problems that come with taking 120 samples of beer in one day!). Nonetheless, it was necessary to be able to examine the brew with known probabilities. To overcome this problem, William Sealy Gosset (1943) developed measures to test hypotheses using sample sizes of less than 120, known as **t-tests**. The only differences between a *t*-test and a *Z* test are that in a *t*-test:

- *s* is used instead of σ (which was actually the case in the operational formula for a *Z* test as shown above).
- $N - 1$ is used instead of *N* in the denominator (which some books use in the formula for a *Z* test).
- The critical value is adjusted for the sample size.

Note here that the reason for using a *t*-test rather than a *Z* test is to account for small sample sizes; therefore, we compensate the most for the smallest sample sizes. Look at Tables B-1 and B-3 in the appendix. At the 0.01 level, for a sample of 5, the difference between the critical value of a *t*-test and that of a *Z* test is 4.03 to 2.58. If the sample gets large enough to approximate one used for a *Z* test, however (in this case, with sample sizes greater than 120), the critical values would become the same—2.58.

The *t*-test was developed by W. S. Gosset in 1908 and published in 1943 as a means of conducting hypothesis tests with small samples. Gosset wrote under the name "Student," so the *t*-test is often called *Student's t*, and the distribution associated with it, **Student's *t* Distribution**. You have surely noticed that the *t* in *t*-test is lowercase whereas the *Z* in *Z* test is uppercase. This is used to denote the difference in working with large samples (theoretically, closer to the population) and small samples, similar to the difference between Greek and Roman notation with populations and samples, respectively.

Assumptions of a t-test

A *t*-test has the same assumptions and goals as a *Z* test. First, it assumes that the data is at interval level, although a *t*-test is robust for this assumption, as discussed below. A *t*-test also assumes a normal distribution, although the actual shape of the *t*-distribution is not exactly normal, especially at smaller sample sizes. A *t*-test also assumes that a probability sample has been used to select the sample elements. Finally, a *t*-test assumes that the observations are independent, as with a *Z* test.

As discussed above, there are differences between a *t* distribution and a *Z* distribution in that the shape of the *t* distribution is flatter (more platykurtic) than the *Z* distribution. When sample sizes are small, a *t* distribution is much more platykurtic than a *Z* distribution. Look at **Figure 17-5**, where a platykurtic curve is superimposed over a normal curve. Notice that even though they both have the same mean, the standard

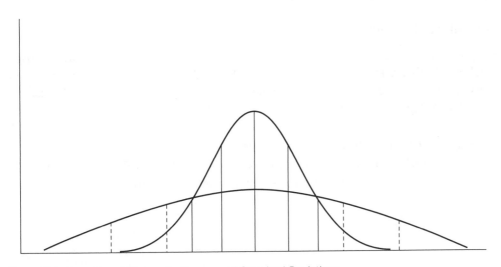

Figure 17-5 Normal and Platykurtic Curves and Standard Deviations

deviations of the normal curve (the solid vertical lines) are much closer to the mean of both distributions than either of the standard deviations (the dashed lines) of the platykurtic curve. This means that for inferential analysis, the critical value of t at a given significance level will be greater than the comparable value of Z at that significance level, making it more difficult to reject the null hypothesis. For example, look at Table B-3 in the appendix. We know that with a normal curve (a Z distribution), the value associated with a 0.05 significance level (5% of the values of the normal curve in the tail) for a two-tailed test is 1.96. With a sample size of 2 ($df = 1$) for a t distribution, this same area under the normal curve occurs at 12.70 (comparable to where the standard deviations are in relation to each other in Figure 17-5). As the sample size increases, though, the shape of the t distribution gets taller, approximating the Z distribution. The critical values of t and Z also get closer together. Look again at Table B-3. For a two-tailed t-test with the degrees of freedom at 120, the critical value is 1.98, not very far from the 1.96 used with a Z test. Also notice that the next line shows the degrees of freedom at 8. This is because it can be argued that the t and Z distributions get close enough in values above $N = 120$ that the distributions are interchangeable, although the critical value of t does not actually match the value of Z until the degrees of freedom reach 500. If the t distribution (the platykurtic distribution) in Figure 17-5 had a sample size of 125, as in the example of juvenile attention spans, it would be very close to a Z distribution: close enough, in fact, that a Z test could be used, as was the case above.

As you can see from the foregoing discussion, the issue of sample size is critical with t-tests; but how small a sample can be used and still achieve reliable results? It is generally accepted that even with t-tests, a sample size of at least 30 should be used. Although not a definite rule, it is often assumed that larger samples are more likely to be normal in distribution. The smaller the sample size, the more likely it is that the distribution will not be normal. As discussed in Chapter 10, the central limit theorem can usually be relied upon to make assumptions of normality as long as the sample has an N of at least 30. Below an N of 30, it is very unlikely that the distribution is normal. This does not mean that a t-test cannot be used with sample sizes smaller than 30. In fact, a t-test can be used with a sample size of 2. Interpretations below a sample size of 30, however, should be more cautious.

Calculation and Example

Suppose the sample of juveniles taken above was only 20. A Z test could not be used reliably with that data, although a t-test could be used. The same steps as those used in the hypothesis-testing process up to determining the obtained value could also be used. These steps are discussed below.

Step 1: Determining the Research Hypothesis

The research hypothesis is the same as with the Z test: to determine if juvenile delinquents have a shorter attention span than other juveniles their age.

Step 2: Developing the Null Hypothesis

The null hypothesis is also the same; there is no statistically significant difference in the mean attention span between juvenile delinquents and other juveniles.

Step 3: Drawing the Sample

Here the attention span survey instrument would be administered to the population of students in the high school, but the sample of juvenile delinquents would be only 20.

Step 4: Selecting the Test

As with the *Z* test, a one-tailed *t*-test would be used, with the critical value set at 0.05.

Step 5: Calculating the Obtained Value

For the sake of comparison, assume that the same results were obtained as with the *Z* test: a mean attention span level of 84 for the population and a mean score of 78 with a standard deviation of 16 for the sample of juvenile delinquents. This would be comparing a small sample to the population of the high school. This information can then be used to calculate the obtained value using a *t*-test. Because, in this case, a sample mean score of 78 was obtained from the juvenile delinquents, the null hypothesis would be that H_o: $\mu = 78$, and the research hypothesis would be that H_r: $\mu = 78$. The formula for the obtained value using a *t*-test is

$$t = \frac{\overline{X} - \mu}{s/\sqrt{N - 1}}$$

Notice in this formula that the denominator has changed. This is because we can no longer have as much confidence that we are obtaining an accurate estimate of the value of the population standard error. With large samples, there was confidence that even a sample standard deviation would be a sound estimate of the population standard error. With smaller samples, though, that assumption cannot be made. As such, the denominator of the formula must be changed to show that we are estimating the standard error and to make an adjustment for a larger error in the calculations.

Including the value of $N - 1$ in the denominator of this formula has the same effect as having larger critical values. Because we are making more biased estimates of the mean and standard deviation of the population, we are less sure of our results, especially as the sample size decreases. As a result, precautions are taken by making it more difficult to reject the null hypothesis (more difficult to commit a Type I error). This is done by making the critical value larger than for *Z* (and increasingly larger as the sample size decreases). This is accomplished by reducing the value of the denominator by 1. This makes the ratio obtained by the calculations smaller; thus making it more difficult to reject the null hypothesis. The value of $N - 1$ also represents the degrees of freedom for this test. This is important when determining the critical value.

Using the formula above, the obtained value can be calculated as

$$
\begin{aligned}
t &= \frac{\overline{X} - \mu}{s/\sqrt{N - 1}} \\
&= \frac{78 - 84}{16/\sqrt{20 - 1}} \\
&= \frac{78 - 84}{16/4.34} \\
&= \frac{78 - 84}{3.67} \\
&= \frac{-6}{3.67} \\
&= -1.63
\end{aligned}
$$

Step 6: Obtaining the Critical Value

Because this is a one-tailed test at the 0.05 level, Table B-3 is used to determine the critical level. For a one-tailed test at the 0.05 level, using 19 degrees of freedom ($N - 1$), the critical value would be 1.729. As you can tell from the calculations, we did not obtain as large a value for t as when calculating Z, and the critical value for this sample size is larger than for the Z test. Both of these work to make it harder to reject the null hypothesis in this case.

Step 7: Making a Decision

Here, the obtained value is 1.63 and the critical value is 1.729. Using the decision criteria, we would fail to reject the null hypothesis at the 0.05 level and conclude that there is no statistically significant difference between the attention spans of "normal" juveniles and delinquents. This is a complete reverse of the Z test, even though only the sample size changed. This shows how researchers must be more confident in calculations where smaller sample sizes are used.

SPSS Analysis for Z tests and t-tests

SPSS includes only t-tests in its analysis. This is probably for several reasons. First, it is possible to calculate a Z score for a given variable or distribution. Because the cutoff values for a Z test are well known (e.g., 1.65, 1.96, 2.58), it would be a simple matter to compare the Z score obtained with a desired critical value. Also, as discussed above, as

TABLE 17-1 One-Sample t-test

One-Sample Statistics

	N	Mean	Std. Deviation	Std. Error Mean
Delinquent Incidents for the Census Tract	45	39.02	40.194	5.992

One-Sample Test

	(Test Value = 0)				95% Confidence Interval of the Difference	
	t	df	Sig. (2-tailed)	Mean Difference	Lower	Upper
Delinquent Incidents for the Census Tract	6.513	44	.000	39.02	26.95	51.10

sample sizes associated with *t*-tests increase, the *t* distribution becomes a closer approximation of a *Z* distribution. There is also very little practical difference between the calculations for a *t*-test and for a *Z* test. Furthermore, as discussed above, the critical values for a *Z* test and a *t*-test become closer at sample sizes above 120. Because one requirement of the *Z* test is a sample size above 120, the critical values will be very close or the same for these two tests. For large sample sizes, therefore (of at least 120), the SPSS *t*-test could be used as a *Z* test.

One-Sample *t*-Test

Table 17-1 is sample output from SPSS showing a **one-sample *t*-test** comparing a single sample to a population. Notice that this output contains two sections. The first section displays descriptive values for the variable that is chosen to be compared to the population. The second set of values shows the results of the *t*-test. The first section of this output contains the same values that could be displayed with a frequency distribution of this variable or with any other descriptive analysis. This underscores the importance of conducting thorough univariate and other descriptive analyses even when conducting inferential tests. This output shows the sample size (45), from which it can be determined that a *t*-test is appropriate. It also shows the mean of the sample (the expected value of the population), which would be the value that we would use in a null or research hypothesis to compare to the theoretical population parameter. Also included is the standard deviation of the sample. This shows a large standard deviation within the sample. The final piece of information is the estimated standard error of the mean, which could be used in hand calculations of the *t* value if something other than the standard SPSS output was desired.

How do you do that?
One-Sample *t*-tests

1. Open a data set.
   a. Start SPSS.
   b. Select *File*, then *Open*, then *Data*.
   c. Select the file you want to open, then select *Open*.
2. Once the data is visible, select *Analyze, Compare Means, One-Sample t-Test*. . . .
3. Select the variable you wish to examine and press the ▶ next to the *Test Variable(s)* window.
4. Click *OK*.
5. An output window should appear containing tables similar to the one-sample *t*-test results displayed in Table 17-1.

The second section of the output in Table 17-1 consists of the results of the *t*-test, which includes several pieces of important information. The first value is that obtained value of the *t*-test (*t*). This is the value that is compared to the critical value in the *t* table if this test is done by hand. The next value in the table is the degrees of freedom. This is calculated as $N - 1$. The next value is the significance of the *t*-test. Instead of using the cutoff values in a table, this output provides the exact significance value for the difference between the obtained and critical values. Incidentally, notice that the significance value is for a two-tailed test. To obtain results for a one-tailed test, simply compare the *t* value to the critical value in a *t* table at 44 degrees of freedom, as was done above. Here, the significance value is 0.000. This means that there is a statistically significant difference between the variables (in this case, between the sample and the population), and the null hypothesis could be rejected. This is the same procedure as with other measures of statistical significance or hypothesis testing. The next value in this output, the *mean difference*, is the mean as stated in the descriptive analysis in the first section of the output. The final portion of the output is the confidence intervals around the mean. These are the boundaries around which we are 95% sure the population parameter (mean) lies. It can be stated with confidence that the population parameter should lie between 26.95 and 51.10, with an expected value of 39.02.

Returning to the issue of one-tailed versus two-tailed tests, there is an additional precaution for *t*-tests. Remember that one of the reasons for using a *t*-test is that there is less assurance of a normal population distribution. It is possible, for example, for the population distribution to be skewed. A two-tailed test can be used and still reach the tails of a skewed distribution. It may be a biased estimate, but it can approximate the population parameters. In conducting a one-tailed test, however, there is the additional risk of misrepresenting the population distribution. When values are estimated using a one-tailed test, it is possible that the sample statistic falls in the skewed portion of the distribution. If this occurs, we may not be in the estimated portion of the curve that we thought we were in, indicating a biased estimate that may not be representative of the population distribution.

Two-Sample *t*-Test

As discussed above, although Z tests and *t*-tests are established and based on calculations of one sample compared to a population, it is rare that researchers actually know (or can even approximate) population parameters. It is more likely that a researcher will draw two samples: one sample that is the subject of the research (usually, an experimental group) and one to "represent" the population (usually, a control group). In this case, the experimental group is generally examined in an attempt to determine if it is different in some way from the control group. If that is the case, the researcher usually concludes that the difference is the result of the treatment or whatever research is being conducted. The difference between the groups is interpreted as a difference between population means. Examining the differences between these two samples is accomplished with a **two-sample *t*-test** (or Z test).

The procedures for conducting a two-sample *t*-test are the same as those for a *t*-test comparing a sample to a population, with the exception of a change in calculating the observed value. The formula for a two-sample *t*-test is

$$t = \frac{\overline{X}_1 - \overline{X}_2}{\sqrt{s_1^2/(N_1 - 1) + s_2^2/(N_2 - 1)}}$$

Essentially, the only difference here is that this formula takes into account the means and standard deviations of the two samples rather than the mean of one sample and the mean and standard deviation of the population. Also, because the main difference between a Z test and a *t*-test is the denominator, this formula can be used for a two-sample Z test simply by replacing the value in the denominator (N instead of $N - 1$).

An example SPSS output for a two-sample *t*-test is shown in **Table 17-2**. Here, those juveniles who reported being a member of a gang are compared to those reporting that they were not in a gang in relation to whether the juvenile had been arrested.

As with a one-sample *t*-test, the first section of the output contains descriptive values for both samples. This section provides information on the different sample sizes. More important, the output provides information on the differences in the means and standard deviations between the two groups. As with the one-sample *t*-test, this output also shows the expected standard error of the mean for both samples.

TABLE 17-2 Two-Sample *t*-test

Group Statistics

	Are there people in your neighborhood	N	Mean	Std. Deviation	Std. Error Mean
Recoded ARREST varible to represent yes or no	Yes	178	1.90	.295	.022
	No	104	1.94	.234	.023

Independent Samples Test

		Levene's Test for Equality of Variances		*t*-test for Equality of Means					95% Confidence Interval of the Difference	
		F	Sig.	t	df	Sig. (2-tailed)	Mean Difference	Std. Error Difference	Lower	Upper
Recoded ARREST varible to represent yes or no	Equal variances assumed	5.203	.023	−1.118	280	.265	−.04	.034	−.104	.029
	Equal variances not assumed			−1.186	254.782	.237	−.04	.032	−.101	.025

> ### How do you do that?
> ### Two-Sample *t*-tests
>
> 1. Open a data set.
>   a. Start SPSS.
>   b. Select *File*, then *Open*, then *Data*.
>   c. Select the file you want to open, then select *Open*.
> 2. Once the data is visible, select *Analyze, Compare Means, Independent-Samples* t-*Test. . . .*
> 3. Select the variable you wish to examine and press the ▶ next to the *Test Variable(s)* window.
> 4. Select the variable you want to compare with and press the ▶ next to the *Grouping Variable* window.
> 5. Click on the *Define Groups*.
> 6. Insert the two pronged numeric coding associated with that variable in each box.
> 7. Click *Continue*.
> 8. Click *OK*.
> 9. An output window should appear containing tables similar to the two-sample *t*-test results displayed in Table 17-2.

The requirement for independent samples also applies here; although it is more typical for independence to be maintained in a two-sample test because researchers are often physically drawing two samples. There is also a requirement for equality of variances with a two-sample test. The equality of variances is examined in SPSS with Levene's Test for Equality of Variances. This is the same test that will be used in ANOVA, and for the same reason. In this case, Levene's Test has a significant value for the equality of variances (0.023), so the *t*-test may proceed. If Levene's Test is not significant, the *t*-test should be reconsidered.

The output for a two-sample *t*-test is essentially the same as that for a one-sample test. The value of *t* is the obtained *t* value for the two samples. Here the value is -1.118. This value could be compared to the critical value in a *t* table for the same degrees of freedom. A difference between one-sample and two-sample tests is in the calculation of the degrees of freedom. Because there are two samples, degrees of freedom must be calculated for each sample. The formula for calculating degrees of freedom for a two-sample *t*-test is $N_1 + N_2 - 2$. In this case, it would make the degrees of freedom $178 + 104 - 2 = 280$. The value given under the *Sig* (2-tailed) column is the significance level for the null hypothesis. With a 0.265 value in this case, we would fail to reject the null hypothesis. Because the null hypothesis could not be rejected, the confidence intervals may be of value. Instead of testing whether the two means are different, the confidence intervals attempt to determine the actual location of the parameter value. This output also contains values for the mean difference between the samples and for the standard error of the difference.

A final note about *t*-tests. Although a *t*-test is used in preference to a *Z* test because we are less able to invoke the central limit theorem, it is important to remember that the central limit theorem has not been abandoned. Without the central limit theorem, it would be impossible to determine that the values used in a *t*-test are approximations, however gross, of the population parameters. It is more a case of degradation in the ability to use the central limit theorem than its abandonment.

■ Chi-square Test for Independence

The tests above have all been for use with parametric data. There are many times, however, when the data available is nonparametric. As discussed in Chapter 8, with nonnormal data, a **Chi-square Test of Independence** is most appropriate for testing the null hypothesis. This was covered extensively in Chapter 7 and is not repeated here. Some additional information as it applies specifically to hypothesis testing is discussed, however.

When using Chi-square to test the null hypothesis, we are testing the hypothesis that there is no difference (a value of zero) between two variables: that they are independent of each other rather than interrelated. In this case a Chi-square Test for Independence would be used rather than one of the other forms of Chi-square. To do this, we compare the observed values to the expected values. If the two variables are not independent, the observed values should equal or be very close to the expected values (the model of no association). If, however, the observed frequencies are different from the expected frequencies, the conclusion may be that the variables are different in some way. The null hypothesis in this case is that the variables are not independent. The relationship between the observed and expected values can be examined to determine if this is, in fact, true.

Two points may be brought out about the Chi-square table that were not possible prior to the discussion in this section on inferential analyses. First, the value of Chi-square for one degree of freedom is the square of *Z*. Notice in Table B-2 that the value of Chi-square for one degree of freedom at the 0.05 level is 3.84. This is 1.96^2. Essentially, what is occurring here is that the distribution of Chi-square is very skewed for small samples. In fact, at one degree of freedom, a Chi-square distribution is so skewed that it represents one-half of a normal distribution. Hypothesis testing at this point is analogous to conducting a one-tailed *Z* test because you only have one-half of the normal curve to work with. As the degrees of freedom increase for Chi-square, the distribution begins to approximate normality. At a value of 12 degrees of freedom, the distribution becomes closer to a normal curve; and by a value of 30 degrees of freedom, it can be said that the Chi-square distribution approximates normality.

This represents a significant step in the analysis of categorical data. As discussed in Chapter 8, Chi-square is extremely sensitive to large sample sizes, such that Chi-square tests with large sample sizes may reject a null hypothesis that is true (thereby committing a Type I error). As discussed above, however, Chi-square begins to approximate a normal curve at 30 *df*.

This statistical property is the foundation of the robustness of *t*-test for use with categorical data. Although there is a great debate concerning the ability to use a *t*-test with categorical data, the discussion above supports an argument that *t*-tests are appropriate, even for categorical data, with sample sizes above 30. The reason that a *t*-test is often not used with this kind of data is the fear of violating the assumptions of the *t*-test (interval level data and a normal distribution, specifically). As shown above, however, a Chi-square distribution with more than 30 degrees of freedom begins to approximate a normal curve. Additionally, the *t*-test is fairly robust to violation of the level of data; so use of a *t*-test with nonparametric data having sample sizes greater than 30 may be justified.

There are two final comments that should be stressed concerning Chi-square. First, the value of Chi-square only tests the null hypothesis; it is not a measure of the strength of association for the data. If the obtained value of chi-square is quite large, that only means that we may be more confident in rejecting the null hypothesis. It does not mean that there is a strong association between the two variables.

■ Conclusion

This chapter has examined hypothesis testing. In this chapter, you should have learned the difference between descriptive and inferential analyses, and you should understand the importance of each type in conducting research. Inferential analysis is built upon the foundations of the central limit theorem, the normal curve, and sampling distributions. Inferential analysis is further clarified and brought to practice with the differences between one-tailed and two-tailed tests, Type I and Type II errors, and statistical power. All of these feed into the process of hypothesis testing and give researchers the ability to conduct research on samples of all types and sizes, to make inferences between the samples or to compare a sample to a population. *Z* tests, *t*-tests, and Chi-square provide hypothesis testing tools that allow analysis of nominal level to ratio level data; and facilitate comparison of one, two, or more samples. The next chapter will focus on the last type of inferential statistic, the *F*-test. A discussion of the analysis of variance (ANOVA) will facilitate the *F*-test, as the *F*-test is one of the key statistics found in ANOVA output.

■ Key Terms

Chi-Square Test for Independence
nonparametric test
one-sample *t*-test
parametric tests

Student's *t* distribution
t-test
two-sample *t*-test
Z test

■ Summary of Equations

Z Test (population)

$$Z = \frac{\overline{X} - \mu}{\sigma/\sqrt{N}}$$

Z Test (sample)

$$Z = \frac{\overline{X} - \mu}{s/\sqrt{N}}$$

Variance (population)

$$\sigma^2 = \frac{\Sigma(X - \overline{X})^2}{N}$$

Variance (sample)

$$s^2 = \frac{\Sigma(X - \overline{X})^2}{N - 1}$$

t-test (obtained value)

$$t = \frac{\overline{X} - \mu}{s/\sqrt{N - 1}}$$

t-test (two-sample)

$$t = \frac{\overline{X}_1 - \overline{X}_2}{\sqrt{s_1^2/(N_1 - 1) + s_2^2/(N_2 - 1)}}$$

■ Exercises

1. A researcher at a counseling clinic wants to evaluate the effectiveness of the program. To do this, she gathers data on 10 people who have been assigned to counseling by the court. She also gathers a control group of people who have the same characteristics and who are receiving counseling at the clinic but who have never been arrested. The data gathered by the researcher follows.

Number of Visits for Professional Counseling			
Sample 1: Criminals		Sample 2: Noncriminals	
7	6	4	4
7	5	2	4
7	6	5	3
8	7	1	5
8	7	1	4

a. Find the mean number of counseling visits for the criminals and noncriminals.
b. Find the standard deviation for the criminals and noncriminals.
c. Using sample 1, find the probability (area under the normal curve) of finding a criminal with between 4 and 6 visits.
d. Assume that the standard deviation in sample 1 is the population standard deviation. Find the standard error of the mean and the 95% confidence interval.
e. For sample 2, find the standard error of the sample mean and the 99% confidence interval.
f. State the null hypothesis to test the difference between the two samples.
g. State the research hypothesis to test for *any* difference between the two groups.
h. Test the hypothesis at the 0.01 level of a difference between the criminals as a sample and noncriminals as the population.

2. A group of social scientists wanted to determine if police officers have a higher IQ than the normal public. To test this theory, they took a sample of 140 police officers and administered an IQ test. The mean score for officers was 110. Knowing that the average IQ in the general public is 100 with a standard deviation of 15, test to see if these researchers were right. Use the steps in the research process to outline your answer.

3. What would the result be if the researchers would have taken only a sample of 30 officers?

4. Find one journal article each that uses a Z test, a t-test, and Chi-square.

   a. Compare the procedures in the article to those in this chapter.

   b. What was different or the same?

   c. Attempt to determine the critical value and significance level using (1) the observed values from the journal article, and (2) the appropriate tables in the appendix.

References

Gosset, W. S. (1943). *"Student's" Collected Papers*, E. S. Pearson and J. Wishart (eds.). London: Biometrika Office, University College (collection of 1908 papers).

Laplace, P. S. (1966). *Celestial Mechanics*. New York: Chelsea (translation of 1823 work).

Note

1. Note that some textbooks use the value of $\sqrt{N-1}$ in this formula. This essentially makes the formula for a Z test and a t-test identical, the only difference being the critical value used for different sample sizes in the Z and t tables.

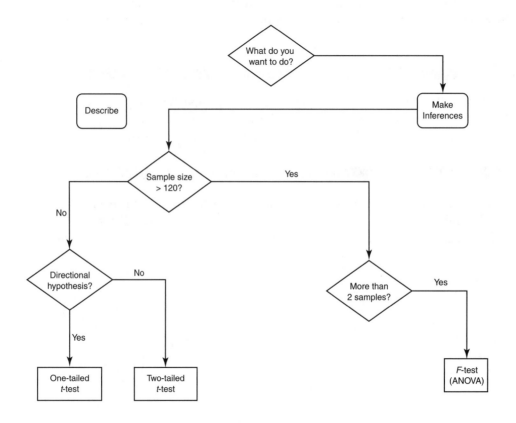

Analysis of Variance (ANOVA)

In Chapter 17, we introduced you to methods of comparing two samples taken from a population. But what happens if you have three or more samples you would like to compare? It is possible to examine the relationship between three or more samples using a Z test or a t-test. There are at least two problems with this, however. First, it would require an enormous amount of effort to calculate all possible combinations of hypothesis tests that would be required to completely examine the relationship among the variables. Also, as discussed earlier, each time a sample is drawn to conduct a hypothesis test, error is present (hopefully not more than 5%). Conducting Z tests or t-tests on each of the combination of samples would greatly increase the possibility of committing a Type I error. It is advisable, then, to use an F-test when more than two samples are to be examined. The **F-test** is drawn from an F distribution, named for R. A. Fisher (1925), who developed the distribution and test.

The F-test was developed as a part of Fisher's work in creating and advancing the **Analysis of Variance**, also known as **ANOVA**, as a statistical procedure. In its purest sense, ANOVA is a descriptive analysis procedure. It is similar to regression or other multivariate procedures that examine a specific set of data. The inclusion of an F-test, however, allows ANOVA to be used as an inferential analysis procedure. In this chapter we explore how ANOVA allows researchers to be able to analyze more than two samples derived from a population and to infer information about these samples back to that population. In this chapter, we conclude the discussion on inferential statistics by exploring ANOVA and the F-test.

■ ANOVA

In conducting research, comparisons must sometimes be made among three or more groups. As noted above, this can be done with Z tests or t-tests, but that would require testing each pair of possible groups and then combining the results for hypothesis testing—which is not only potentially time consuming, but very tedious. This would also be a problem because it would greatly increase the possibility of a Type I error (rejecting a null hypothesis that is true) because the different tests would not be completely independent and could double measure some differences. There would also be the problem of finding some pairs of groups that were significant and others that were not. Then a decision would have to be made concerning which to believe. Luckily, multiple Z or t-tests are not required to explain more than two groups because the F-test allows researchers to make comparisons between three or more groups fairly easily and accurately.

This procedure is called Analysis of Variance (ANOVA). ANOVA is a test of the measure of dispersion between group means. This test allows simultaneous comparison of the means of more than two samples. Specifically, ANOVA tests the null hypothesis that several group means are equal ($H_0: \overline{X}_1 = \overline{X}_2 = \overline{X}_3 = \cdots$). It does this by examining the variability of data within each group and the variability of data between group means.

You may wonder why this procedure is called Analysis of Variance if we are examining the means. The reason is that we are actually examining the variation around the mean, which is the variance, in this case the variance within and between samples.

ANOVA was developed by R. A. Fisher, with some of his first efforts appearing in Fisher and Mackenzie (1923). Fisher also sent his formulas and ideas to Student, who began to refine them (see Student, 1923). Student is often credited with developing the terms *variance* and *analysis of variance*, but he has pointed out that Fisher (1918) was actually the person who coined both terms. Fisher essentially established ANOVA as one of the standards for experimental analysis in his 1935 book on experimental design.

ANOVA tests a null hypothesis, as discussed previously, and the results are typically used to state that the means in the sample drawn have some relation to the means expected in the population. This allows conclusions to be drawn about populations (all parole officers in the United States) based on sample data (officers drawn at random from three regions).

There are actually several types of ANOVA procedures. The first is **One-Way ANOVA**. This is essentially a bivariate procedure in which there is a single dependent variable with different categories. When there are two independent variables, a two-way ANOVA would be utilized. For a multivariate model (more than two independent variables), a *Simple Factorial ANOVA* (often called **MANOVA**; found under GLM in SPSS) is used. Finally, with a continuous dependent variable and a categorical independent variable, use ANCOVA (analysis of covariance; see the discussion of dummy variables in Chapter 12). For simplicity of explanation, a One-Way ANOVA is used for the example here.

Assumptions

There are several important assumptions for ANOVA analyses. These are generally the same as those for other inferential statistics reviewed in Chapter 16.

The first assumption is that the data should be random samples from a normal population. This assumption is important because the goal of the procedure will be to make inferences from the samples to the population. If the samples are not random (if there is systematic error) or if the population is not normal (such that the central limit theorem cannot be used), it makes the inferences suspect at best and probably impossible.

The second assumption is that the population variances are equal for all groups of variables. This seems rather severe because we do not know the population parameters (or we could simply examine them with descriptive statistics). It is, however, possible to measure the extent to which the variances are equal in the population. This is accomplished through a **Levene Test** (Levene, 1960), which addresses the null hypothesis that all variances are not equal. If the significance value for the Levene Test is significant, you can reject this null hypothesis and assume that the variances are equal. If this value is not significant, you should reconsider using ANOVA. The results of a Levene Test for the data to be used in the ANOVA analysis for this chapter are shown in **Table 18-1**. Notice here that the Levene Test is significant, so we could reject the null hypothesis that all the variances are not equal and continue with our interpretation of the ANOVA results.

It is also important for a One-Way ANOVA that the variables are independent. You should not use ANOVA in a time series or similar design where the same variables are observed over time. This is because it makes the test think it has more degrees of freedom than are actually present in the data; thus negatively influencing the results. In the event of time series data, it is better advised to utilize some form of time series analysis.

Finally, an often overlooked assumption of ANOVA is that the categories of analysis are the only ones in the population or the only ones of interest. Because you are attempting to examine the difference in means in a sample and infer this variance to a population, you would not want your sample categories to be only a portion of the possible categories. For example, you would not want to do an analysis on crime and choose only one or two crimes to examine in the ANOVA. You would typically want to use all crime categories for the analysis. The exception here is if theory dictates the use of only a portion of the categories, such as studying only violent crimes. Even here, though, you have defined the population as violent crimes.

Calculation and Interpretation

A *t*-test for two samples represents a hypothesis test of the means of the two samples divided by the standard deviations of the two samples. A hypothesis test of the sample variances, though, could use an F distribution to measure the differences. Although the calculations for an *F*-test are more complicated and more suited to use of computers, the procedures for using an *F*-test as an inferential analysis procedure are the same as with a Z test or *t*-test. These procedures are outlined in the accompanying box.

TABLE 18-1 **Test of Homogeneity of Variances**

How long have you lived at this address?

Levene Statistic	df1	df2	Sig.
7.416	2	269	.001

How do you do that?
ANOVA

1. Open a data set.
   a. Start SPSS.
   b. Select *File*, then *Open*, then *Data*.
   c. Select the file you want to open, then select *Open*.
2. Once the data is visible, select *Analyze*, then *Compare Means*, then One-Way ANOVA.
3. Select the dependent variable you wish to examine and press the ▶ next to the *Dependent List* window.
4. Select the independent variable that you want to use and press the ▶ next to the *Factor* window.
5. Click on *Options*.
6. Place a check by the *Descriptive* box and the *Homogeneity of Variance Test* box.
7. Click *Continue*.
8. Click on *Post Hoc. . . .*
9. Place a check by the *Bonferroni* box and click *Continue*.
10. Click *OK*.
11. An output window should appear containing tables similar to the One-Way ANOVA results displayed in Tables 18-1 through 18-5.

Step 1: Determining the Research Hypothesis

The procedure for creating a research hypothesis for an *F*-test is the same as creating one for a *Z* test or *t*-test. The only difference is that the research hypothesis may be more complex because there are more samples to deal with.

Step 2: Developing the Null Hypothesis

This is also the same procedure as with other hypothesis tests but with one possible modification. For an *F*-test, researchers are examining whether the variances are equal or not. The inferential assumption, then, is that the variances are equal (that there is no statistically significant difference between the variances); an alternative assertion is that the means of the different samples are equal. Either way, the null hypothesis is stated in such a way that the assumption is that the samples were drawn from the same population.

Step 3: Drawing the Sample

This is also the same as with other hypothesis tests. The only difference is that the researcher will be drawing more than two samples.

Step 4: Selecting the Test

An *F*-test is the choice when there are more than two samples to be examined. An *F*-test is different from other hypothesis tests, however, in that there is no issue of one- versus two-tailed tests. Because more than two samples are used, it is only possible to examine a hypothesis of whether or not there is a difference between the groups. As such, there is only one type of *F*-test associated with this type of hypothesis test: one measuring the possible magnitude of the difference between the groups.

Step 5: Examining the Univariate Statistics for Each of the Variables

This is important for ANOVA because it allows an examination of the central tendency and dispersion of the variables. Because ANOVA uses the means and variance to make comparisons, knowing these measures for the variables can provide a preliminary understanding of what is expected in the analysis. Box and whisker plots, discussed in Chapter 3, are also beneficial here because they provide an indication of the measure of central tendency and the spread of the values about that measure. These show, visually, what the ANOVA analysis may produce.

The example in this chapter will explore the variance between samples of people who feel safe in their neighborhood and how long the same people have lived in their neighborhood. The variables used in this example are ADDRESS (how long have you lived at this address) and WALK_EVE (how often do you walk in your neighborhood in the evening). The univariate statistics for these variables are shown in **Table 18-2**.

It is also important here to examine the combined descriptive statistics. This is essentially the univariate measures for the combined categories of the two variables under observation. This is necessary because there may be nothing unusual about each of the variables, but their joint measures may be out of line. For example, there is nothing unusual about a person stating that he or she considered moving. There is also nothing special about a person who has some fear of crime in his or her neighborhood. It would be of interest, however, to know that a person had considered moving because of his or her fear of crime. This bivariate analysis of the mean, standard error, and so on, is therefore important. The combined descriptives for the variables above, as well as in a typical ANOVA, are shown in **Table 18-3**.

TABLE 18-2   Univariate Measures for ADDRESS and WALK_EVE

Descriptive Statistics

	N	Mean	Std. Dev.	Variance	Skewness	Kurtosis
How long have you lived at this address?	337	13.36	13.704	187.790	1.313	1.397
How often do you walk in your neighborhood in the evening?	278	1.60	.655	.428	.633	−.612
Valid N (listwise)	272					

TABLE 18-3   **Combined Descriptives (Univariate Values) for ADDRESS and WALK_EVE**

Descriptives

How long have you lived at this address?

	N	Mean	Std. Deviation	Std. Error	95% Confidence Interval for Mean Lower Bound	Upper Bound	Minimum	Maximum
Never	132	13.70	13.649	1.188	11.35	16.05	0	51
Occasionally	114	11.10	11.774	1.103	8.91	13.28	0	50
Often	26	7.54	6.754	1.325	4.81	10.27	0	25
Total	272	12.02	12.479	0.757	10.53	13.51	0	51

Step 6: Examining the Variability of the Groups

This is a two-part procedure. The first part examines the variability of data around the mean in each group. This is the **sum of squares within groups** (SS_w), which is the sum of the squared deviations between each score and the mean (just as in multiple regression). This measures the variability within each category. The second part of the procedure examines the variability between the group means. This analysis produces the **sum of squares between groups** (SS_b), which is the sum of the squared deviations between each sample mean and the total mean for all observed values.

Each of these analyses has its own formulas and analysis procedures. The formula for SS_w is

$$SS_w = \Sigma(n_i - 1)s_i^2$$

where n is the total for each of the categories and s_i^2 is the variance for each of the categories. This totals (sums) the variance of each category. The formula for SS_b is

$$SS_b = \Sigma n_i(\overline{X}_i - \overline{X})^2$$

where n is the total for each category, $\overline{X}_i$ is the mean of each category, and $\overline{X}$ is the total mean for all categories. This, in effect, subtracts the variance of each sample mean from the total mean, squares it, multiplies it by the number of samples, and then sums this value.

Step 7: Calculating the Degrees of Freedom for the *F*-test

The calculation for the *F*-test is somewhat different than other hypothesis tests in that the degrees of freedom must be taken into account in calculating the obtained value as well as when determining the critical value. The degrees of freedom for an *F*-test must take into account all of the variability within the model. To do this, you must account for variation both between groups and within groups. The formula for calculating the degrees of freedom for an F-test is

$$df_{total} = df_{within} + df_{between}$$
$$= (N - k) + (k - 1)$$

where N is the number of cases and k is the number of categories.

Step 8: Calculating the Obtained Value

Now that the sums of squares and degrees of freedom have been calculated, we can test to see if the samples are different. It stands to reason that if different samples are being examined, they should be different from (have more variation around the total mean) the individual samples themselves. If people are classified into groups based on their environment, history, and so on, people in the same groups should be similar (little variation around the mean), but there should be some sharp differences between groups.

This difference is examined with the **F-ratio**. The F-ratio is a measure of the size of the difference between groups relative to the size of the variation within groups. Remember that the null hypothesis is stating that there is no statistically significant difference between the two groups. If that is true, the measure of the within-groups variability (SS_w) should be the same, or close to the same, as the between-groups variability (SS_b). Examining the ratio of these two values should determine the closeness of the two groups. The F-ratio does the same thing as the t-ratio but is used for multiple groups. The formula for the F-ratio is[1]

$$F = \frac{\text{variance between samples}}{\text{variance within groups}}$$

As you can see, this formula is simply the sum of squares between groups (SS_b) divided by the sum of squares within groups (SS_w). This is essentially a comparison of the differences in the means of the different groups (SS_b), taking into account the variation assumed within each sample. Only when SS_b is larger than SS_w can the null hypothesis that there is a true difference between the groups be rejected. If there is no difference at all between the groups, the result of this calculation would be 1. If there is little difference between the groups, the value will be close to 1 or less than 1. As with other tests of the null hypothesis, the larger the value of F, the more likely it is that the null hypothesis will be rejected.

Step 9: Obtaining the Critical Value

Like all other hypothesis tests, the use of an F-test requires a determination of the degrees of freedom and the critical value to be used. For an F-test, the degrees of freedom are calculated as described above. Because both types of degrees of freedom are necessary, a separate table is required for each significance level used for an F-test. Only a 0.05 table is shown in the appendix.

As shown in Table B-4 in the Appendix, both types of degrees of freedom for an F-test are used in the table to establish the critical value. The degrees of freedom between ($k - 1$) determines the degrees of freedom across the top of the table and the degrees

TABLE 18-4   Output of One-Way ANOVA

ANOVA

How long have you lived at this address?

	Sum of Squares	df	Mean Square	F	Sig.
Between groups	990.629	2	495.315	3.233	.041
Within groups	41210.279	269	153.198		
Total	42200.908	271			

of freedom within $(N - k)$ determines the degrees of freedom down the side of the table. The degrees of freedom between is the value that determines the 5% area in the right side of the tail. The degrees of freedom within is the value that establishes the 1% area in the right side of the tail. If the ratios of these two values were equal, this value would be 1. Because the objective is to find more variation between than within, however, all the values in Table B-4 are greater than 1,[2] and larger values represent a value farther into the tail. With these two values, the critical value of F can be obtained by determining the point at which the degrees of freedom within intersect with the degrees of freedom between. As with other hypothesis tests, however, determining the critical value in this way is generally not required. The F value is displayed along with its significance value in all computer statistics programs (see Table 18-4, for example), so a table is not necessary to determine the critical value.

Step 10: Decision Making

A decision can be made about whether to reject or fail to reject the null hypothesis that the variances (or means) of the samples are equal. This step is the same as all other hypothesis tests. If the obtained value is greater than the critical value, the null hypothesis is rejected; otherwise, we fail to reject the null hypothesis. Using output from SPSS, if the value of F is significant (less than 0.05), the null hypothesis can be rejected and a conclusion made that the groups were drawn from different populations. If the value is not significant, you would fail to reject the null hypothesis and conclude that the groups are equal, and the data can be considered as being drawn from the same population. As shown in **Table 18-4**, the significance value is less than 0.05 (0.041), so the null hypothesis that there is no difference between the means of the groups can be rejected.

The only caution is that the criticisms against hypothesis testing are voiced even more loudly for an F-test. Here, even though the null hypothesis that there is no difference between the groups may be rejected, we cannot say which of the groups are different, only that one of them is different from the others. This issue is overcome with the post hoc tests discussed next.

■ Post Hoc Tests

As with other hypothesis tests, you can determine if there is reason to believe the groups did or did not come from the same population, but little else. It would also be beneficial to know which groups are different in their means. This can be established through additional tests associated with ANOVA, called **post hoc** or **multiple comparison tests**.

Post Hoc

There are a number of tests that can be used to examine the difference of means within groups. One of the oldest and easiest is the **Bonferroni test** (from a 1936 work never translated from Italian). Other tests that are also popular include Tukey's HSD (Honestly Significant Difference) and the Scheff test. The discussion here is limited to the Bonferroni test, as it is the most common post hoc test utilized. This test makes comparisons among all of the means and then standardizes the result by dividing by the number of pairs (or tests). The output from a Bonferroni test is shown in **Table 18-5**. Notice that the groups are repeated. This is because all combinations are taken into account, but the reality is that, for example, the category Often on the first variable combined with Never on the second variable is the same as Never–Often.

This table presents an interesting situation for ANOVA, but one that is not altogether uncommon. As discussed above, the null hypothesis could be rejected that there are no differences between the means of the different groups. This value was close to 0.05, however. In the Bonferroni test, there are no significant differences between any of the groups. The closest group was Never–Often (0.063). This situation may be a result of the fact that the sample sizes were different for each category (an assumption often put forth for ANOVA but rarely attainable in social science or field research). Although the null hypothesis can be rejected, it is not possible to make any conclusions about the differences between any of the groups being studied.

It should also be mentioned that post hoc tests should not be approached haphazardly. Although many researchers conduct post hoc tests as an automatic follow-up to ANOVA, there should be some specific relationships you are attempting to explore if you do so rather than simply testing to see what might pop up. If you do not care which of the categories are causing the difference, post hoc tests may not be necessary, other than as a check to ensure some intercategory differences when the F is close to nonsignificant. If you have hypothesized a priori about certain relationships, however, post hoc tests are a logical next step in your analysis.

TABLE 18-5   Bonferroni Test: Dependent Variable, ADDRESS

Multiple Comparisons

Dependent Variable: How long have you lived at this address?
Bonferroni

(I) How often do you walk in your neighborhood in the evening?	(J) How often do you walk in your neighborhood in the evening?	Mean Difference (I − J)	Std. Error	Sig.	95% Confidence Interval Lower Bound	Upper Bound
Never	Occasionally	2.60	1.583	.305	−1.21	6.41
	Often	6.16	2.656	.063	−.24	12.56
Occasionally	Never	−2.60	1.583	.305	−6.41	1.21
	Often	3.56	2.690	.561	−2.92	10.04
Often	Never	−6.16	2.656	.063	−12.56	.24
	Occasionally	−3.56	2.690	.561	−10.04	2.92

■ Conclusion

This discussion of ANOVA completes the process of statistical analysis. With the tools of descriptive and inferential statistics at his or her disposal, a researcher can analyze any data source. The key to statistical analysis is determining what type of data is available and matching that with the appropriate procedure. Error is always present in any analysis, and statistical analysis alone cannot provide a researcher with proof of absolute causality. This can only be accomplished with a combination of theory, methodology, and statistical analysis. It is this combination of research techniques that is the subject of the final chapter of this book.

■ Key Terms

Analysis of Variance (ANOVA)
Bonferroni test
F-ratio
F-test
Levene test

MANOVA
One-Way ANOVA
post hoc or multiple comparison tests
sum of squares between groups
sum of squares within groups

■ Summary of Equations

Sum of Squares within Groups

$$SS_w = \Sigma(n_i - 1)s_i^2$$

Sum of Squares between Groups

$$SS_b = \Sigma n_i(\overline{X}_i - \overline{X})^2$$

Degrees of Freedom (*F*-test)

$$df_{total} = df_{within} + df_{between}$$

F-Ratio

$$F = \frac{\text{variance between samples}}{\text{variance within groups}}$$

■ Exercises

1. After studying the difference between police officers and "normal" people, a researcher wondered what the relationship in IQ was between different members of the criminal justice system. The researcher proposed that police officers would have a higher IQ than correctional officers and that correctional officers would have a higher IQ than lawyers. The researcher then took a sample of 150 each of these groups and found the following: $SS_{between} = 40$ and $SS_{within} = 12.93$.
   a. Use an *F*-test to determine if the researcher's theory was right.
   b. What does this tell us about the difference in IQ between the members of the criminal justice system (police officers, correctional officers, and lawyers)?

2. Find one journal article that uses an *F*-test (ANOVA).
   a. Compare the use of ANOVA in the article to the information in this chapter.
   b. What was different and the same?
   c. Attempt to determine the critical value and significance level using (1) the observed values from the journal article, and (2) the appropriate tables in the appendix.
3. Design a research project as you did in Chapter 1 and discuss how ANOVA could be used to analyze the data. Discuss what limitations or problems you would expect to encounter.

References

Fisher, R. A. (1918). The correlation between relatives on the supposition of Mendelian inheritance. *Transcripts of the Royal Society of Edinburgh*, 52:399–433.

Fisher, R. A. (1925). *Statistical Methods for Research Workers*, 11th ed. Edinburgh: Oliver & Boyd.

Fisher, R. A. (1935). *The Design of Experiments*. Edinburgh: Oliver & Boyd.

Fisher, R. A., and W. A. Mackenzie (1923). The manurial response of different potato varieties. *Journal of Agriculture Science*, 13:311–320.

Levene, H. (1960). Robust tests for equality of variances. In I. Olkin, S. Ghurye, W. Hoeffding, W. Madow, and H. Mann (eds.), *Contributions to Probability and Statistics*. Stanford, CA: Stanford University Press, pp. 278–292.

Student (1923). On testing varieties of cereals. *Biometrika*, 15:271–294.

For Further Reading

Girden, E. R. (1992). *ANOVA: Repeated Measures*. Thousand Oaks, CA: Sage Publications.

Iversen, G. R., and H. Norpoth (1987). *Analysis of Variance*. Thousand Oaks, CA: Sage Publications.

Notes

1. Mathematically, an *F*-ratio is actually a ratio of the χ^2 values of the variables, each divided by their degrees of freedom.
2. It is interesting to note the relationship between an *F* distribution and a *t* distribution. By examining the *F* and *t* values in each of the tables in the appendix, you can see that the values of *F* are closely related to the values of *t*. In fact, each value of *F* equals the *t* values squared.

Putting It
All Together

Over the course of this book, topics have included:

- The variety of possible statistical procedures
- How the type of data affects the statistical procedure used
- The mathematical foundations of statistical procedures through their formulas
- How to analyze and interpret statistical output

The goal of this final chapter is to look at how these are related, how statistics fits into the research process along with theory and methodology, and how all of these are applied in *practical* scientific inquiry.

What you should have noticed by now is how so much of the material in this book is interrelated. Look at how much of Chapters 8 and 12 is included in Chapter 17. Also realize that some of the important information about Chi-square, which was discussed most fully in Chapter 8, could not even be presented until after the material in Chapters 15 and 16. The foundations of statistical analysis are very closely interrelated—so much so that you must almost take a statistics course before you can really understand a statistics course.

The reason the basics of statistical analysis are often repeated in graduate courses (and even repeated in doctoral courses) is because academic departments rarely have room in the curriculum to require enough courses in statistical analysis to facilitate a full understanding of the material. As you can see after finishing this book, you really need the understanding that comes from learning about inferential analysis to understand descriptive analysis; and you need the understanding of descriptive analysis before you can fully understand inferential analysis.

This same type of interrelation spills over into the relationship between statistics, theory, and methodology. It is to this relationship that the discussion now returns.

■ The Relationship Between Statistics, Methodology, and Theory

Don't let the statistical tail wag the conceptual dog.

Never be fooled into thinking that statistics can answer all of the questions of research. As discussed at the beginning of this book, statistical analysis is only one part of the complex work of scientific discovery. Good statistical analysis must be based on a sound and carefully planned methodology; and sound methodology is driven by careful construction of a theoretical/conceptual model.

That does not mean that theory and methodology can stand on their own, however. Theory without some statistical analysis[1] is little more than armchair philosophizing. You have surely read articles in which the author was putting forth a lot of ideas with little or no empirical support, and you came away thinking, "That was OK, but I don't see how it fits in with reality." These articles have no support for their conclusions. Using the examples from Chapter 1, it is like building a house without a foundation. It may stand for awhile, but the least bit of wind will probably tumble it; and even the natural shift of the ground will bring it down. So it is with theories without statistical support. They may sound as if they would hold true, but they are often easily refuted because the author has nothing to back up his or her arguments, and the natural shifting of human behavior may change enough that the theory is no longer useful—and we do not know why because there was never a foundation for the arguments.

The same can be said for methodology without statistical analysis or qualitative support. Methodology is brought to life with analysis. Having a good theory with sound methodology and no analysis is tantamount to having a great idea for an invention but never telling anyone. Without statistical analysis, many forms of methodology cannot be accomplished.

It is also essential to understand that researchers cannot wait until the analysis stage of a research project to begin to think about what procedures to use. Waiting until this point will probably result in the researcher having a very difficult time finding a good match between the data and an appropriate statistical procedure. For example, the researcher may have some data that is nominal, some that is ordinal, and some that is interval. Although this can be overcome (especially by using nominal level analyses), it is not making the best use of the data. Sometimes a researcher cannot avoid collecting data at different levels. Through careful conceptualization and operationalization, however, a researcher can often collect data that is at the appropriate level or is compatible for statistical analysis. Some even propose that researchers construct dummy tables (empty tables with only the headings) to see what the data will look like in certain bivariate analyses.

From the very outset of the research, therefore, a researcher should be cognizant of the statistical analyses to be used. When developing the primary and research questions, the researcher should be aware of what the data might look like and what analysis procedures would be most appropriate. This will ensure that the best analyses are used for the research to be conducted.

Understanding the statistical procedure at the outset of the research is also important in determining the sample size to be used. For example, to use a Z test requires a sample size of 120 or more. Additionally, as discussed in Chapter 16, the issue of statistical power is determined in part by sample size. It is important, then, to have a full understanding of sample size requirements for particular statistical procedures early in the research process.

■ Describe It or Make Inferences

Statistics is just another way of never having to say you are sorry.

One of the biggest questions that arises in deciding on a statistical analysis procedure for research is whether to use descriptive or inferential analyses. This, of course, assumes that the researcher is not attempting to use descriptive analyses and make inferences from them, as happens all too often (see the section on abuses of statistics, below). The issue of descriptive versus inferential analyses is appropriate and should be a conscious decision made early in the research process.

Even though inferential analyses are central to many fields, there are those in the social sciences who argue that the data available in fields such as criminal justice and criminology are not appropriate for inferential analyses. Labovitz (1970, 1971) argues that there is very little in social science data that supports an argument that the population parameter can be estimated with any real confidence. As discussed in the section on inferential analyses, he proposes that we cannot even *estimate* the population parameters for social science data sufficiently to use them in inferential analyses. Furthermore, Labovitz argues that even if the population parameter could be estimated mathematically such that inferential formulas could be used, we do not have a good enough grasp on the characteristics of the population to be able to make sound inferences from a sample to the population. Selvin (1957) also argues that inferential analyses should not be used with surveys. He proposes that surveys are not a sufficient method of sampling and that the data is insufficient to estimate population parameters.

Given these arguments, should inferential analyses be abandoned for criminal justice and criminology research in favor of using only descriptive analyses? That is a decision that only you can make for your own research; however, abandoning inferential analyses is probably a little strong. The line between making inferences to a population and describing the data drawn is actually much finer than many would like to admit. In fact, many readers of research do not notice the difference. For example, if there is published research on trial judges in the northwest, in the purest sense of statistical analysis the results can be used in one of two ways: Either they can be used as a descriptive analysis of the judges chosen, or if the proper sampling methodology has been followed, inferences can be made to all judges in the northwest. Most readers will not interpret the findings this way, however. A large number of readers will infer (either explicitly or unconsciously) the results of the analysis to all judges in the United States. Is this technically correct? No, but most will make this leap of inference even if the researcher did not.

As the editor of a journal, one of the authors has had to referee between authors who wished to report only descriptive analyses and reviewers who wanted to see significance tests (e.g., adding a significance value to Lambda or Pearson's *r*). One group explicitly did not want to make inferences with the data; the other group used tests of significance as an estimate of what they might expect in the population. Who was wrong in this case? Probably neither group. Experienced researchers do not let the results of statistical analyses themselves sway their thinking. The results do not make the decision, the researcher makes the decision. Using the old statistical adage that statistics are like a lamppost for a drunk—used more for support than illumination—most statisticians simply use the results of analyses to guide their theoretical decisions. In this case, the group that wanted to see the significance results perhaps did not want to make inferences beyond what the data allowed, but they would have liked the additional information to support or not support a decision to generalize the results.

Although they are of value, inferential analyses should not be elevated to the place of prominence that they have in many disciplines, however. As discussed in earlier chapters, inferential analyses are only as good as the univariate and bivariate (and sometimes multivariate) analyses upon which the conclusions can be drawn. In essence, inferences are based only on the observation of the relationships that exist in the sample. As stated repeatedly, rejection of the null hypothesis means only that there is a small chance that there is no relationship between the variables studied and that whatever is happening with the sample may happen in the population. As such, the description of what is happening in the sample is at least as important as making the inference, and maybe more.

In the end, it will be up to you to decide when and if to use descriptive or inferential analyses, and the most important decision may not be which to use. The most important decision should be what procedure is most appropriate for the data and research being conducted. A correct decision here will provide you with the proper tool to make the most of the results of the research. Like any tool, however, statistics can be misused. It is the abuses and misuses of statistics that should be avoided.

■ Abuses of Statistics

Prediction is very difficult, especially when you are talking about the future.

—Neils Bohr (also attributed to Yogi Berra)

The greatest abuse of statistics is probably the one discussed above: researchers employing descriptive analyses and trying to use them to make inferences. This is not the same as when significance results are included to assist in examining whether a relationship exists between two variables. The abuse suggested here occurs when researchers use purely descriptive analyses and make explicit references to the population.

There is also the problem of using the wrong analysis procedure for the data available or for the level of data. It is acceptable to use certain lower-level data with certain higher-order analyses, such as using dichotomized nominal level data with Pearson's r. Often, though, using lower-level data is a violation of the assumptions of the procedure. Even when it is not an explicit violation of the assumptions, any departures from the assumptions of a statistical procedure should be thoroughly analyzed and justified. For example, if you wanted to use Pearson's r and the data was in a Likert scale (1 to 5, high to low), it is accepted by some to do so; but you would need to make sure that there is a good underlying continuum such that the data can be said to be somewhat continuous, and you should justify your use of Pearson's r with this data.

Another abuse of statistics is simply ignoring the assumptions of the statistical procedure. As in the example above, there are people who conduct correlations (Pearson's r) with any and all levels of data. They simply ignore the levels of measurement requirements. Others will use regression when more than one of the assumptions are so violated as to make the analysis practically worthless. This is not to say, however, that all the assumptions must be adhered to strictly. Before using a statistical procedure when the assumptions are violated, the researcher should know how robust the procedure is to the violation. If the procedure is robust for the violation, it may be acceptable to use the procedure in that manner.

The final abuse of statistics is using the latest fad just to use it. Statistics is an evolving field. There are new procedures and improvements in old procedures added almost daily. Indeed, our statistical techniques have surpassed researchers' sampling abilities. With every new advancement, researchers are better able to analyze data and draw conclusions. Many researchers get wrapped up in what is current, however. They will use a new statistical procedure with data that is perfectly suited for an existing procedure just to attempt to appear current. This is not necessarily wrong; but it is an abuse of statistics because the researcher is not using the most appropriate statistical procedure, and for the wrong reason.

■ When You Are On Your Own

In the real world, we really don't know what we are doing. To close your mind to the possibility that the data are telling you something that you didn't anticipate is just asking to miss the interesting discoveries. Many of the most important advances in science have come from the insightful observation of something unexpected in the data.

—Paul Velleman

One thing that may have bothered you about this book is that there were not always direct and steadfast rules. A lot of statistical analysis is knowing what applies (differently) in certain situations, knowing what assumptions may be violated, and justifying the procedure or interpretation. That is the nature of statistical analysis. Although there are a lot of rules and procedures, the majority of them probably can be violated under certain (sometimes many) conditions. Also, the key to statistical analysis is in the analysis. Anyone can be taught the formulas; but you earn your money as a statistician/social scientist through your ability to interpret the findings.

When you are on your own, you will have a lot of latitude in how you work and in what procedures you use. Whether you choose a descriptive or inferential analysis procedure is almost entirely up to you: depending, of course, on the type of data. Also, the particular statistical procedure you choose is your decision. The rules that have been laid out in this book are designed to provide you with guidelines to use in making your selections. Additionally, the example analyses and interpretations were included to provide you with illustrations of how one might analyze or interpret particular findings from particular data. That does not mean that you must follow these guidelines or examples to the letter. There is a great deal of room for "artistic license" in statistical analysis. You may decide that the data requires a particular statistical procedure and that the results should be interpreted one way, and someone else may argue something completely different. Your job at that point (and any time you conduct research) will be to argue your point successfully. Your success will come from your ability to articulate your position.

When you stay within the guidelines presented in this book, you will have behind you the support of hundreds of years of statistical analysis. This does not mean that you may not draw the ire of a reviewer or reader, but it does mean that you have something to fall back on. Furthermore, we have attempted to provide you with the original works and writings where possible, so that if you are challenged, you can invoke the words of the person who developed the statistical procedure to support your arguments. This does not mean that you cannot disagree with what has been written here, nor that you cannot depart from the guidelines presented; it only means that these are the more accepted attitudes about statistical analysis and that they are supported by a wider portion of social science researchers. If you choose to depart from the guidelines and discussion in this book, you only have to be sure of your arguments and have support for them.

■ Conclusion

We want to end with a short discussion of how you can benefit from the information contained here. There are four ways that statistics in general, and the material in this book, can affect you, depending on where you are in your academic career and what you want to do with your life. If you are a graduating senior and want to be unemployed the rest of your life, you will probably not be affected very much by this material, except that you will know more about how the world around you works (such as election polls). This probably is not your goal.

If you are graduating and you plan to go to work, especially in the criminal justice field, you may see this material again. Do not be caught in the trap of thinking "I'm going to be a cop. I'll never need this stuff!" There are many police officers who upon promotion have found themselves in the research and planning division of a police department. Also, it is not uncommon for a person who never thought he or she would work in the research area to find a great state job with an agency such as a crime information center or other criminal justice data center. Finally, even in the most unexpected places, you can be moved into a job that requires you to conduct statistical analyses, interpret research, and write grants simply because "you've got a college education." Do not underestimate the future; it is better to be open-minded and prepared than closed-minded and behind the power curve when an opportunity arises.

If you still have some time to go before obtaining your bachelor's degree, the material in this book can help you read the articles and books that you will have to read in the rest of your courses and actually understand what they are talking about. If you have always skipped the sections on statistics and findings, you can now read them and get a better understanding of what the authors did. You may not be able to pick apart an article's statistical methods, but you can read it and find what the authors used, for example, a regression or hypothesis test. You can then add that information in a discussion or paper and have more complete information—and impress your professors if you do it right!

If you are going on to a graduate program (or if this book was used in a graduate course), this material provides you with a basis of understanding that will be essential to fully comprehend the material and complete a graduate program successfully. If you really get serious and go on to a Ph.D. program (or if this book is used in a Ph.D. program), the book will be invaluable for the same reasons as with a master's program, and you will need it because, eventually, you will be conducting research and writing articles.

The focus of this book was on the practical theory and application of statistics. If the goal was met, you now have a better understanding of how statistics works, you have seen the problems and pitfalls that you can expect when you begin to conduct research on your own, and you have an understanding of what will be expected when the output starts rolling out of a printer. Good luck, and have some fun with research!

References

Labovitz, S. (1970). The nonutility of significance tests: The significance of significance tests reconsidered. *Pacific Sociological Review*, 13(Summer):141–148.

Labovitz, S. (1971). The zone of rejection: Negative thoughts on statistical inference. *Pacific Sociological Review*, 14(4):373–381.

Selvin, H. (1957). A critique of tests of significance in survey research. *American Sociological Review*, 22(5):519–527.

Note

1. This, of course, includes qualitative methods and other sound methodological ways of supporting arguments. It does not imply that a theory must have quantitative support to be useful.

Appendix A: Math Review and Practice Test

Although there are many similarities between statistics and math (numbers, formulas, etc.) there are also some fundamental differences. As described in Chapter 1 and in other places discussing the history of statistics, it is founded and based in math; however, just as a person does not have to know how to build an internal combustion engine or understand the electronics of a stereo to drive a car, a researcher does not have to have the complex mathematical skills needed to develop statistical procedures to be able to conduct statistical analyses. Especially since computers now complete most of the calculations necessary for statistical analysis, it is more important for you to understand the theory behind a procedure, to understand how it works with data and how data affects it, and what the analyses mean, rather than to understand the mathematical workings. As such, most of the statistical procedures covered in this book can be understood using simple mathematics.

This review is intended to assist you in recalling the math skills that you will need to use in working through the formulas in this book and in working some of the problems and exercises. In addition to basic addition, subtraction, multiplication, and division, this review covers squaring and square roots, negative numbers, and the order of operations within formulas.

Following the review is a practice exercise that will allow you to test your knowledge of these basic math skills. If you can pass this practical exercise, you can complete any formula in this book with a little thought and diligence.

A. Addition, Subtraction, and Negative Numbers

Adding and subtracting numbers should not be a problem. Sometimes, though, a review of adding and subtracting negative numbers can be a benefit. When adding and subtracting negative and positive numbers, sum (Σ) all of the positive numbers, then sum (Σ) all of the negative numbers. Then subtract the smaller total from the larger and take the sign of the larger total.

Examples

$4 + 9 = 13$	$-4 + -9 = -13$	$-4 + -9 + 5 + 7 = -1$
$4 - 9 = -5$	$-4 - 9 = -13$	$-4 - 9 - 5 - 7 = -25$
$4 + (-9) = -5$	$-4 - (-9) = 5$	$-4 + (-9) + (-5) + (-7) = -25$

B. Multiplication, Division, and Negative Numbers

In multiplying and dividing numbers with the same sign, the product (multiplication) or quotient (division) will be positive. When multiplying or dividing numbers with different signs, the product or quotient will be negative. There are many ways multiplication can be symbolized, including $(x)(y)$, $x * y$, $x \times y$ and $x \cdot y$. In this book, multiplication will most always be symbolized either by $(x)(y)$ or by $x \times y$, and division will be noted by x/y rather than $x \div y$.

Examples

$$4 \times 9 = 36 \qquad 4/9 = 0.44$$
$$4 \times -9 = -36 \qquad 4/-9 = -0.44$$
$$-4 \times -9 = 36 \qquad -4/-9 = 0.44$$

C. Statistical Addition: Summation

As discussed in Chapter 1, the Greek letter Σ means to sum or add numbers. This can be undertaken with numbers in a row or in a column. You will see and have to complete the summation of numbers in many of the formulas and exercises in the text. An example of this is:

$$\Sigma (1 + 2 + 3 + 4 + 5) = 15$$

The same procedure is used in summing numbers in columns. In the example below, the numbers in the X and Y columns are simply added top to bottom. This points out an important issue. Often, numbers in columns will be added top to bottom. In some instances, however, such as with the median, the numbers will be added from the lowest number to the highest, which may be from bottom to top. You should be careful to examine the data and attempt to determine which is the lowest value (or where the data appears to begin). In this case, for example, the fact that A is at the top of the distribution and E is at the bottom suggests that the numbers are to be added from top to bottom. The distribution could have been reversed, however, and the A placed at the bottom. This would have suggested that the numbers be added from bottom to top.

Subject	X	Y
A	1	5
B	2	6
C	3	7
D	4	8
E	5	9
ΣX	15	ΣY 35

D. Exponents and Roots

Remember that exponents and roots are really nothing more than multiplication and division. You simply multiply the number times itself a specific number of times (the exponent) or divide it by itself a specific number of times (the root).

Examples

$$2^2 = 4 \qquad \sqrt{2} = 1.41$$
$$2^3 = 8 \qquad \sqrt[3]{2} = 1.26$$
$$2^4 = 16 \qquad \sqrt{25} = 5$$

E. Order of Operations

In more complicated formulas, it is necessary to undertake several different functions. In these cases, it is necessary to follow the rules for the order of operations. In general, operations should be undertaken in the following order:

1. Complete the operations within parentheses or brackets (this may require several steps).
2. Complete the exponents and roots.
3. Multiply and divide.
4. Add and subtract.
5. Move left to right if the operations are the same.

Examples

$$5 - 8 \times (6 - 2) \times 5^2$$
$$= 5 - 8 \times (4) \times 5^2$$
$$= 5 - 8 \times 4 \times 25$$
$$= 5 - 32 \times 25$$
$$= 5 - 1600$$
$$= -1595$$

$$(4 - 2) \times (8 - 9) \times \sqrt{25}$$
$$= (2) \times (-1) \times \sqrt{25}$$
$$= 2 \times -1 \times 5$$
$$= -2 \times 5$$
$$= -10$$

F. Putting It All Together: Simple Math for Statistical Analysis

The operations above will be used in various statistical formulas. Let's now look at a real example from statistical analysis to show how these operations work together. One of the more complicated formulas in descriptive statistics is that of the *variance,* whose formula is

$$s^2 = \frac{\Sigma x^2 - \dfrac{(\Sigma x)^2}{N}}{N}$$

This is actually quite a simple formula, though, when it is broken down and the rules discussed above are applied. All that is actually needed to address this formula is two sets of numbers: N, Σx^2, and $(\Sigma x)^2$. The ordering is important, though, because Σx^2 and $(\Sigma x)^2$ are not the same. The expression Σx^2, called the *sum of the x squares,* is obtained by squaring each score and then adding all the squared numbers. The second expression, $(\Sigma x)^2$, called the *sum of the x's squared,* is obtained by summing all the x's and then squaring the sum. Look at the steps used to calculate the variance for the following small data set.

Subject		x		x^2
A		2		4
B		3		9
C		4		16
D		5		25
E		6		36
	Σx	20	Σx^2	90

Plugging these values into the formula above would produce the following:

$$s^2 = \frac{90 - \frac{(20)^2}{5}}{5}$$

Now it is simply a matter of using the basic rules of math discussed above to solve the equation:

$$s^2 = \frac{90 - \frac{400}{5}}{5} = \frac{90 - 80}{5} = \frac{10}{5} = 2$$

Notice that the one place where this seemed to deviate from the rules is that the top portion of the formula was completed first and then the final division was undertaken. When you have major divisions as contained in the formula, you should always perform all the calculations on the top and bottom of the formula before dividing the top by the bottom.

G. Practice Test

Now that you have refreshed your knowledge with a review of basic math, you can test that knowledge with a practice test. This will show you where you need some practice and also that you can undertake statistical analyses without fear.

1. $-4 + 11 + 7 - 8 + 2 + 3 - 10 =$

2. $(4)^2 =$

3. $\sqrt{280} =$

4. $3(4) + 8(4 + 1) - 20 =$

5. $4^2 =$

6. $(0.25)^4 =$

7. $(\sqrt{17} + 9)^2 =$

8. $56/-13 =$

9. $(-9)(-10) =$

10. $1.96 + 5/7(34) =$

11. $\sqrt{49}/\sqrt{7} =$

12. $20/4(0.50)/\sqrt{25} =$

For the following values of x:

$$1, -2, 2, 3, 4, -4, 5, 6, 7, 7, -8$$

13. Σx

14. $(\Sigma x)^2$

15. Σx^2

Appendix B: Statistical Tables

TABLE B-1 Area Under the Normal Curve

a Z	b Area between X and Z	c Area beyond Z	a Z	b Area between X and Z	c Area beyond Z	a Z	b Area between X and Z	c Area beyond Z	a Z	b Area between X and Z	c Area beyond Z
0.00	0.0000	0.5000	0.45	0.1736	0.3264	0.90	0.3159	0.1841	1.35	0.4115	0.0885
0.01	0.0040	0.4960	0.46	0.1772	0.3228	0.91	0.3186	0.1814	1.36	0.4131	0.0869
0.02	0.0080	0.4920	0.47	0.1808	0.3192	0.92	0.3212	0.1788	1.37	0.4147	0.0853
0.03	0.0120	0.4880	0.48	0.1844	0.3156	0.93	0.3238	0.1762	1.38	0.4162	0.0838
0.04	0.0160	0.4840	0.49	0.1879	0.3121	0.94	0.3264	0.1736	1.39	0.4177	0.0823
0.05	0.0199	0.4801	0.50	0.1915	0.3085	0.95	0.3289	0.1711	1.40	0.4192	0.0808
0.06	0.0239	0.4761	0.51	0.1950	0.3050	0.96	0.3315	0.1685	1.41	0.4207	0.0793
0.07	0.0279	0.4721	0.52	0.1985	0.3015	0.97	0.3340	0.1660	1.42	0.4222	0.0778
0.08	0.0319	0.4681	0.53	0.2019	0.2981	0.98	0.3365	0.1635	1.43	0.4236	0.0764
0.09	0.0359	0.4641	0.54	0.2054	0.2946	0.99	0.3389	0.1611	1.44	0.4251	0.0749
0.10	0.0398	0.4602	0.55	0.2088	0.2912	1.00	0.3413	0.1587	1.45	0.4265	0.0735
0.11	0.0438	0.4562	0.56	0.2123	0.2877	1.01	0.3438	0.1562	1.46	0.4279	0.0721
0.12	0.0478	0.4522	0.57	0.2157	0.2843	1.02	0.3461	0.1539	1.47	0.4292	0.0708
0.13	0.0517	0.4483	0.58	0.2190	0.2810	1.03	0.3485	0.1515	1.48	0.4306	0.0694
0.14	0.0557	0.4443	0.59	0.2224	0.2776	1.04	0.3508	0.1492	1.49	0.4319	0.0681
0.15	0.0596	0.4404	0.60	0.2257	0.2743	1.05	0.3531	0.1469	1.50	0.4332	0.0668
0.16	0.0636	0.4364	0.61	0.2291	0.2709	1.06	0.3554	0.1446	1.51	0.4345	0.0655
0.17	0.0675	0.4325	0.62	0.2324	0.2676	1.07	0.3577	0.1423	1.52	0.4357	0.0643
0.18	0.0714	0.4286	0.63	0.2357	0.2643	1.08	0.3599	0.1401	1.53	0.437	0.0630
0.19	0.0753	0.4247	0.64	0.2389	0.2611	1.09	0.3621	0.1379	1.54	0.4382	0.0618
0.20	0.0793	0.4207	0.65	0.2422	0.2578	1.10	0.3643	0.1357	1.55	0.4394	0.0606
0.21	0.0832	0.4168	0.66	0.2454	0.2546	1.11	0.3665	0.1335	1.56	0.4406	0.0594
0.22	0.0871	0.4129	0.67	0.2486	0.2514	1.12	0.3686	0.1314	1.57	0.4418	0.0582
0.23	0.0910	0.4090	0.68	0.2517	0.2483	1.13	0.3708	0.1292	1.58	0.4429	0.0571
0.24	0.0948	0.4052	0.69	0.2549	0.2451	1.14	0.3729	0.1271	1.59	0.4441	0.0559
0.25	0.0987	0.4013	0.70	0.2580	0.2420	1.15	0.3749	0.1251	1.60	0.4452	0.0548
0.26	0.1026	0.3974	0.71	0.2611	0.2389	1.16	0.377	0.1230	1.61	0.4463	0.0537
0.27	0.1064	0.3936	0.72	0.2642	0.2358	1.17	0.379	0.1210	1.62	0.4474	0.0526
0.28	0.1103	0.3897	0.73	0.2673	0.2327	1.18	0.381	0.1190	1.63	0.4484	0.0516
0.29	0.1141	0.3859	0.74	0.2704	0.2296	1.19	0.383	0.1170	1.64	0.4495	0.0505
0.30	0.1179	0.3821	0.75	0.2734	0.2266	1.20	0.3849	0.1151	1.65	0.4505	0.0495
0.31	0.1217	0.3783	0.76	0.2764	0.2236	1.21	0.3869	0.1131	1.66	0.4515	0.0485
0.32	0.1255	0.3745	0.77	0.2794	0.2206	1.22	0.3888	0.1112	1.67	0.4525	0.0475
0.33	0.1293	0.3707	0.78	0.2823	0.2177	1.23	0.3907	0.1093	1.68	0.4535	0.0465
0.34	0.1331	0.3669	0.79	0.2852	0.2148	1.24	0.3925	0.1075	1.69	0.4545	0.0455
0.35	0.1368	0.3632	0.80	0.2881	0.2119	1.25	0.3944	0.1056	1.70	0.4554	0.0446
0.36	0.1406	0.3594	0.81	0.2910	0.2090	1.26	0.3962	0.1038	1.71	0.4564	0.0436
0.37	0.1443	0.3557	0.82	0.2939	0.2061	1.27	0.398	0.1020	1.72	0.4573	0.0427
0.38	0.1480	0.3520	0.83	0.2967	0.2033	1.28	0.3997	0.1003	1.73	0.4582	0.0418
0.39	0.1517	0.3483	0.84	0.2995	0.2005	1.29	0.4015	0.0985	1.74	0.4591	0.0409
0.40	0.1554	0.3446	0.85	0.3023	0.1977	1.30	0.4032	0.0968	1.75	0.4599	0.0401
0.41	0.1591	0.3409	0.86	0.3051	0.1949	1.31	0.4049	0.0951	1.76	0.4608	0.0392
0.42	0.1628	0.3372	0.87	0.3078	0.1922	1.32	0.4066	0.0934	1.77	0.4616	0.0384
0.43	0.1664	0.3336	0.88	0.3106	0.1894	1.33	0.4082	0.0918	1.78	0.4625	0.0375
0.44	0.1700	0.3300	0.89	0.3133	0.1867	1.34	0.4099	0.0901	1.79	0.4633	0.0367

TABLE B-1   Area Under the Normal Curve (*continued*)

a	b	c	a	b	c	a	b	c	a	b	c
	Area	Area		Area	Area		Area	Area		Area	Area
	between	beyond		between	beyond		between	beyond		between	beyond
Z	X and Z	Z	Z	X and Z	Z	Z	X and Z	Z	Z	X and Z	Z
1.80	0.4641	0.0359	2.25	0.4878	0.0122	2.70	0.4965	0.0035	3.15	0.4992	0.0008
1.81	0.4649	0.0351	2.26	0.4881	0.0119	2.71	0.4966	0.0034	3.16	0.4992	0.0008
1.82	0.4656	0.0344	2.27	0.4884	0.0116	2.72	0.4967	0.0033	3.17	0.4992	0.0008
1.83	0.4664	0.0336	2.28	0.4887	0.0113	2.73	0.4968	0.0032	3.18	0.4993	0.0007
1.84	0.4671	0.0329	2.29	0.4890	0.0110	2.74	0.4969	0.0031	3.19	0.4993	0.0007
1.85	0.4678	0.0322	2.30	0.4893	0.0107	2.75	0.4970	0.0030	3.20	0.4993	0.0007
1.86	0.4686	0.0314	2.31	0.4896	0.0104	2.76	0.4971	0.0029	3.21	0.4993	0.0007
1.87	0.4693	0.0307	2.32	0.4898	0.0102	2.77	0.4972	0.0028	3.22	0.4994	0.0006
1.88	0.4699	0.0301	2.33	0.4901	0.0099	2.78	0.4973	0.0027	3.23	0.4994	0.0006
1.89	0.4706	0.0294	2.34	0.4904	0.0096	2.79	0.4974	0.0026	3.24	0.4994	0.0006
1.90	0.4713	0.0287	2.35	0.4906	0.0094	2.80	0.4974	0.0026	3.25	0.4994	0.0006
1.91	0.4719	0.0281	2.36	0.4909	0.0091	2.81	0.4975	0.0025	3.30	0.4995	0.0005
1.92	0.4726	0.0274	2.37	0.4911	0.0089	2.82	0.4976	0.0024	3.35	0.4996	0.0004
1.93	0.4732	0.0268	2.38	0.4913	0.0087	2.83	0.4977	0.0023	3.40	0.4997	0.0003
1.94	0.4738	0.0262	2.39	0.4916	0.0084	2.84	0.4977	0.0023	3.45	0.4997	0.0003
1.95	0.4744	0.0256	2.40	0.4918	0.0082	2.85	0.4978	0.0022	3.50	0.4998	0.0002
1.96	0.4750	0.0250	2.41	0.4920	0.0080	2.86	0.4979	0.0021	3.60	0.4998	0.0002
1.97	0.4756	0.0244	2.42	0.4922	0.0078	2.87	0.4979	0.0021	3.70	0.4999	0.0001
1.98	0.4761	0.0239	2.43	0.4925	0.0075	2.88	0.4980	0.0020	3.80	0.4999	0.0001
1.99	0.4767	0.0233	2.44	0.4927	0.0073	2.89	0.4981	0.0019	3.90	0.49995	0.00005
2.00	0.4772	0.0228	2.45	0.4929	0.0071	2.90	0.4981	0.0019	4.00	0.49997	0.00003
2.01	0.4778	0.0222	2.46	0.4931	0.0069	2.91	0.4982	0.0018	4.50	0.4999966	0.0000034
2.02	0.4783	0.0217	2.47	0.4932	0.0068	2.92	0.4982	0.0018	5.00	0.4999997	0.0000003
2.03	0.4788	0.0212	2.48	0.4934	0.0066	2.93	0.4983	0.0017	5.50	0.4999999	0.0000001
2.04	0.4793	0.0207	2.49	0.4936	0.0064	2.94	0.4984	0.0016			
2.05	0.4798	0.0202	2.50	0.4938	0.0062	2.95	0.4984	0.0016			
2.06	0.4803	0.0197	2.51	0.4940	0.0060	2.96	0.4985	0.0015			
2.07	0.4808	0.0192	2.52	0.4941	0.0059	2.97	0.4985	0.0015			
2.08	0.4812	0.0188	2.53	0.4943	0.0057	2.98	0.4986	0.0014			
2.09	0.4817	0.0183	2.54	0.4945	0.0055	2.99	0.4986	0.0014			
2.10	0.4821	0.0179	2.55	0.4946	0.0054	3.00	0.4987	0.0013			
2.11	0.4826	0.0174	2.56	0.4948	0.0052	3.01	0.4987	0.0013			
2.12	0.4830	0.0170	2.57	0.4949	0.0051	3.02	0.4987	0.0013			
2.13	0.4834	0.0166	2.58	0.4951	0.0049	3.03	0.4988	0.0012			
2.14	0.4838	0.0162	2.59	0.4952	0.0048	3.04	0.4988	0.0012			
2.15	0.4842	0.0158	2.60	0.4953	0.0047	3.05	0.4989	0.0011			
2.16	0.4846	0.0154	2.61	0.4955	0.0045	3.06	0.4989	0.0011			
2.17	0.4850	0.0150	2.62	0.4956	0.0044	3.07	0.4989	0.0011			
2.18	0.4854	0.0146	2.63	0.4957	0.0043	3.08	0.4990	0.0010			
2.19	0.4857	0.0143	2.64	0.4959	0.0041	3.09	0.4990	0.0010			
2.20	0.4861	0.0139	2.65	0.4960	0.0040	3.10	0.4990	0.0010			
2.21	0.4864	0.0136	2.66	0.4961	0.0039	3.11	0.4991	0.0009			
2.22	0.4868	0.0132	2.67	0.4962	0.0038	3.12	0.4991	0.0009			
2.23	0.4871	0.0129	2.68	0.4963	0.0037	3.13	0.4991	0.0009			
2.24	0.4875	0.0125	2.69	0.4964	0.0036	3.14	0.4992	0.0008			

TABLE B-2 **Values of Chi-Square**

	Probability (Top Row) and Significance (Bottom Row)					
df	0.999 0.0001	0.99 .01	0.95 .05	0.90 .10	0.80 .20	0.70 .30
1	10.827	6.635	3.841	2.706	1.642	1.074
2	13.815	9.210	5.991	4.605	3.219	2.408
3	16.268	11.345	7.815	6.251	4.624	3.665
4	18.465	13.277	9.488	7.779	5.989	4.878
5	20.517	15.086	11.070	9.236	7.289	6.064
6	22.457	16.812	12.592	10.645	8.558	7.231
7	24.322	18.475	14.067	12.017	9.803	8.383
8	26.125	20.090	15.507	13.362	11.030	9.524
9	27.877	21.666	16.919	14.684	12.242	10.656
10	29.588	23.209	18.307	15.987	13.442	11.781
11	31.264	24.725	19.675	17.275	14.631	12.899
12	32.909	26.217	21.026	18.549	15.812	14.011
13	34.528	27.688	22.362	19.812	16.985	15.119
14	36.123	29.141	23.685	21.064	18.151	16.222
15	37.697	30.578	24.996	22.307	19.311	17.322
16	39.252	32.000	26.296	23.542	20.465	18.418
17	40.790	33.409	27.587	24.769	21.615	19.511
18	42.312	34.805	28.869	25.989	22.760	20.601
19	43.820	36.191	30.144	27.204	23.900	21.689
20	45.315	37.566	31.410	28.412	25.038	22.775
21	46.797	38.932	32.671	29.615	26.171	23.858
22	48.268	40.289	33.924	30.813	27.301	24.939
23	49.728	41.638	35.172	32.007	28.429	26.018
24	51.179	42.980	36.415	33.196	29.553	27.096
25	52.620	44.314	37.652	34.382	30.675	28.172
26	54.052	45.642	38.885	35.563	31.795	29.246
27	55.476	46.963	40.113	36.741	32.912	30.319
28	56.893	48.278	41.337	37.916	34.027	31.391
29	58.302	49.588	42.557	39.087	35.139	32.461
30	59.703	50.892	43.773	40.256	36.250	33.530

TABLE B-3   Student's *t* Distribution

	Level of Significance for One-Tailed Test					
	.10	.05	.025	.01	.005	.0005
	Level of Significance for Two-Tailed Test					
df	.20	.10	.05	.02	.01	.001
1	3.078	6.314	12.706	31.821	63.657	636.62
2	1.886	2.920	4.303	6.965	9.925	31.598
3	1.638	2.353	3.182	4.541	5.841	12.941
4	1.533	2.132	2.776	3.747	4.604	8.610
5	1.476	2.015	2.571	3.365	4.032	6.859
6	1.440	1.943	2.447	3.143	3.707	5.959
7	1.415	1.895	2.365	2.998	3.499	5.405
8	1.397	1.860	2.306	2.896	3.355	5.041
9	1.383	1.833	2.262	2.821	3.250	4.781
10	1.372	1.812	2.228	2.764	3.169	4.587
11	1.363	1.796	2.201	2.718	3.106	4.437
12	1.356	1.782	2.179	2.681	3.055	4.318
13	1.350	1.771	2.160	2.650	3.012	4.221
14	1.345	1.761	2.145	2.624	2.977	4.140
15	1.341	1.753	2.131	2.602	2.947	4.073
16	1.337	1.746	2.120	2.583	2.921	4.015
17	1.333	1.740	2.110	2.567	2.898	3.965
18	1.330	1.734	2.101	2.552	2.878	3.922
19	1.328	1.729	2.093	2.539	2.861	3.883
20	1.325	1.725	2.086	2.528	2.845	3.850
21	1.323	1.721	2.080	2.518	2.831	3.819
22	1.321	1.717	2.074	2.508	2.819	3.792
23	1.319	1.714	2.069	2.500	2.807	3.767
24	1.318	1.711	2.064	2.492	2.797	3.745
25	1.316	1.708	2.060	2.485	2.787	3.725
26	1.315	1.706	2.056	2.479	2.779	3.707
27	1.314	1.703	2.052	2.473	2.771	3.690
28	1.313	1.701	2.048	2.467	2.763	3.674
29	1.311	1.699	2.045	2.462	2.756	3.659
30	1.310	1.697	2.042	2.457	2.750	3.646
40	1.303	1.684	2.021	2.423	2.704	3.551
60	1.296	1.671	2.000	2.390	2.660	3.460
120	1.289	1.658	1.980	2.358	2.617	3.373
∞	1.282	1.645	1.960	2.326	2.576	3.291

TABLE B-4   **Distribution of F; $p = .05$**

$df_{between}$	1	2	3	4	5	6	8
df_{within}							
1	161.4	199.5	215.7	224.6	230.2	234	238.9
2	18.51	19.00	19.16	19.25	19.3	19.33	19.37
3	10.13	9.55	9.28	9.12	9.01	8.94	8.85
4	7.71	6.94	6.59	6.39	6.26	6.16	6.04
5	6.61	5.79	5.41	5.19	5.05	4.95	4.82
6	5.99	5.14	4.76	4.53	4.39	4.28	4.15
7	5.59	4.74	4.35	4.12	3.97	3.87	3.73
8	5.32	4.46	4.07	3.84	3.69	3.58	3.44
9	5.12	4.26	3.86	3.63	3.48	3.37	3.23
10	4.96	4.10	3.71	3.48	3.33	3.22	3.07
11	4.84	3.98	3.59	3.36	3.20	3.09	2.95
12	4.75	3.89	3.49	3.26	3.11	3.00	2.85
13	4.67	3.81	3.41	3.18	3.03	2.92	2.77
14	4.60	3.74	3.34	3.11	2.96	2.85	2.70
15	4.54	3.68	3.29	3.06	2.90	2.79	2.64
16	4.49	3.63	3.24	3.01	2.85	2.74	2.59
17	4.45	3.59	3.20	2.96	2.81	2.70	2.55
18	4.41	3.55	3.16	2.93	2.77	2.66	2.51
19	4.38	3.52	3.13	2.90	2.74	2.63	2.48
20	4.35	3.49	3.10	2.87	2.71	2.60	2.45
21	4.32	3.47	3.07	2.84	2.68	2.57	2.42
22	4.30	3.44	3.05	2.82	2.66	2.55	2.40
23	4.28	3.42	3.03	2.80	2.64	2.53	2.37
24	4.26	3.40	3.01	2.78	2.62	2.51	2.36
25	4.24	3.39	2.99	2.76	2.60	2.49	2.34
26	4.23	3.37	2.98	27.4	2.59	2.47	2.32
27	4.21	3.35	2.96	2.73	2.57	2.46	2.31
28	4.20	3.34	2.95	2.71	2.56	2.45	2.29
29	4.18	3.33	2.93	2.70	2.55	2.43	2.28
30	4.17	3.32	2.92	2.69	2.53	2.42	2.27
40	4.08	3.23	2.84	2.61	2.45	2.34	2.18
60	4.00	3.05	2.76	2.53	2.37	2.25	2.10
80	3.96	3.11	2.72	2.48	2.33	2.21	2.05
120	3.92	3.07	2.68	2.45	2.29	2.17	2.02
∞	3.84	3.00	2.60	2.37	2.21	2.10	1.94

TABLE B-4   Distribution of F; p = .05 (continued)

$df_{between}$	1	2	3	4	5	6	8
df_{within}							
1	241.9	245.9	248	250.1	251. 1	252.2	253.3	254.3
2	19.4	19.43	19.45	19.46	19.47	19.48	19.49	16.5
3	8.79	8.70	8.66	8.62	8.59	8.57	8.55	8.53
4	5.96	5.86	5.80	5.75	5.72	5.69	5.66	5.63
5	4.74	4.62	4.56	4.50	4.46	4.43	4.40	4.36
6	4.06	3.94	3.87	3.81	3.77	3.74	3.70	3.67
7	3.64	3.51	3.44	3.38	3.34	3.30	3.27	3.23
8	3.35	3.22	3.15	3.08	3.04	3.01	2.97	2.93
9	3.14	3.01	2.94	2.86	2.83	2.79	2.75	2.71
10	2.98	2.85	2.77	2.70	2.66	2.62	2.58	2.54
11	2.85	2.72	2.65	2.57	2.53	2.49	2.45	2.40
12	2.75	2.62	2.54	2.47	2.43	2.38	2.35	2.30
13	2.67	2.53	2.46	2.38	2.34	2.30	2.25	2.21
14	2.60	2.46	2.39	2.31	2.27	2.22	2.18	2.13
15	2.54	2.40	2.33	2.25	2.20	2.16	2.11	2.07
16	2.49	2.35	2.28	2.19	2.15	2.11	2.06	2.01
17	2.45	2.31	2.23	2.15	2.10	2.06	2.01	1.96
18	2.41	2.27	2.19	2.11	2.06	2.02	1.97	1.92
19	2.38	2.23	2.16	2.07	2.03	1.98	1.93	1.88
20	2.35	2.20	2.12	2.04	1.99	1.95	1.90	1.84
21	2.32	2.18	2.10	2.01	1.96	1.92	1.87	1.81
22	2.30	2.15	2.07	1.98	1.94	1.89	1.84	1.78
23	2.27	2.13	2.05	1.96	1.91	1.86	1.81	1.76
24	2.25	2.11	2.03	1.94	1.89	1.84	1.79	1.73
25	2.24	2.09	2.01	1.92	1.87	1.82	1.77	1.71
26	2.22	2.07	1.99	1.90	1.85	1.80	1.75	1.69
27	2.20	2.06	1.97	1.88	1.84	1.79	1.73	1.67
28	2.19	2.04	1.96	1.87	1.82	1.77	1.71	1.65
29	2.18	2.03	1.94	1.85	1.81	1.75	1.70	1.64
30	2.16	2.01	1.93	1.84	1.79	1.74	1.68	1.62
40	2.08	1.92	1.84	1.74	1.69	1.64	1.58	1.51
60	1.99	1.84	1.75	1.65	1.59	1.53	1.47	1.39
80	1.95	1.80	1.70	1.60	1.54	1.49	1.41	1.32
120	1.91	1.75	1.66	1.55	1.50	1.43	1.35	1.25
∞	1.83	1.67	1.57	1.46	1.39	1.32	1.22	1.00

Appendix C: The Greek Alphabet

Letter Name	Uppercase	Lowercase	English Letter
Alpha	A	α	a
Beta	B	β	b
Gamma	Γ	γ	g
Delta	Δ	δ	d
Epsilon	E	ϵ	ê
Zeta	Z	ζ	z
Eta	H	η	e
Theta	Θ	θ	th
Iota	I	ι	i
Kappa	K	κ	k, c
Lambda	Λ	λ	l
Mu	M	μ	m
Nu	N	ν	n
Xi	Ξ	ξ	x
Omicron	O	o	ô
Pi	Π	π	p
Rho	P	ρ	r
Sigma	Σ	σ	s
Tau	T	τ	th
Upsilon	Υ	υ	y
Phi	Φ	ϕ	ph
Chi	X	χ	ch
Psi	Ψ	ψ	ps
Omega	Ω	ω	o

Appendix D: Variables in Data Sets

AR_SENTENCING

cident	Court Identifier		
delta	Delta vs. Other Counties		
	Value		Label
	0		Other counties
	1		Delta counties
warrant	Warrant Status		
	Value		Label
	0		No warrant issued
	1		Arrest warrant issued
trial	Type of Trial		
	Value		Label
	0		Plead guilty
	1		Went to Trial
violent	Violent Offense		
	Value		Label
	0		Non-violent crime=0
	1		Violent crime=1
drug	Drug Offense		
	Value		Label
	0		Non-drug crime=0
	1		Drug crime=1
chargety	Felony		
	Value		Label
	0		All other charges
	1		Felony charges

AR_SENTENCING (*continued*)

offserio   Offense Seriousness

nocounts   Total Number of Counts

status   Supervision Status

Value	Label
0	Not on probation/parole
1	On probation/parole

race   Race

Value	Label
0	White=0
1	Non-White=1

sex   Sex

Value	Label
0	Male
1	Female

age   Age

crhistot   Criminal History Total

sentotmo   Sentence in Months

GANG_MEM

sex   Sex

Value	Label
0	Male
1	Female
9	No Answer

race	Race	
	Value	Label
	1	White/Anglo, Non-Hispanic
	2	Black/African American
	3	Hispanic/Latino
	4	American Indian/Native American
	5	Asian/Pacific Islander/Oriental
	6	Other
	7	Mixed
	9	No Answer

age	Age	
	Value	Label
	11	11 or Younger
	18	18 or Older
	99	No Answer

f_school	Father's Schooling	
	Value	Label
	1	Grade school or less
	2	Some high school
	3	Completed high school
	4	Some college
	5	Completed college
	6	More than college
	7	Don't Know
	9	No Answer

m_school	Mother's Schooling	
	Value	Label
	1	Grade school or less
	2	Some high school
	3	Completed high school
	4	Some college
	5	Completed college
	6	More than college
	7	Don't Know
	9	No Answer

GANG_MEM (*continued*)

parents   Do your parents know where you are?

Value	Label
1	Strongly disagree
2	Disagree
3	Neither agree or disagree
4	Agree
5	Strongly agree
9	No Answer

gang_mem   Ever been a gang member?

Value	Label
1	No
2	Yes
9	No Answer

POISSONEX

viol   Cleveland Police Dept Violent Crime Count

totpop90   Total Population 1990

totpop00   Total Population 2000

ownocc90   Owner Occupied Housing 1990

ownocc00   Owner Occupied Housing 2000

perblack   Percent Black 1990

pervac90   Vacancy Rate 1990

povrate9   Poverty Rate 1990

LIST OF VARIABLES IN EACH DATA SET

CENSUS

TRACT   CENSUS TRACT NUMBER

WM4   WHITE MALES 0-4

WM5   WHITE MALES 5-17

WM18   WHITE MALES 18-44

WM45   WHITE MALES 45-64

WM62   WHITE MALES 62 AND OLDER

WF4   WHITE FEMALES 0-4

WF5   WHITE FEMALES 5-17

WF18   WHITE FEMALES 18-44

WF45   WHITE FEMALES 45-64

WF62   WHITE FEMALE 62 AND OLDER

BM4   BLACK MALE 0-4

BM5   BLACK MALE 5-17

BM18   BLACK MALE 18-44

BM45   BLACK MALE 45-64

BM62   BLACK MALE 62 AND OLDER

OTHER4   OTHER RACE 0-4

OTHER5   OTHER RACE 5-17

OTHER18   OTHER RACE 18-44

OTHER45   OTHER RACE 45-64

OTHER62   OTHER RACE 62 AND OLDER

MNOMARID MALE NEVER MARRIED

MMARRIED MALE MARRIED

MSEPARAT MALE SEPARATED

MWIDOW   MALE WIDOW

MDIVORCE MALE DIVORCED

FNOMARID FEMALE NEVER MARRIED

FMARRIED FEMALE MARRIED

FSEPARAT FEMALE SEPARATED

FWIDOW   FEMALE WIDOW

FDIVORCE FEMALE DIVORCED

MALE1   1 MALE HOUSEHOLDER

FEMALE1   1 FEMALE HOUSEHOLDER

MARCHILD MARRIED WITH CHILDREN

MNOKIDS   MARRIED WITH NO RELATED CHILDREN

MHOSHEAD SINGLE MALE HEAD OF HOUSEHOLD

FHOSHEAD FEMALE HEAD OF HOUSEHOLD

OWNOCCUP OWNER OCCUPIED HOUSING UNITS

RENTOCCP RENTER OCCUPIED HOUSING UNITS

VACRENT VACANT HOUSING UNITS FOR RENT

VACSALE VACANT HOUSING UNITS FOR SALE ONLY

VACSOLD VACANT HOUSING UNITS RENTED OR SOLD BUT NOT OCCUPIED

OTHERVAC OTHER VACANT HOUSING UNITS

OWNROOM ROOMS PER UNIT FOR OWNER OCCUPIED HOUSING

RENTROOM OCCUPIED ROOMS PER UNIT FOR RENTER OCCUPIED HOUSING

VACRNTRM NUMBER OF ROOMS PER UNIT FOR VACANT FOR RENT HOUSING

VACSALRM NUMBER OF ROOMS PER UNIT FOR VACANT FOR SALE ONLY

VACSLDRM NUMBER OF ROOMS PER UNIT FOR RENTED/SOLD NOT OCCUPIED

VACOTHRM NUMBER OF ROOMS PER UNIT FOR OTHER VACANT HOUSING

WHITEOWN WHITE OWNER OCCUPIED HOUSING UNITS

WHITERNT WHITE RENTER OCCUPIED HOUSING UNITS

BLACKOWN BLACK OWNER OCCUPIED HOUSING UNITS

BLACKRNT BLACK RENTER OCCUPIED HOUSING UNITS

OTHEROWN OTHER RACE OWNER OCCUPIED HOUSING UNITS

OTHERRNT OTHER RACE RENTER OCCUPIED HOUSING UNITS

OWN15   15-24 YEAR OLD OWNER OCCUPIED HOUSING UNITS

RENT15   15-24 YEAR OLD RENTER OCCUPIED HOUSING UNITS

OWN25   25-44 YEAR OLD OWNER OCCUPIED HOUSING UNITS

RENT25   25-44 YEAR OLD RENTER OCCUPIED HOUSING UNITS

OWN45   45-64 YEAR OLD OWNER OCCUPIED HOUSING UNITS

RENT45   45-64 YEAR OLD RENTER OCCUPIED HOUSING UNITS

OWN65   65 YEARS OLD AND OLDER OWNER OCCUPIED HOUSING UNITS

RENT65   65 YEARS OLD AND OLDER RENTER OCCUPIED HOUSING UNITS

OWN1   1 PERSON PER UNIT OWNER OCCUPIED HOUSING UNITS

OWN2   2 PERSON PER UNIT OWNER OCCUPIED HOUSING UNITS

OWN3   3 PERSON PER UNIT OWNER OCCUPIED HOUSING UNITS

OWN4   4 PERSON PER UNIT OWNER OCCUPIED HOUSING UNITS

OWN5   5 PERSON PER UNIT OWNER OCCUPIED HOUSING UNITS

OWN6   6 PERSON PER UNIT OWNER OCCUPIED HOUSING UNITS

OWN7   7 OR MORE PERSON PER UNIT OWNER OCCUPIED HOUSING UNITS

RENT1   1 PERSON PER UNIT RENTER OCCUPIED HOUSING UNIT

RENT2   2 PERSON PER UNIT RENTER OCCUPIED HOUSING UNITS

RENT3   3 PERSON PER UNIT RENTER OCCUPIED HOUSING UNITS

RENT4   4 PERSON PER UNIT RENTER OCCUPIED HOUSING UNITS

RENT5   5 PERSON PER UNIT RENTER OCCUPIED HOUSING UNITS

RENT6   6 PERSON PER UNIT RENTER OCCUPIED HOUSING UNITS

RENT7   7 OR MORE PERSONS PER UNIT FOR RENTER OCCUPIED HOUSING UNITS

OWNPERAV AVG PERSONS PER UNIT FOR OWNER OCCUPIED HOUSING UNITS

RNTPERAV AVG NUMBER OF PERSONS PER UNIT FOR RENTER OCCUPIED HOUSE

OWN.5   NO. OF OWNER OCCUPIED HOUSING UNITS < .5 PERSONS PER ROOM

OWN.51   NO OF OWNER OCCUPIED HOUSING UNITS .51-1.00 PER ROOM

OWN1.00 OWNER OCCUPIED HOUSING UNITS WITH 1.00-1.50 PERSONS PER ROOM

OWN1.51 OWNER OCCUPIED HOUSING UNITS WITH 1.51-2.00 PERSONS PER ROOM

OWN2.00 OWNER OCCUPIED HOUSING UNITS WITH >2.01 PERSONS/ROOM

VACRENT2 VACANT HOUSING UNITS FOR RENTS LESS THAN 2 MONTHS

VACRENT6 VACANT HOUSING UNITS FOR RENT 2-6 MONTHS

VACRENT7 VACANT HOUSING UNITS FOR RENT MORE THAN 6 MONTHS

VACSALE2 VACANT HOUSING UNITS FOR SALE ONLY LESS THAN 2 MONTHS

VACSALE6 VACANT HOUSING UNITS FOR SALE ONLY 2-6 MONTHS

VACSALE7 VACANT HOUSING UNITS FOR SALE ONLY MORE THAN 6 MONTHS

OTHVAC2 OTHER HOUSING UNITS VACANT LESS THAN 2 MONTHS

OTHRVAC6 OTHER HOUSING UNITS VACANT 2-6 MONTHS

OTHRVAC7 OTHER HOUSING UNITS VACANT MORE THAN 6 MONTHS

ROOM1   HOUSING UNITS WITH 1 ROOM

ROOM2   HOUSING UNITS WITH 2 ROOMS

ROOM3   HOUSING UNITS WITH 3 ROOMS

ROOM4   HOUSING UNITS WITH 4 ROOMS

ROOM5   HOUSING UNITS WITH 5 ROOMS

ROOM6   HOUSING UNITS WITH 6 ROOMS

ROOM7   HOUSING UNITS WITH 7 ROOMS

ROOM8   HOUSING UNITS WITH 8 ROOMS

ROOM10   HOUSING UNITS WITH MORE THAN 9 ROOMS

BOARDED   BOARDED UP HOUSING UNITS

NOBOARDS NOT BOARDED UP HOUSING UNITS

VAL35K   OWNER OCCUPIED HOUSING UNITS VALUED LESS THAN $35,000

VAL60K   OWNER OCCUPIED HOUSING UNITS VALUED $40,000-$60,000

VAL125K   OWNER OCCUPIED HOUSING UNITS VALUED $75,000-$125,000

VAL500K   OWNER OCCUPIED HOUSING UNITS VALUED $150,000 OR MORE

RENT150   CONTRACT RENT LESS THAN $150

RENT300   CONTRACT RENT LESS THAN $350

RENT550   CONTRACT RENT $350-$550

RENT1000 CONTRACT RENT $550 OR MORE

NORENT   NO CASH RENTAL HOUSING UNITS

WHITEVAL AVERAGE VALUE OF WHITE OWNER OCCUPIED HOUSING UNITS

BLACKVAL AVERAGE VALUE OF BLACK OWNER OCCUPIED HOUSING UNITS

OTHERVAL AVERAGE VALUE OF OTHER RACE OWNER OCCUPIED HOUSING

WHITRENT AVERAGE CONTRACT RENT FOR WHITE RENTER HOUSING UNITS

BLAKRENT AVERAGE CONTRACT RENT FOR BLACK RENTER HOUSING UNITS

OTHRENT   AVERAGE CONTRACT RENT FOR OTHER RACE RENTER HOUSING

DETACH1O 1 UNIT DETACHED OWNER OCCUPIED

ATTACH1O 1 UNIT ATTACHED OWNER OCCUPIED

CONDO2   2 UNIT OWNER OCCUPIED HOUSING UNIT

CONDO3   3-4 UNIT OWNER OCCUPIED HOUSING UNIT

CONDO5   5-19 UNIT OWNER OCCUPIED HOUSING UNIT

CONDO20   20-49 UNIT OWNER OCCUPIED HOUSING UNIT

CONDO50   50 UNIT OR MORE HOUSING UNIT

TRAILERO OWNED MOBILE HOME

DETACH1R 1 UNIT DETACHED RENTER OCCUPIED HOUSING UNIT

ATTACH1R 1 UNIT ATTACHED RENTER OCCUPIED HOUSING UNIT

APPART2   2 UNIT APARTMENT

APPART3   3-4 UNIT APARTMENT

APPART5   5-19 UNIT APARTMENT

APPART20 20-49 UNIT APARTMENT

APPART50 50 UNIT OR MORE APARTMENT

TRAILERR RENTED MOBILE HOME

RENTASKD AVERAGE RENT ASKED

PRICEASK AVERAGE PRICE ASKED FOR HOMES FOR SALE

BF4   BLACK FEMALE 0-4

BF5   BLACK FEMALE 5-17

BF189   BLACK FEMALE 18-44

BF45   BLACK FEMALE 45-64

BF62   BLACK FEMALE 62 OR OLDER

OTHERF4   OTHER RACE FEMALE 0-4

OTHERF5   OTHER RACE 5-17

OTHERF18 OTHER RACE 18-64

OTHERF62 OTHER RACE 62 YEARS OLD AND OLDER

RENT.5   .5 OR LESS PERSONS PER ROOM IN RENTER OCCUPIED HOUSING UNITS

RENT.51   .51-1.00 PERSONS PER ROOM IN RENTER OCCUPIED HOUSING UNITS

RENT1.00 1.01-1.50 PERSONS PER ROOM IN RENTER OCCUPIED HOUSING UNITS

RENT1.51 1.51-2.00 PERSONS PER UNIT IN RENTER OCCUPIED HOUSING UNITS

RENT2.01 2.01 OR MORE PERSONS PER ROOM IN RENTER OCCUPIED HOUSING

OTHERF45 OTHER RACE FEMALE 45

DELINQ   DELINQUENT INCIDENTS FOR THE CENSUS TRACT

POP80   POPULATION IN 1980

POP90   POPULATION IN 1990

WELFARE   NUMBER OF FAMILIES RECEIVING SOCIAL ASSISTANCE

MEDRENT   MEDIAN RENTAL VALUE

BFAMILY   BLACK COUPLE FAMILIES

BMHOUSE   BLACK MALE HEAD OF HOUSEHOLD

BFHOUSE   BLACK FEMALE HEAD OF HOUSEHOLD

WELFAREN NUMBER OF PEOPLE RECEIVING SOCIAL ASSISTANCE

POPCHANG POPULATION CHANGE 1980-1990

BLAKHEAD BLACK HEAD OF HOUSEHOLD

JUVENILE PROPORTION OF JUVENILES IN THE POPULATION

LOGWEL   LOG TRANSFORMED VARIABLE OF WELFARE

SINOWN   SINE TRANSFORMED VARIABLE OF OWNOCCUP

MODPOP   MOD10 TRANSFORMED VARIABLE OF POPCHANGE

GANG

DATE   Date of Appearance

AGE   Age of juvenile

SEX   Sex of juvenile
Value   Label
    1   Male
    2   Female

RACE   Race of Juvenile
Value   Label
    1   White
    2   Black
    3   Other

CHILDREN   Do you have any children of your own?
Value   Label
    1   Yes
    2   No

PARENTS   What parents do you live with?
Value   Label
    1   Both
    2   Neither
    3   Mother
    4   Father
    5   Grandparents

MO_AGE   How old is your mother?
Value   Label
    1   20-30
    2   30-40
    3   Over 40

FA_AGE   How old is your father?

Value   Label

    1   20-30

    2   30-40

    3   Over 40

MO_HS   Did your mother graduate high school?

Value   Label

    1   Yes

    2   No

FA_HS   Did your father graduate high school?

Value   Label

    1   Yes

    2   No

MO_CRIME   Has your mother ever been convicted of a crime?

Value   Label

    1   Yes

    2   No

FA_CRIME   Has your father ever been convicted of a crime?

Value   Label

    1   Yes

    2   No

MO_JAIL   Has your mother ever been in jail or prison?

Value   Label

    1   Yes

    2   No

FA_JAIL   Has your father ever been in jail or prison?

Value   Label

    1   Yes

    2   No

MO_BOOZE Does your mother drink alcohol?

Value   Label

1   Yes

2   No

FA_BOOZE Does your father drink alcohol?

Value   Label

1   Yes

2   No

MA_DRUG   Does your mother use drugs?

Value   Label

1   Yes

2   No

FA_DRUG   Does your father use drugs?

Value   Label

1   Yes

2   No

P_FIGHT   Have your parents ever hit one another in an argument?

Value   Label

1   Yes

2   No

SIBS   How many brothers and sisters do you have?

HOM_SIBS How many siblings live with you?

HOME   What type of house do you live in?

Value   Label

1   House

2   Duplex

3   Trailer

4   Apartment

5   Other

OWN_HOME Do your parents own their home?

Value   Label

    1   Yes

    2   No

TENURE   How long have you lived at your current address (in months)

DETER   How many homes are in your neighborhood boarded up or have broken windows?

SCHOOL   Are you still in school?

Value   Label

    1   Yes

    2   No

GRADE   School grade

CLUBS   Are you involved in any clubs or sports in school?

Value   Label

    1   Yes

    2   No

CURFEW   Do you have a curfew at home?

Value   Label

    1   Yes

    2   No

TIME_CUR What is the time of your curfew?

RESTRICT Are there times when you are required to be at home?

Value   Label

    1   Yes

    2   No

TIM_REST When are you required to be at home (combinations represent multiple responses)?

Value   Label

1   School nights

2   Until homework is done

3   Meals

4   Other

QUALITY   How many hours a week do you spend with your parents (not counting sleeping)?

CHURCH   Do you ever go to church?

Value   Label

1   Yes

2   No

SKIPPED   How often have you skipped school?

Value   Label

1   Never

2   Once a school year

3   Once a 9 weeks

4   Once a month

5   Twice a month

6   Once a week

7   More than once a week

BOOZE   What kind of alcohol have you consumed (combinations represent multiple responses)?

Value   Label

1   Never consumed alcohol

2   Beer

3   Wine

4   Mixed drinks

BOZE_TIM How often do you consume any type of alcohol?

Value   Label

    1   Once

    2   Once a year

    3   A few times a year

    4   Once a month

    5   A few times a month

    6   Weekly or more

BOZE_PL   Where did you get the alcohol (combinations represent multiple responses)?

Value   Label

    1   Bought it yourself

    2   Had someone else buy it

    3   At a party/in a group

    4   Got it from your parents

DRUGS   Have you ever used illegal drugs (combinations represent multiple responses)?

Value   Label

    1   Never

    2   Marijuana

    3   Cocaine

    4   LSD

    5   Pills

    6   Huffing

    7   Crack

    8   Herion

    9   Other

DRUG_PL   Where did you get the drugs (combinations represent multiple responses)?

Value   Label

    1   Friend

    2   At a party/in a group

    3   Dealer

    4   Parent

    5   Other relative

    6   Other

DRUG_TIM How often have you used drugs?

Value   Label

    1   Once

    2   Once a year

    3   A few times a year

    4   Once a month

    5   A few times a month

    6   Weekly or more

LAW_DRUG Have you ever been in trouble with the law over drugs or alcohol)?

Value   Label

    1   Yes

    2   No

SCH_DRUG Have you ever been in trouble at school because of drugs or alcohol)?

Value   Label

    1   Yes

    2   No

SKIP_DRG Have you ever skipped school because you had been drinking or using drugs?

Value   Label

    1   Yes

    2   No

GRND_DRG Have you ever been grounded because of drugs or alcohol?

Value   Label

    1   Yes

    2   No

OUT_DRUG Have your parents ever threatened to throw you out of the house because of drugs?

Value   Label

    1   Yes

    2   No

GUNS   Do you think guns are becoming a problem in your neighborhood?

Value   Label

    1   Yes

    2   No

GUN_NEIG Have you ever seen anyone in your neighborhood with a gun?

Value   Label

    1   Yes

    2   No

SCH_GUN   Have you ever seen anyone at your school with a gun?

Value   Label

    1   Yes

    2   No

GRUP_GUN Have you ever been in a group where someone was carrying a gun?

Value   Label

    1   Yes

    2   No

OUT_GUN   Have you ever carried a gun out with you when you went out at night?

Value   Label

    1   Yes

    2   No

GUN_WHER Where did you get the gun?

Value   Label

    1   Bought it yourself

    2   Got it from a friend

    3   Had someone else buy it

    4   Got it from your parents

    5   Other

GUN_REG   Do you regularly carry a gun with you?

Value   Label

    1   Yes

    2   No

GANGS   Are there people in your neighborhood or school who say they are in a gang?

Value   Label

    1   Yes

    2   No

U_GANG   Are you in a gang?

Value   Label

    1   Yes

    2   No

TIME_GAN How long (in years) have you been in a gang?

FIGHT   Have you ever gotten into a fight with a gang member?

Value   Label

    1   Yes

    2   No

ARREST   How many times have you been arrested?

U_CRIME   How many times have you committed a crime but not been caught

U_CONVIC How many times have you been convicted of a crime?

U_GUN_CM Have any of your arrests involved a firearm?

Value   Label

    1   Yes

    2   No

SKIPPEDR Recoded SKIPPED variable to yes and no

Value   Label

    1   Yes

    2   No

MUNDANR Scale measure of mundane activities

DELINQR   Scale measure of delinquent activities

POTENTLR Scale measure of potentially delinquent activities

GANGR   Scale measure of gang activities

DRUGR   Recoded DRUGS variable to represent yes or no

Value   Label

    1   Yes

    2   No

BOOZER   Recoded BOOZE variable to represent yes or no

Value   Label

    1   Yes

    2   No

UCONVICR Recoded U_CONVIC variable to represent yes or no

Value   Label

    1   Yes

    2   No

ARRESTR   Recoded ARREST varible to represent yes or no

Value   Label

    1   Yes

    2   No

LR_COP

WALK_DAY How Often Do You Walk In Your Neighborhood During the Day

Value   Label

    1   Never

    2   Occasionally

    3   Often

WALK_EVE How often do you walk in your neighborhood in the evening?

Value   Label

    1   Never

    2   Occasionally

    3   Often

WALK_NIT How often do you walk in your neighborhood at night?

Value   Label

    1   Never

    2   Occasionally

    3   Often

ASSOC   Do you participate in any neighborhood associations?

Value   Label

    0   None

    1   Yes, No Other Response

    2   Block Club

    3   Church

    4   Athletics

    5   Neighborhood Association

    6   Community Agency (United Way)

    7   Other

VICTIM   Have you been a victim of any of these crimes in the last 6 months

Value   Label

    1   Burglary

    2   Vandalism

    3   Car Theft

    4   Robbery

    5   Assault

    6   Other

CRIME_PL Did the crime take place in your neighborhood?

Value   Label

    1   Yes

    2   No

CRIME_TM What time did the crime take place?

Value   Label

    1   7 A.M. to 3 P.M.

    2   3 P.M. to 11 P.M.

    3   11 P.M. to 7 A.M.

WITNESS   Were there any witnesses to the crime?

Value   Label

    1   Yes

    2   No

NEIGH_HL Did any of your neighbors help you?

Value   Label

    1   Yes

    2   No

HOOD_HOM Do you think the criminal was from your neighborhood?

Value   Label

    1   Yes

    2   No

REPORT   Where did you report the crime?

Value   Label

    0   None

    1   Yes, No Other Response

    2   Main Police Department

    3   Substation

    4   Local Patrol Officer

TIME_OK   How satisfied were you with the time it took the police to respond?

Value   Label

    1   Very Satisfied

    2   Somewhat Satisfied

    3   Not At All Satisfied

RESP_OK   How satisfied were you with the quality of the police response?

Value   Label

    1   Very Satisfied

    2   Somewhat Satisfied

    3   Not At All Satisfied

US_SAFE   How safe is the U.S.?

Value   Label

1   Becoming Safer

2   Not Changing

3   Becoming Less Safe

LR_SAFE   How serious is crime in this city compared with other cities?

Value   Label

1   Very Serious

2   About the Same

3   Less Crime In Little Rock

MOVE   Is crime such a problem that you have considered moving in the past 12 months?

Value   Label

1   Yes

2   No

NEIGH_SF Is your neighborhood safety changing?

Value   Label

1   Becoming Safer

2   Not Changing

3   Becoming Less Safe

ENVN_PRB How big a problem is the environment?

Value   Label

1   No Problem

2   Small Problem

3   Big Problem

PARK_PRB How big a problem is parking?

Value   Label

1   No Problem

2   Small Problem

3   Big Problem

SHOP_PRB How big a problem is inadequate shopping?

Value   Label

1   No Problem

2   Small Problem

3   Big Problem

CRIM_PRB How big a problem is crime?

Value   Label

    1   No Problem

    2   Small Problem

    3   Big Problem

BUS_PRB   How big a problem is public transportation?

Value   Label

    1   No Problem

    2   Small Problem

    3   Big Problem

SCHL_PRB How big a problem is public schools?

Value   Label

    1   No Problem

    2   Small Problem

    3   Big Problem

HOMY_BUS How big a problem are neighbors?

Value   Label

    1   No Problem

    2   Small Problem

    3   Big Problem

JOB_PRB   How big a problem is unemployment?

Value   Label

    1   No Problem

    2   Small Problem

    3   Big Problem

PART_FER How much has fear of crime affected your social activities in your neighborhood?

Value   Label

    1   Great Effect

    2   Small Effect

    3   No Effect

DAY_FER   How much has fear of crime effected your decision to walk during the day?

Value   Label

1   Great Effect

2   Small Effect

3   No Effect

EVE_FER   How much has fear of crime affected your decision to walk during the evening?

Value   Label

1   Great Effect

2   Small Effect

3   No Effect

NITE_FER How much has fear of crime affected your decision to walk at night?

Value   Label

1   Great Effect

2   Small Effect

3   No Effect

KIDS_FER How much has fear of crime affected supervision of your children

Value   Label

1   Great Effect

2   Small Effect

3   No Effect

REC_FER   How much has fear of crime affected your recreation?

Value   Label

1   Great Effect

2   Small Effect

3   No Effect

COP_ACCT Police officers should be accountable to the public for their actions.

Value   Label

1   Strongly Agree

2   Agree

3   Disagree

COP_PREV Police officers should concentrate their efforts on crime prevention.

Value   Label

     1   Strongly Agree

     2   Agree

     3   Disagree

COP_SPOT Police officers should be able to recognize area residents.

Value   Label

     1   Strongly Agree

     2   Agree

     3   Disagree

COP_RPT   Police officers should encourage people to report crimes.

Value   Label

     1   Strongly Agree

     2   Agree

     3   Disagree

COP_JUV   Police officers should provide guidance to juveniles.

Value   Label

     1   Strongly Agree

     2   Agree

     3   Disagree

COP_SAFE Police officers should try to increase personal safety.

Value   Label

     1   Strongly Agree

     2   Agree

     3   Disagree

COP_SCHL Police officers should work with schools to deter crime.

Value   Label

     1   Strongly Agree

     2   Agree

     3   Disagree

COP_VICT Police officers should recognize the needs of victims.

Value   Label

1   Strongly Agree

2   Agree

3   Disagree

COP_CONT What contact have you had with the police in the last 12 months?

Value   Label

1   Acquaintance With Police

2   Emergency Assistance From Police

3   Complaint Made To Police

4   Witness Questioned By Police

5   Arrested By Police

6   Other

7   No Contact

COP_OK   In your neighborhood, how well do you think the police performing their duties?

Value   Label

1   Very Well

2   Average

3   Below Average

4   Not At All

IMPROVE   To what extent does the police department need improvement?

Value   Label

1   Many Changes Needed

2   Some Changes Needed

3   No Changes Needed

IMP_NO   To what extent does the police department need to increase the number of officers?

Value   Label

1   Very Important

2   Important

3   Not Important

IMP_QUAL To what extent does the police department need to increase officer qualifications?

Value   Label

    1   Very Important

    2   Important

    3   Not Important

IMP_TRNG To what extent does the police department need to improve training?

Value   Label

    1   Very Important

    2   Important

    3   Not Important

IMP_TIME To what extent does the police department need to improve response time?

Value   Label

    1   Very Important

    2   Important

    3   Not Important

IMP_PATR To what extent does the police department need to improve patrol

Value   Label

    1   Very Important

    2   Important

    3   Not Important

IMP_FOLW To what extent does the police department need to follow up on reports of crime?

Value   Label

    1   Very Important

    2   Important

    3   Not Important

IMP_PCR   To what extent does the police department need to improve community relations?

Value   Label

    1   Very Important

    2   Important

    3   Not Important

IMP_MINR To what extent does the police department need to improve minority relations?

Value   Label

1   Very Important

2   Important

3   Not Important

LR_TH_YR How has the police performance changed in the last year?

Value   Label

1   Better

2   Same

3   Worse

JOB   What is your occupation

Value   Label

1   Professional (Principal, Banker)

2   Clerical/Technical

3   Blue Collar (Factory Worker, Sales Person)

4   Retired

5   Housewife

6   Other (Part-time, Self-employed)

7   Unemployed

ADDRESS   How long have you lived at this address?

HOME   Do you own or rent your home

Value   Label

1   Own

2   Rent

NEIGH_HM How long have you lived in this neighborhood?

LR_HOME   How long have you lived in this city?

KID_HOME What type of town did you live in when you were growing up?

Value   Label

1   Rural Area

2   Small City

3   Suburb Of A Large City

4   A City As Large As Little Rock

AGE   Age of respondent

SEX   Sex of respondent

Value   Label

    1   Male

    2   Female

RACE   Race of respondent

Value   Label

    1   Black

    2   White

    3   Other

MARRIED   Marital status of respondent

Value   Label

    1   Single

    2   Married

    3   Widowed

    4   Separated

    5   Divorced

KIDS   How many children do you have?

KIDS_LIV How many kids live with you?

ORGS   What local groups do you belong to?

Value   Label

    1   Church

    2   Volunteer Group

    3   Social Club

    4   Neighborhood Association

    5   Veteran's Association

    6   Labor Union

    7   Other

VICT_Y_N Recoded VICTIM to yes and no

Value   Label

     1   Yes

     2   No

EDUCATIO What is your highest level of education?

Value   Label

     1   Less than High School

     2   GED

     3   High School Graduate

     4   Some College

     5   Collegel Graduate

     6   Post Graduate

Index

Italicized page locators indicate a figure; tables are noted with a *t*.